中国湿地保护地管理者操作手册

Operational Handbook for Wetland Protected Area Managers in China

〔英〕约翰·马敬能（John MacKinnon）著
卢和芬
赖鹏飞 译

科学出版社
北京

内 容 简 介

本书是一本为中国湿地保护地管理者编写的实用手册。分为 8 部分：第 1 部分湿地简介；第 2 部分湿地基本知识和理论；第 3 部分湿地保护地的规划；第 4 部分栖息地管理；第 5 部分物种管理；第 6 部分主流化与传播；第 7 部分运行；第 8 部分监测和报告。书后附录了相关国际组织、湿地生物多样性公约 / 计划、最佳实践指南、湿地野外作业建议、湿地管理适用的生态原则、特定指标监控的具体内容、选择适当的统计测试方法、巡护和监测报告样表、改善中国湿地保护所需要的改革方案与实践变更汇总等。由全球环境基金“增强湿地保护地子系统管理有效性，保护具有全球意义的生物多样性”项目支持出版。

本书可供湿地保护管理人员、研究人员、规划设计人员及其他有关人士参阅。

图书在版编目（CIP）数据

中国湿地保护地管理者操作手册 = Operational Handbook for Wetland Protected Area Managers in China：汉英对照 /（英）约翰 · 马敬能（John MacKinnon），卢和芬著；赖鹏飞译. —北京：科学出版社，2020.11
ISBN 978-7-03-064139-7

Ⅰ. ①中… Ⅱ. ①约…②卢… ③赖… Ⅲ. ①沼泽化地 - 自然资源保护 - 中国 - 手册 - 汉、英 Ⅳ. ① P942.078-62

中国版本图书馆 CIP 数据核字（2020）第 010205 号

责任编辑：张会格　白　雪 / 责任校对：严　娜
责任印制：吴兆东 / 封面设计：北京图阅盛世文化传媒有限公司

科学出版社 出版
北京东黄城根北街 16 号
邮政编码：100717
http://www.sciencep.com

北京九州迅驰传媒文化有限公司印刷
科学出版社发行　各地新华书店经销
*
2020 年 11 月第　一　版　开本：787×1092　1/16
2025 年　1 月第三次印刷　印张：24 1/4
字数：575 000
定价：298.00 元

前　言

本手册是一本专为湿地保护地从业人员编写的指南，着眼于保护好居住在这些栖息地的珍贵野生动物，加强这些栖息地为人类提供的重要生态服务。

湿地包括沼泽、池塘、湖泊、河流和浅海水域等类型。这里不仅栖息着多种水生生物，还是许多鸟类特别是重要的迁徙水鸟和滨鸟的栖息和繁殖场所。它不仅与上下游关系密切，同时在水源、渔业、水力发电、航运，以及净化水质等方面发挥作用。

本手册更好地解读了“湿地”的含义，描述了中国不同类型湿地的分布，并强调了湿地对于生物多样性及人类自身环境福祉的重要性。

与管理机构或企业不同的是，因湿地的情况更加复杂多样，需要管理者有丰富的知识和能力，才能兼顾不同的管理目标并更好地应对由气候、季节和生物变化等带来的不同状况。

受多种因素影响，中国的湿地面积有所减少，保持、恢复甚至创造新的湿地成为当务之急。

本手册将理论与实践相结合，内容涵盖了湿地保护地边界划分、水位控制、最佳植被管理，以及保护目标陆生和水生物种等方面。还就如何进行游客管理、在不同层面开展监测，以及提供教育和娱乐服务等提供了建议和指导。

中国已经将湿地保护确定为国家战略，建立了不同类型的保护地，并给予不同程度的保护，湿地管理体系和机制逐渐完整，然而湿地管理者的能力仍需进一步提高，希望本手册可以在这方面发挥作用。

Preface

This handbook is written as a specific guide for the managers of China' s wetlands and wetland protected areas from the perspective of conserving the precious wildlife that occupy these habitats and enhancing the important ecological services that such habitats deliver to the human population.

Wetlands range from marshes, ponds, lakes, rivers and shallow seas. They harbour many aquatic creatures but also a rich bird fauna, especially for important seasonal migrating waterfowl and shorebirds. But wetlands are complex being linked to upstream and downstream impacts, multiple uses—sources of water and fisheries, providers of hydropower, routes of transport but also serving as drains and sewers.

The handbook clarifies what we mean by the term wetland; describes where there are different types of wetlands in China and demonstrates the importance of such wetlands for both biodiversity and our own environmental well-being.

Wetlands are very complicated. Managing a wetland is not like administering a business or department. It requires a lot of knowledge and understanding help the manager balance the different objectives of management and to respond to the different conditions that climate, season and biological changes may present.

Overall China has lost much of its original wetland area and there is a huge need to retain, restore and even create new wetlands.

The manual offers a combination of theoretical guidance and practical details covering aspects of boundaries, maintaining suitable water levels, management and encouragement of optimal vegetation and conservation of target terrestrial and aquatic species. It further gives advice and instruction in how to manage visitors, how to monitor different aspects of each site and how to provide educational and recreational roles.

Protection of wetlands has been placed high on the national agenda. Many hundreds of important wetland sites have been listed and given degrees of protection, yet there is little specific training offered to wetlands managers. This manual hopes to fill some of this gap.

目　　录

Contents

第1部分　湿地简介

1.1　定　　义

什么是生物多样性?

生物多样性是我们所生活的地球上各种各样的生物，包括构成数百万不同物种及其变种的基因、物种本身及这些物种通过与其他物种及其物理环境发生相互作用所构成的可正常发挥功能的生态系统。

什么是湿地?

湿地是指所有永久性或定期被浅水所淹没的土地，包括所有湖泊和江河、地下蓄水层、沼泽、湿草地、泥炭地、绿洲、河口、三角洲和滩涂、红树林和其他沿海地区、珊瑚礁、水深不超过6m的海水，以及鱼塘、稻田、水库和盐田等所有人工湿地。

湿地评估：确定湿地的现状和威胁因素，为通过监测活动收集有关湿地的更多具体信息奠定基础。

湿地监测：针对湿地评估活动的假设，收集有关湿地的具体信息，用于湿地管理。

1.2　湿地的重要性何在?

湿地可以通过经济和非经济方式促进人类福祉。重要的湿地功能包括养分循环、泥沙保留、维护重要生物群落、保护重要遗传资源、蓄水和净水、固碳、防洪和保护海岸地区。湿地也可作为放牧区，提供持续的动植物产品，具有娱乐和景观价值，并能提高土地价值（表1.1）。

表1.1　湿地提供的或来自湿地的生态系统服务

服务	功能和范例
供给服务	
食物	产出鱼类、野生动物、水果和谷物
淡水*	储存和保留水分；提供家庭生活用水、灌溉用水和饮用水
纤维和燃料	产出木材、薪柴、泥炭和饲草
生物化学品	从生物群中提取药物和其他物质
遗传物质	提供抵抗植物病原体的基因及培育观赏物种等

续表

服务	功能和范例
调节服务	
调节气候	温室气体的源和汇；影响局地和区域性气温、降水及其他气候过程
调节水（水文状况）	地下水的补给和排放
净化水和废弃物处理	保留、恢复和消除过多的养分和其他污染物
预防侵蚀	土壤保持，保存沉积物
调控自然灾害	防洪、抵御风暴
授粉	为授粉者提供栖息地
文化服务	
精神和灵感	灵感的源泉；很多宗教都很重视湿地生态系统各个方面的精神和宗教价值
休闲娱乐	提供休闲活动的机会
美学	人们普遍能从湿地生态系统的各个方面发现其美学价值
教育	提供正规和非正规教育和培训的机会
支持服务	
土壤形成	保留沉积物、富集有机物
养分循环	养分的储存、再循环、加工和获取

* 在千年生态系统评估（MA）中，淡水被看作供给服务，但在其他情况下淡水也常常被看作调节服务

湿地提供的上述惠益许多直接进入市场，其他一些惠益则由全社会免费享受，不进入市场，但对其经济价值仍可分别进行估算，并且价值通常非常巨大。另外，也可开展全球性的价值估算。Costanza 等 1997 年指出，全球湿地生态服务每年的总经济价值约为 19 万亿美元。此后的 2014 年，Costanza 等利用最新湿地价值和面积开展的评估，在 2011 年将该数据提升至 50 万亿美元。各个湿地类型的单位价值见表 1.2。

表 1.2　湿地的生态系统服务价值（全球平均值）[美元 / （$hm^2 \cdot a$）]

湿地类型	生态系统服务（2007 年）	生态系统服务（2011 年）	主要功能
河口	31 509	28 916	迁徙鸟类、可食用软体动物和甲壳动物的停留地，鱼种场
海草 / 海藻床	26 226	28 916	抑制波浪，为人类提供食物，丰富了生物多样性
珊瑚礁	8 384	352 249	保护海岸、发展渔业和旅游业
潮沼 / 红树林	13 785	193 843	保护海岸、发展渔业、截获土壤、净水、丰富生物多样性
沼泽 / 泛滥平原	27 021	25 681	净水、蓄水、丰富生物多样性、提供产品
湖泊 / 江河	11 727	12 512	引水、蓄水、提供水能、发展渔业、丰富生物多样性

珊瑚和红树林生态服务的估值得到提高，其主要原因是在 2004 年印度尼西亚和 2011 年日本的大型海啸事件（图 1.1）后人们更好地意识到了珊瑚和红树林在保护海岸方面的作用。

图 1.1　2011 年席卷日本内陆地区的海啸

湿地的经济价值总体要大大超过热带森林［5382 美元 / ($hm^2 \cdot a$)］、温带森林［3137 美元 / ($hm^2 \cdot a$)］或农田［5567 美元 / ($hm^2 \cdot a$)］。值得注意的是，在湿地上植树没有任何经济价值，相反只会直接降低湿地的生态系统价值。

湿地的这些巨大价值可能难以解释或把握。有时，将这些价值转化为决策者更加易于理解的指标可能会更好。

水库等价物

某个面积为 $200km^2$ 的森林集水区每年可能提供价值 6000 万美元的生态系统服务。不过，对于当地的规划者和决策者来说，该数据很难理解，这是因为此类数据通常不会被纳入普通的支出和生产账目中。如果将这种生态系统服务以更为有形的方式来表示，比如说这相当于容量 50 亿 m^3 的水库，可能更加易于理解。

污水处理厂等价物

沼泽湿地的主要功能之一是其在净化水源方面提供的重要服务。沼泽植被可以将金属和其他污染物捕获到土壤中，并使沉积物沉淀下来。它们的作用相当于污水处理厂。由太阳提供能量的植物和细菌可以分解污染物，净化污水。

我们可以计算出湿地在净化水源服务方面的效率，并将之转化为人造污水处理厂的对等规模和数量，从而估算出建造和维护人造污水处理厂的经济成本。按照这种方式来体现湿地的价值，可以更好地让当地规划者和决策者认识到保持这些健康湿地不被开发和不被排干的必要性。

娱乐等价物

湿地的部分价值，如风景、娱乐、精神和文化价值更加难以用纯货币的方式进行计算。如果某个湿地提供的一项服务为娱乐服务，我们可以根据游客数量或旅游收入来计算其潜在的经济收益。不过，其大部分的价值为非经济价值，体现为人类幸福、福祉和健康的形式。例如，将湿地的娱乐功能按照相当于10座城市公园的价值来体现，可能让政府规划部门更加易于理解。

1.3 生物多样性如何促进湿地功能？

湿地的生物多样性可以通过以下多种方式造福于人类。

- 我们可以从湿地中获得许多可再生资源，如鱼类、甲壳动物、软体动物、木材、薪柴、植物和栅栏柱、动物饲料、可食用植物、野味肉、菌类和蜂蜜等。
- 生物多样性高可增强湿地功能的效率和健康状况，大大改善湿地生态系统服务的供给状况，包括形成土壤、养分循环、蓄水和净化水源、固碳、防洪、抵御风暴等。
- 我们可以从湿地中获得其他许多惠益，包括景区和娱乐区、生活场地、文化圣地和遗产地等。
- 生物多样性可作为环境健康状况的指标及危险变动状况的预警系统。

1.4 湿 地 调 查

湿地调查包括收集和（或）整理有关湿地的核心信息，以用于开展湿地管理，包括提供用于开展特定的湿地评估和监测活动的信息库。湿地调查可在各种尺度，如从项目点到流域再到全国尺度上开展。《湿地公约》提供的有关湿地调查的详细指南（http://www.ramsar.org/sites/default/files/documents/pdf/lib/hbk4-15.pdf.）如下。

该指南手册建议针对湿地调查采取13个步骤。

- **明确目标**——湿地调查的目标是什么？
- **审阅现有信息**——审阅已出版的、未出版的地图、遥感影像、照片、此前的调查结果等。
- **审查可能采用的调查方法**——分析哪些方法能够提供为实现既定目标所必需的数据。
- **确定调查的尺度和分辨率**—— 依据湿地的面积及可提供基础图的类型。
- **整合成核心基准数据库**——无须所有已提供的信息，只选择与目标最为相关的关键元素。
- **选择适宜的栖息地类型**——利用或适当调整现有的植被类型、土壤类型、土地单元或可从遥感影像中明显识别出的类型。
- **选择最适宜的调查方法**——调查方法有数十种，选择最符合当地情况和人力状况等的调查方法。
- **制订数据收集规范和管理体系**——确定必需的数据集及其元数据库，规划数据流和数据共享协议。

- **制订采集不同数据的时间表**——规划需要采集哪些数据集及间隔时间和负责人员。
- **评估各种方法和对策的成本效益比**——需要更多的信息，但由于时间、资金、人力和设备的限制，必须对信息有所选择。
- **制订报告规程**——在报告之前，需要对原始数据进行整理、分析、解析，并与其他数据整合。
- **制订评审和评估流程**——确定用来核查和批准不同数据集的规程，考虑哪些人可以获取哪些数据集。
- **规划试验性研究**——开展试验，看评审和评估流程是否切实可行和足以实现目标。在开展全面调查前，对流程进行必要的修改。

国家林业局①现已组织了两次全国湿地资源调查。第一次调查于2003年结束，对单块面积在100hm^{2}以上的所有湿地开展了调查，调查了其湿地类型、所有权和法律地位、生物和物理特征。2009-2013年，开展了第二次全国湿地资源调查，本次调查更为详细；单块面积8hm^{2}以上的湿地均纳入调查范围，并且绘制了所有湿地地图。调查结果显示，全国湿地总面积5360.26万hm^2，占国土面积的比率为5.58%。与第一次调查相比，湿地面积减少了339.63万hm^2（第一次调查甚至未包括许多面积较小的湿地）。其中自然湿地4667.47万hm^2，占全国湿地总面积的87.08%。不过，自然湿地也减少了337.62万hm^2。在湿地的分布上，青海、黑龙江、西藏、内蒙古的湿地面积最大。

1.5 湿地的多种用途

中国许多具有重要生物学意义的湿地已被列为国家级或省级保护区。不过，这并不意味着湿地管理的目的仅仅是自然保护。所有湿地都具有多种用途，可向社会其他部门提供各种服务。国家林业局颁布了有关湿地保护管理的规章，但其他部门，如渔业、砂石疏浚、水利、污染排放治理、航运等部门也颁布了自己的规章，因此不同目的、使用权、所有权和规章之间相互重叠、错综复杂。

合作体制

鉴于不同部门之间规章的大量重叠（参见2.3节），同时，目的之间往往存在冲突，这种多用途土地（绝大部分湿地都是如此）要实现协调利用，不同人群、机构和部门之间必须开展深入的合作和协调。创建这种合作体制是一个非常复杂和微妙的过程，必须依托现有的合作体制。例如，现有的灌溉体系已存在数十年，这些体系制订了相关的规则，确定了哪些人可以何时、何地从整个供水体系中取多少水

① 2018年3月，根据第十三届全国人民代表大会第一次会议批准的国务院机构改革方案，将国家林业局的职责整合，组建国家林业和草原局。

量。类似的合作体制也适用于那些对共享渔业资源、森林资源、放牧权和其他共有资源的使用。

1.6 湿地的动态属性

从本质上讲，湿地是一种随机而动的栖息地。它们总是不断变化，从未保持稳定。湖泊形成后，如果充满沉淀物，就可能变成陆地。海岸线不断变化，海平面快速上升。冰川冬季扩张，夏季退缩，但在气候变化的影响下，目前所有冰川都在退缩。随着冰川的逐一融化，融水形成的溪流也在干涸。河流改道，河道的切割日益加深。不过，偶尔发生的暴雨和洪水会阻塞泛滥平原的排水系统，形成新的湿地。一旦出现洼地或排水渠道遭到阻塞，新的湿地就会形成，湿地植物和动物大批生长和繁殖，并在自然界中不断演变。为湿地创造适宜条件的因素可能是自然、地质或地理事件，也同样可能是人类活动导致的结果。农业和工程活动可能加快或减缓排水，从而形成、破坏或至少显著改变湿地的条件和性质。因此，世界上许多大型湿地实际上是人工湖泊、水库或排水量被人为减少的陆地。湿地的自然成因不如自然物种的组成重要，而人为活动却能够形成、破坏、保护或改变湿地以确保我们获得所需的服务。湿地管理的成功秘诀就是，了解必须从某个湿地获得哪些服务，并开展良好的管理，确保湿地提供这些服务，或者保护需要保护的生物多样性。一方面，生物多样性保护本身是一个可使我们生活的地球变得更为丰富多彩的高尚目标；另一方面，人类为了生存，总是会把自身的发展需求置于野生动植物的需求之上。不过，我们必须同时意识到，生物多样性最为重要的服务之一，是确保湿地的生态健康，并体现湿地的健康状况。

第2部分　湿地基本知识和理论

2.1　中国湿地的分布和类型

湿地包括浅水（通常深度不足6m）中永久性或周期性被淹没的土地。这些不同的湿地类型见表2.1。

表2.1　中国的湿地类型

类型	描述	分布	重要性
沼泽	排水不畅、带有草本植物或灌木等植被的潮湿地区，间杂有多条溪流和水塘	全国各地相对平坦、排水不畅的地区，包括位于高原的许多地区	非常重要，通常含泥炭的储水区。许多水鸟和部分重要哺乳动物的繁殖栖息地
草甸	开阔的草地和草本植物地区，会周期性地被水淹没，通常用于放牧而非农耕	大型河流的冲积泛滥平原，通常地处河流旁或湖泊周围	可作为野生动物吃草的地方和家畜的放牧区。许多鸟类和哺乳动物等的栖息地，对于水生污染物的净化至关重要
湖泊	淡水或部分含盐的大型天然水体（高原上）	中国各地	重要的渔业、水鸟繁殖和越冬地，天然储水地。可开展人员运送，拥有部分重要的水生哺乳动物
水库	利用大坝蓄积的人工湖泊，用于蓄水和灌溉	中国各地	通常相对贫瘠，但可能提供重要的渔业资源，为水鸟和其他水生生物提供支持
池塘	流动缓慢的小型天然或人工水体	中国各地	支持硅藻、蜻蜓、两栖动物、小鱼和其他池塘生物群
江河和溪流	从流域上游流向海洋的天然水系	中国各地，但湿润地区相对更多	水生生物分布的至关重要的连接。鱼类、甲壳动物、鸟类、水獭等的栖息地，对水的氧化作用至关重要
沟渠和运河	引入来自天然水系中的水或将水引回至天然水系中的人工水道。部分沟渠用于将湿地的水排干开垦为农业用地	中国各地	许多沟渠受到污染，相对贫瘠，但可与天然江河和溪流一样支持丰富的生物群
海岸线	拥有环礁湖和珊瑚礁的岩质、沙质、泥质海滩和悬崖峭壁	所有沿海地区，有许多种栖息地类型	很多生境类型维系了不同的生物群。南部的珊瑚栖息地拥有丰富的生物群，有利于渔业和海岸保护，也提供了娱乐机会
潮间带	由泥沙、沙洲、淡水、海水构成的地带。有多种不同的植被类型	所有沿海地区	对涉禽、软体动物、蟹和鱼类产卵至关重要
红树林	由树木及能够抵御海水经常性泛滥的相关植物组成的植被类型	受潮汐作用影响的热带冲积地	支持甲壳动物、鱼类、蠕虫、软体动物、涉禽、林冠鸟、昆虫、哺乳动物等极其丰富的生物群

2.2　中国的湿地保护地系统

目前，中国的湿地保护地主要包括两种类型：湿地自然保护区和湿地公园。建立湿地自然保护区的主要目的是保护各种物种和生态系统，而建立湿地公园的主要目的是保护湿地风景区，供游客参观享受。第二次全国湿地资源调查结果显示，中国现有湿地自然保护区 577 处，湿地公园 468 处，共有 2324.32 万 hm^2 湿地受到保护，较 2003 年增长了 525.94 万 hm^2。目前正计划改革中国的保护地体系。很快，随着采用世界自然保护联盟（IUCN）的国际保护地分类体系，中国的湿地管理对策可能将拥有更大程度的灵活性。

2.2.1　国际保护地分类法和中国国内的保护地分类法

世界自然保护联盟国际保护地分类体系见表 2.2。

表 2.2　世界自然保护联盟国际保护地分类体系

<table>
<tr><th colspan="2">七类法（世界自然保护联盟）</th><th colspan="2">四类法（建议中国使用）</th></tr>
<tr><td>Ⅰa</td><td>严格自然保护区</td><td rowspan="2">Ⅰ类</td><td rowspan="2">严格保护类</td></tr>
<tr><td>Ⅰb</td><td>原野保护地</td></tr>
<tr><td>Ⅱ</td><td>国家公园</td><td rowspan="3">Ⅲ类</td><td rowspan="3">自然展示类</td></tr>
<tr><td>Ⅲ</td><td>自然纪念物</td></tr>
<tr><td>Ⅳ</td><td>陆地 / 海洋景观保护地</td></tr>
<tr><td>Ⅴ</td><td>栖息地和物种管理地</td><td>Ⅱ类</td><td>干预保护类</td></tr>
<tr><td>Ⅵ</td><td>自然资源可持续利用保护地</td><td>Ⅳ类</td><td>限制利用类</td></tr>
</table>

2.3　湿地法律法规

目前，不同部门已各自颁布了有关湿地和水路使用的法律法规。受国务院的委托，国家林业局负责全国湿地保护（包括野生动植物、湿地保护地、流域保护等），但其他部门的法规同样适用于全国湿地保护工作（表 2.3）。

表 2.3　其他主要负责机构

部门	职能与职责
环境保护部	污染问题、生物多样性保护问题的标准和总体协调
水利部	大坝、水库管理、调水和抽水及防洪
农业部	渔业，农业中的化学品使用，芦苇采割
国家海洋局	海洋渔业、海水养殖业、滨海湿地和海洋自然保护区
交通运输部	轮船、船只、河道和桥梁
国土资源部	挖取砂石，用作施工材料

注：2018 年 3 月，根据第十三届全国人民代表大会第一次会议批准的国务院机构改革方案，将环境保护部的职责整合，组建生态环境部；将农业部的职责整合，组建农业农村部；将国家海洋局、国土资源部的职责整合，组建自然资源部，自然资源部对外保留国家海洋局牌子

这种职权的交叉重叠，导致不同部门之间出现许多矛盾。通常，农业部门负责发放捕捞和芦苇采割许可证；水利部门负责水闸控制和取水总量控制；环保部门负责污染和水道排放监控；交通部门负责航道和交通运输的组织管理。所有这些活动都可能在同一处湿地内开展。甚至，自然保护区内也存在不同的利益冲突，保护区管理局必须保持游客参观、资源利用和物种保护等不同目标之间的平衡。

由某个部门颁发的法规可能仅在其自身负责的领域内有效。拥有不同职权的其他部门可能认为无须遵守这些法规。因此，林业部门可能规定不能在其保护区内开展开矿活动，但国土资源部门可能会签发勘探许可证，并在整个陆地景观中开展勘探工作。

1994 年颁布的《中华人民共和国自然保护区条例》规定每个自然保护区可以分为三个区，即受到严格保护的核心区，相对不能进行人为改变的缓冲区，以及可以进行人为改变的实验区。该条例遵照“人与生物圈”（MAB）计划的设计，但该计划的目的是研究人与环境之间的关系，而其本身并非为了保护自然。该条例令人困惑不解的一点在于：在许多国际自然保护计划中，缓冲区通常是经过高度人为改变的外部区，而非受到良好保护的内部区，而中国的区划正好反过来。此外，该条例对自然保护区的设计或管理对策的规定方面缺乏灵活性。因此，尽管事实上中国的许多自然保护区面临高强度的人为利用和改变状况，但没有适宜的分区，自然保护区也只有一种类型。

有关专家和部门已多次提出必须全面修订自然保护区条例，建立更多可允许的区划和基于世界自然保护联盟的分类体系进行保护地分类，但迄今仍未达成一致意见。与此同时，以其他名称命名的保护地（如森林公园、湿地公园、国家公园、国家重点风景名胜区等）条例也在制定过程之中，但是由不同级别（国家、省级、县级等）的不同部门未得到协调一致的方式来开展的。

2.4　中国湿地面临的威胁

中国湿地面临以下许多威胁。

2.4.1　栖息地的丧失

湿地的丧失不仅会导致引人入胜的自然遗产的丧失，还会对数亿人口的生命、健康、安全和福祉构成威胁，造成土地和财产数万亿美元的损失，并危及我们同样赖以生存的海洋的健康状况。

湖泊：依据 2003 年公布的首次全国湿地资源调查数据与第二次全国湿地资源调查数据，按可比条件分析，近 10 年来中国湿地发生了显著变化，湖泊湿地面积减少了 58.91 万 hm^2，减少率为 7.05%。《全国水资源综合规划》的数据显示，1950 年以来的半个世纪，全国面积大于 10km^2 的 635 个湖泊中，有 231 个湖泊发生不同程度的萎缩，湖泊萎缩总面积约 1.38 万 km^2（含干涸面积 0.43 万 km^2）。

沼泽和陆生湿地：与2003年公布的首次全国湿地资源调查数据相比，全国天然陆域湿地面积共计减少约1350万 hm^2，减少了28%。

潮间带：亚洲潮间带遭到破坏，使其成为全球生物多样性丧失速度最快的地区。这些潮间带不仅是数百种数百万只水鸟的重要栖息地，同时也是濒危海龟的筑巢产卵地、亚洲海豹的繁殖地，以及成千上万只甲壳动物、蠕虫和软体动物的栖息地。大量全球受威物种的生存依赖于这些潮间带栖息地的最明显证据来源于鸟类，尤其是水鸟，其中全球受威状况最为严重的水鸟包括涉禽（如琵鹭、鹤类）、水禽及海鸟等。

潮间带及其沙洲、海滩和红树林可提供许多重要的生态服务，而我们在那些资金实力雄厚的投资者所兜售的短期经济活动项目的诱惑下，只顾追求短期的发展目标，而忽略了这些重要的生态服务。尽管所涉及的潮间带开发地区总面积很小，但这些潮间带非常脆弱，并在快速消失。

永久冻土：中国东北和西北的最北部地区拥有许多含有泥炭的永久冻土湿地。在这里，地下冰层阻止了排水，并阻碍了树木的生长，从而形成生物特征独特的湿地类型。泥炭地本身是一个大型的碳存储区，必须始终保持湿润，才能避免分解和释放碳氧化物和甲烷等温室气体。因为人类活动的干扰和全球变暖的影响，这些永久冻土区的面积正在快速减小。

2.4.2　污染

中国工业、农业和建筑行业的发展，使得数百万人口脱贫，但同时也带来了严重的污染。

- 自20世纪80年代以来，采用化学品控制野草和病虫害的活动逐渐引入，并得到了当地政府的支持。农民通过采用这些农药，提高了农业产量，但同时也有越来越多的水生物种受到了负面影响。
- 近来发布的一份政府报告显示，农业活动占中国化学需氧量（水中有机化合物主要评测指标）的43.7%，占磷排放量的67%和氮排放量的57%。

2.4.3　引水

中国绝大部分地区缺水，对水的需求非常迫切。人们引用天然水道中的水，作为农业、工业和家庭用水。甚至小型溪流和森林覆盖的山地中的水源也被引入管道或水渠，将水从自然生态系统中引走。

大型引水工程包括南水北调引水工程。同时，成千上万个水坝阻隔了大型江河的流动，将它们隔断为许多梯级水库。

这些引水工程导致许多湿地水资源的丧失，同时也阻碍了许多水生动物（如鱼类、两栖动物等）的自由流动。为完成其生命周期，这些水生动物必须到达水温较冷的河流上游进行繁殖。

澜沧江 - 湄公河沿岸，特别是澜沧江 - 湄公河次区域上游地区将修建大量水坝，用于发电和灌溉目的。这些大坝将对部分洄游鱼类，如鲶鱼 *Pangasius sanitwongsei* 的生存构成很大的威胁。由于大坝对整个河流系统的平衡状况具有很大影响，要解决上述问题不仅仅是新建鱼类通道那么简单。现在，许多科学家都不认同在澜沧江 - 湄公河沿岸修建更多的水坝用于发电和推动该地区经济发展的观点。修建水坝的影响不仅将对西双版纳当地社区造成影响，同时也将对下游国家造成更大的影响。甚至可能出现柬埔寨洞里萨湖的高低水位的长期波动周期遭到干扰的情况。这种水位波动周期形成了世界上最为突出的淡水生态系统之一，拥有很高的生物多样性和生产力。

2.4.4　地下水位的降低

在许多地区，通过孔泵进行工农业引水或者从河流上游引水，已导致总体地下水位的降低。

1980-2008 年，中国的浅层地下水开采量从 557 亿 m^3 增至 1081 亿 m^3。北方地区增加的地下水开采量占中国增加的地下水总开采量的 90%。许多地区的实际地下水开采量超过了地下水开采量的最高允许限度，导致地下水位的持续下降，形成许多环境和地质问题。地下水的过度开采问题在北部地区的平原地带尤为集中。

此外，种植部分极需水分的树种（如柳树、杨树等），也可能导致当地地下水位的降低，并与湿地争夺急需的水资源。

2.4.5　盐化作用

淡水水体受到高盐分污染时，就会发生盐化作用。盐分可能来自海水、内陆土壤或者灌溉水中的盐。

河口是流经河流的淡水与潮汐海水相交汇的地方。河口的盐度水平从纯淡水到纯海水及微咸的中间带不等。中间带流经河口，并随着海潮涨落。由于上游用水量或引水量较高导致的淡水流量减少，同时气候变化导致的海平面上升，流经河流的盐分及流入滨海地区、滨海水井和饮用水供水水源地的盐分也将随之增长。

在内陆地区，由于地表水的含盐浓度随着蒸发作用而增加，渗透作用会将盐分从地下伏土层中析出。因此，世界上太阳能非常丰富的干旱地区拥有许多盐沼和盐湖。在中国，青藏高原也拥有许多盐湖。受气候变化的影响，盐湖的面积还在不断扩大之中。

如同许多浮游生物、甲壳动物、软体动物和蠕虫一样，部分高盐耐受性的植物（如红树林和部分藻类）可以在含盐环境下茁壮生长。这些含盐栖息地通常支持其他觅食这些植物的特有生物群，如涉禽、鸭类等。

2.4.6　富营养化

富营养化（图 2.1、图 2.2）是湖泊从周边流域吸收过多养分（氮磷物质）和沉淀物，从而营养度变高和水位变浅的过程。过多的磷素会刺激藻类和部分有毒浮游植物的快速繁殖。藻类可能增加水的浊度，阻塞水道，导致鱼类和鸟类在栖息地窒息等。当水华时，过度的细菌作用会大量消耗氧气，可能导致鱼类和其他生物的死亡，造成危险有毒细菌和病毒的大量传播。

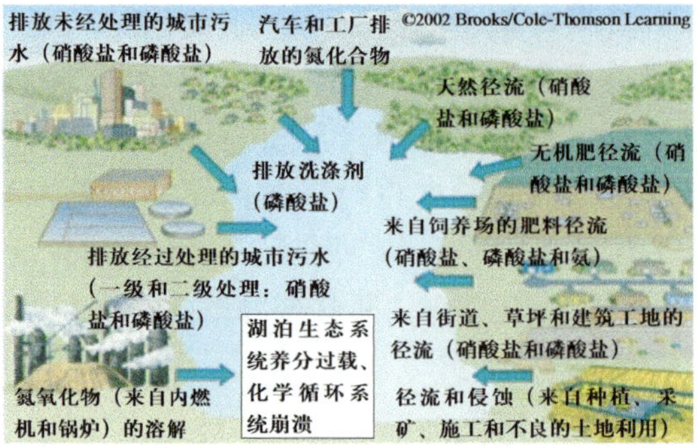

图 2.1　富营养化过程

图 2.2　士兵们正在清理山东沿海中的富营养化海藻

管理不善或失灵的化粪池系统、含磷洗涤剂、含磷化肥、农业活动和动物饲养场可能导致磷、氮和其他养分进入湖泊。在中国，过度施用化肥及污水的不当处理，是造成水体营养过剩的主要根源。营养过剩会造成水体中浮游植物的增加，并往往形成有毒的水华。现在，这些周期性的有毒水华现象越来越多地侵入周边地区。

水体中的氮素过多也会导致沿海水域出现类似的水华现象，破坏鱼类繁殖区和海草床，触发海蜇瘟疫或动物的其他变化。现在，中国东部海岸每年都会受到巨大水华的影响，与此同时，中国几乎所有的淡水湖都出现了富营养化现象。《第一次全国污染源普查公报》政府报告显示，农业活动的主要水污染物排放量总磷占 67%、总氮占 57%。

要将富营养化的湖泊恢复至贫营养的湖泊是一件几乎不可能完成的任务。不过，通过减少水体中的新增养分和沉淀物，可以减缓湖泊富营养化的过程。

当湖岸或溪岸受到侵蚀时，过剩的土壤将被冲刷到水体中，导致水的浊度上升，使得水草和下沉在湖底的沉淀物光合作用的效率低下，从而导致湖泊水体形成淤泥质且更加不稳定。建议对湖岸或溪岸状况开展调查，确定遭受侵蚀影响的地区，以及为改善这些地区需采取的必要措施。最为有效的改善技术是通过种植乡土植物来保护湖岸线。

2.4.7　氧气的丧失

通常，水生动物（如鱼类、水生昆虫、甲壳动物、蠕虫和软体动物等）需要氧气才能呼吸，因此水体中的氧气含量是决定一块湿地能够支持的动物数量的关键因素。水生植物的光合作用及水流的流动或落水形成的物理搅动（如人工曝气装置）会导致氧气释放到水中。与温水相比，冷水可储存更多的氧气。增加水温的活动，如遮阳物的丧失、水流速度的减缓及气候变化往往会降低特定水体中的氧气含量。同时，污染、富营养化、淤积作用或过牧也会导致水生植物遭到破坏甚至死亡。在死物质的分解过程中会消耗氧气。

只有水体始终保持较高的氧气含量，才能保持较高的动物丰度。因此，管理人员应避免或最大限度地减少那些可能降低水体中氧气含量的因素。

2.4.8　外来入侵物种

多个国际机构将外来入侵物种列为除栖息地丧失和遭到破坏，对全球生物多样性威胁最大的因素。

外来入侵物种是指在新的地区站稳脚跟并传播，给当地环境或本地物种造成破坏的非本地物种。

中国尤其容易受到外来入侵物种的影响。这是因为中国现已成为一个全球贸易大国，大量的人员和商品在世界各地流动，为外来入侵物种提供了各种潜在的环境条件，很多的外来入侵物种可以成功地在中国找到立足之地。中国的陆地景观正在经历快速的变化，为许多新物种的入侵提供了大量空间。

由于人们对外来物种的意识和识别物种的能力总体较弱，导致各种新物种的大量传播，最终带来危害。

当外来入侵物种过度扩展，就会出现各种问题，如破坏或者超过本地物种、破坏物理环境或不可持续地消耗有限的养分或资源。

部分外来入侵物种已给当地造成了巨大的经济损失。这些外来物种包括入侵中国南方稻田的福寿螺、对中国北方森林造成破坏的天牛甲虫、阻塞中国许多水道并导致许多湖泊和河道氧气含量减少的凤眼蓝、已成为南方病虫害的红火蚁。同时也出现越来越多的农业杂草，如马缨丹、泽兰等。

包括动植物在内的外来入侵物种的传播，正导致中国湿地的日益退化。部分外来入侵物种的大批繁殖生长纯属偶然，而许多外来物种的引入是有意而为之的，主要用于园艺装饰、作为经济作物或新的食物种类。

过去数十年来，部分水生物种，特别是罗非鱼、南美大盖巨脂鲤和南美福寿螺被引入中国南部地区。这些外来物种的经济或生态影响从未得到认真的评估，可能与部分本地的水生物种形成竞争关系，甚至导致这些本地物种的灭绝。

目前有几百种外来物种已被证实对中国的湿地具有破坏性。这些外来物种的清除或控制往往耗费巨大的人力、物力。表 2.4 列出了导致中国湿地退化的部分主要问题物种。本书 5.9 节详细阐述了控制外来入侵物种的方法。

表 2.4　中国湿地的主要外来入侵物种

中文种名	学名	来源地	中国受影响地区	导致的问题
凤眼蓝	*Eichhornia crassipes*	南美洲	南部和中部地区	浮草会堵塞湖泊和河道，妨碍船只的通行，降低水中氧气的含量，破坏渔业的发展
大薸	*Pistia stratiotes*	南美洲	南部和中部地区	传播速度很快，会阻塞水道，降低水中氧气的含量，导致鱼类和甲壳动物的死亡
空心莲子草	*Alternanthera philoxeroides*	南美洲	绝大部分地区	入侵水生和陆地栖息地，导致其他草本植物窒息而死，要控制空心莲子草代价非常昂贵
互花米草	*Spartina alterniflora*	北美洲	东部和南部沿海地区	浓密的灌木丛，会导致本地滨海植物窒息而死
马缨丹	*Lantana camara*	南美洲	绝大部分地区	灌木丛取代本地物种
仙人掌	*Opuntia*	南美洲	南部地区	灌木丛取代本地物种
巴西含羞草	*Mimosa pudica*	南美洲	南部地区	多刺灌木丛蔓延
斑马贻贝	*Dreissena polymorpha*	通过美国进入俄罗斯	东南沿海地区	阻塞排水渠；禽流感肉毒杆菌中毒的来源
非洲大蜗牛	*Achatina fulica*	东非	南部地区	破坏作物和果园
福寿螺	*Pomacea insularum*	南美洲	南部和中部地区	破坏水生植被和稻田
克氏原螯虾	*Procambarus clarkii*	美国东南部	南部和中部地区	损害防洪堤，并与本地物种进行竞争
莫桑比克口孵非鲫	*Oreochromis mossambica*	南部非洲	东南部和河口地区	破坏底栖底质，吞食其他鱼类及其食物
美洲巨蟾蜍	*Bufo marinus*	通过澳大利亚传入南美洲和中美洲地区	南部地区	吞食本地两栖动物
牛蛙	*Rana catesbeiana*	北美洲	南部和东部地区	数量超过并吞食小型的本地特有两栖动物
水貂	*Martes vison*	北美洲	东北和西北地区	数量超过本地的紫貂和水獭，破坏渔业和松鼠种群等
海狸鼠	*Myocastor coypus*	南美洲	南部和东部地区	大型食草动物，导致栖息地的状况发生转变

2.4.9　过度捕捞和狩猎

捕猎鸟类、哺乳动物、爬行动物和两栖动物是湿地生物多样性面临的一个主要威胁因素。任何可以用来出售或食用的野生动物都是狩猎者或偷猎者的目标，狩猎者或偷猎者有时会使用一些极其短视和贪婪的捕猎方法。中国一半的湿地未得到明确的保护，即便那些被正式确定为自然保护区的湿地也没有得到足够的巡护和保护，因此当地的农民或渔民很可能设置圈套，甚至在一些更加偏远的地区使用枪支来捕猎这些野生动物。

当地的农民和偷猎者经常使用罗网、弹簧陷阱和鸟网来捕获鸟类。部分珍贵的鸟类可能在宠物市场上出售，其他一些鸟类则可能在其他市场上出售，最终被当作食物。

当野生的雁鸭或鹤在农田觅食时，当地的农民有时会撒下有毒的谷物或诱饵来诱杀这些鸟类。中毒的野生鸟类甚至被当作食物出售到市场上，一些人不小心购买到这种鸟类也因此而丧生。

湿地哺乳动物（如水獭、河狸和獐等）也会遭到捕猎。此外，中国对食用蛇、龟和青蛙的需求量也非常巨大。

迁徙物种在其迁徙路线的沿线也经常面临罗网和狩猎者的威胁，可能在远离湿地保护地的地方遭到捕杀。

破坏性“现代化”捕捞方法造成的影响。过去十年里，当地农民开始利用电、炸药和化学毒物来捕鱼。尽管《中华人民共和国渔业法》禁止使用上述这些方法捕鱼，但许多人仍在小型河流及支流中使用上述方法捕鱼，有的人甚至在澜沧江等大型江河和湖泊中捕捞鲶鱼等大型鱼类。

《中华人民共和国渔业法》的颁布及当地政府为保护当地河流所颁布的各种规章制度，可以说地方政府为保护水生资源和控制非法捕捞做出了巨大的努力。不过，由于这些法律和规章制度在农村地区的执行力度较为欠缺，这些非法捕捞方法仅在少数地方得到有效的控制。

2.4.10　风轮机、线缆、建筑物和机场

中国的大型机场也对迁徙鸟类造成巨大危害（图 2.3- 图 2.5）。机场跑道设置鸟网以保护飞机免受飞鸟撞击，每年这些鸟网导致大量飞鸟死亡。而这些死亡的鸟类并不一定是会对飞机造成危害的鸟类，如大型猛禽、天鹅和雁鸭类或大群的鸥类等。其他国家虽然不使用鸟网，但也拥有其他驱鸟技术，如利用猎犬、受过训练的猎鹰、巨大的噪声等。在中国，邻近国际重要湿地，如上海浦东和海南海口的机场，上述问题尤其值得关注。

图 2.3　飞过风轮机的鸟群

图 2.4　被风轮机绞死的鹤

图 2.5　被风轮机绞死的鹰

尽管当地主管部门向人们保证风轮机不会对鸟类造成威胁，但大型风轮机确实会导致大量的大型鸟类死亡。许多国家尽力避免在重要鸟类的筑巢地附近建立风场，同时英国皇家鸟类保护协会（RSPB）发布了专门的指南，控制此类装置对鸟类造成的危害。

灯塔和输电线也会对大型鸟类造成危害。许多国家在输电线上放置易于看见的物体（图 2.6），以警告鸟类不要飞入这些物体。2005 年，鸟类与电缆线互动委员会（APLIC）和美国鱼类及野生动植物管理局（U.S. Fish and Wildlife Service）颁布了《鸟类保护计划（APP）指南》。鸟类保护计划指南提供了相关资源，帮助公用事业部门管理飞鸟 / 输电线问题（图 2.7）。

图 2.6　靠近电网的报警灯

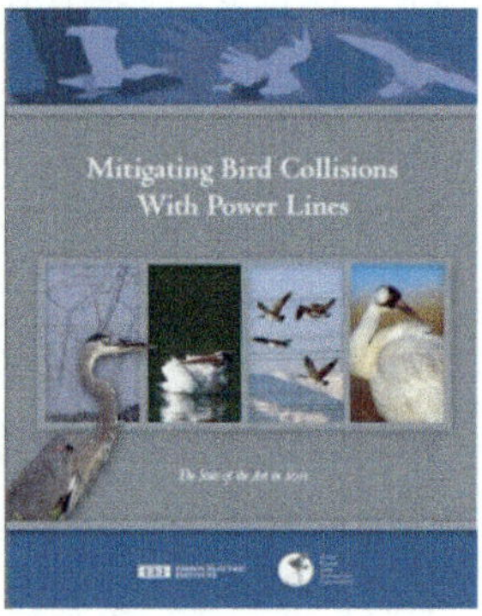

图 2.7　国际电线指南

由于中国许多电线的电缆之间的距离过小，一些大型鸟类可能接触到两条电缆，从而触电而死。必须加大横梁的宽度，以增加电缆之间的距离。

2.4.11　机场和鸟网

高灯塔和拥有大型玻璃窗的大型机场建筑也会导致那些天真幼小的迁徙鸟类死亡。这些鸟儿天真地认为自己可以穿过这些玻璃窗，不幸撞击而死。例如，在鸟类的迁徙季节，每天昆明机场的灯塔和建筑物的玻璃窗会导致数只鸟儿撞击死亡。

不过，更具破坏性的是，中国机场通常使用鸟网来减少喷气式飞机在起降过程中可能遭受鸟类撞击的事故（图 2.8）。

图 2.8　利用鸟网来减少机鸟撞击

中国绝大多数的机场会在主跑道的整个沿线至少设置一排鸟网。在驱鸟方面，鸟网并不十分有效，其他国家都不使用这种方法。在国际上，普遍使用的更为先进和破坏性更小的许多驱鸟方法包括利用猎犬、猎鹰、尖锐发声法和闪光体等。许多机场利用模仿猎鹰的遥控模型飞机或使用小型无人机将鸟从跑道上驱走。

与保护作物的农民、保护鱼池的渔民及想捕获鸟类的非法狩猎者所杀死的鸟类数量相比，跑道上安装的鸟网杀死的鸟类数量可能相对较少。不过，当这些人看到有关部门随意利用这些危险的驱鸟工具时，要控制鸟网的公开使用就非常困难。

2.4.12　气候变化

气候变化是事实，这一点无可争议。数据本身说明了一切。近年来的气温是有史以来最高的。海洋温度不断上升，地表温度也在不断上升，冰川和极地冰原正在融化，海平面不断升高。在全球范围内，极端气候事件日益频繁和严重，天气模式更不稳定，难以预测。

科学证明，大部分的气候变化与大气中二氧化碳浓度的上升有关。目前大气中的二氧化碳含量是过去3万年间最高的，并且绝大部分是由人为活动引起的。《联合国气候变化框架公约》极力希望各国达成国际协议，将碳排放量控制在一定水平上，在21世纪末前将全球平均气温控制在不超过工业革命前的平均气温2℃的水平上。不过，绝大多数的气候科学家对该目标的实现持悲观态度。目前，包括中国的高原和高山地带在内的全球部分地区的气温上升幅度已突破了2℃，超过全球平均水平。

气候变化对湿地及其生物多样性的影响将非常巨大，目前这种影响已经得到显现。事实上，全球变暖所带来的许多后果已不可逆转。今后，即便我们完全停止碳排放，这种后果也将持续下去。

湿地管理者将面临更多的暴风雨、台风、洪灾和旱灾，更多的热浪、寒潮；植被区和物种分布将会出现更多的变化，植物开花结果模式的季节性、昆虫的出现和物种迁徙的时间也会出现更多的变化，湿地管理者必须为之做好规划。在面临此类变化时（图2.9）要保持湿地生态系统的恢复力，将是一个巨大的挑战。

2.4.13　海平面不断上升

随着海洋温度的上升，海洋的面积也在不断扩大。极地冰块的融化进一步导致海平面的升高。科学家普遍估计，到21世纪末海平面的上升幅度将在1.0-2.0m。这将对海岸线、滨海湿地、红树林和小岛的状况带来巨大影响。中国的许多大型城市，如上海、天津和厦门等位于亚洲沿海地区，很容易遭到海水淹没。

海平面上升带来的最大影响将出现在海南、江苏及其长江河口地区。在海南，几乎所有的现有红树林将被海水淹没，不过，可以在现有的高潮线沿线建立新的红树林地区。在长江河口地区，大面积的土地将被海水吞没，部分长江下游湖泊的排水状况可能发生重大转变。

图 2.9　海啸的后续影响

海啸导致村庄被毁，船只被冲上海岸

江苏盐城的大型滨海湿地可能被淹没并相对较早地被冲走。

2.5　必须认识整个生态系统

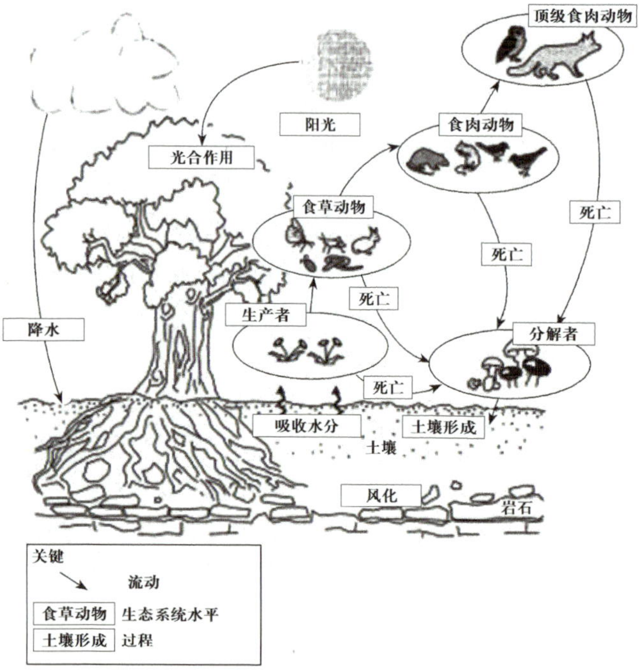

图 2.10　生态系统的能流路径

生态系统是生活在一起并相互适应的物种群落及其物理环境的综合体。每个物种都需要整个生态系统，而整个生态系统也需要所有的组分物种。在任何一个生态系统内，都有能流、食物链和各种营养级（图 2.10）。

绝大多数生态系统的初级能量来自于阳光，通过绿色植物的光合作用过程产生能量。植物叶中的叶绿素将阳光和水、二氧化碳综合在一起，形成糖分并释放氧气。动物在呼吸的过程中会吸入氧气。绝大多数的动物是食草动物，吃食某些绿色植物，并通过消化这些植物获得食物能量。而食肉动物通过吃食其他动物——猎物的肉来获取能量。

我们将食物的不同级别称为金字塔营养级。一个简单的生态系统可能只有三个不同的营养级，而一个复杂的生态系统可能有许多营养级。

除了食用植物外，部分动物可能在生态系统内部发挥其他重要作用。部分物种作为授粉者，另外一些物种则可能作为种子传播媒介。部分物种承担重要的控制或平衡作用。因此，狐狸吃食老鼠，防止老鼠数量过多。老鼠数量过多会吃食所有的谷物种子，导致其自身的食物供应链崩溃。

土壤中看不见的物体包括生态系统的其他重要组分。蠕虫会采集和消化枯死的树叶，帮助建立和提高土壤的肥力。其他分解者，如真菌、等足目动物、线虫和细菌会分解死物质，循环利用养分和基本矿物质，为绿色植物维持土壤肥力，以便绿色植物可以捕获更多的太阳能，从而为整个系统提供能量。

疾病在控制生态系统的平衡方面发挥着重要作用。一个物种越常见，越容易遭受来自同类另外一个物种疾病的传染。因此，疾病和寄生物作为密度制约因子，当某个物种数量过多时，疾病和寄生物对该物种产生的不利影响更大；但在该物种的密度较低时，疾病和寄生物的影响则相对较小。

自然生态系统依赖于所有这些功能，包括初级生产者、基本食草动物、顶级食肉动物、授粉者、种子传播者、分解者、固氮植物和疾病调解者等。一个生态系统越复杂就越稳定，在捕获和转换太阳能方面也更为有效，生产力更高。

湿地管理者必须将其管理的湿地视为一个更大生态系统的组成部分，必须确保周边地区的基本功能得到维护，以维持水质、土壤肥力、植被健康和野生动物的安全，维护整个系统的高生产力和重要生态系统服务的高效提供。

湿地管理者可能必须将保护范围扩大至超过其自身湿地边界限制的其他生态系统，这是因为，湿地可能依赖于来自更大范围的周边景观的授粉者和吃食昆虫者；而湿地显而易见依赖于来自其边界以外地区的水源，并可能需要来自湿地边界以外地区的树木提供保护，保护湿地免于遭受暴风雨的影响，或者作为许多湿地物种的栖息地和筑巢地。我们将这种保护地管理的整体途径称之为“景观”途径。

2.5.1　多样性意味着安全性

由少数几个物种组成的单一生态系统非常脆弱。异常的旱灾、洪灾或疾病和寄生物的出现，可能很容易导致这种单一的生态系统崩溃，使其生态系统服务完全丧失。复杂的生态系统可以保持更好的生态平衡，更有效地利用能源和提供生态系统服务，在出现生态变化时更具有适应力和恢复力。鉴于湿地生态系统的动态属性及当前快速变化的气候条件，维持湿地的高生物多样性是保持湿地健康和安全的最佳方法。

上述理论同样适用于人工生态系统，如人造林和农场等。单一种植风险很大，而多元化种植更加安全。例如，2008 年春，一场罕见的暴风雪席卷了中国南方地区，给当地的竹林、森林和农作物造成了巨大的破坏。一些单一的竹林几乎全部遭到毁坏。与单一种植的人工竹林相比，农民在混交林系统中种植的竹林所遭受的损失要小得多。

2.6　迁徙物种的特殊需要

全球或地区性的大部分迁徙鸟类会在繁殖地以外地区，在时间安排和目的地可预测的前提下进行周期性的、有规律的移动。

迁徙水鸟是指在生态上依赖于湿地的迁徙鸟类。迁徙水鸟包括涉禽（如鹤类、鹭类、朱鹮）、雁鸭类、鹈鹕和海鸟（如潜鸟、鸬鹚、鸥类、军舰鸟、热带鸟等）及其他水鸟群。

大量的迁徙水鸟在迁徙过程中，特别是在跨越大型生态屏障前，经常会在停歇地聚集，重新补充能量。因此，湿地停留栖息地的丧失，可能对迁徙水鸟的成功迁飞和生存造成重大影响。

图 2.11　内蒙古向上游洄游的鱼类

洄游鱼类竞相游往河流上游，希望更快到达冷水繁殖地

对于那些未被专门认定为湿地物种的其他迁徙鸟类，含有湿地或本身被认定为湿地的区域也可能是它们重要的停留地、繁殖地或越冬地。

除鸟类外，迁徙物种还包括：迁徙哺乳动物，如鲸、海豚、驯鹿、蝙蝠等；洄游鱼类（图 2.11），如鲑鱼、鲟鱼、鳝鱼、鲨鱼、金枪鱼和许多远洋物种；可能与特定的湿地管理需求相关的两栖动物甚至昆虫等。

2.7　对连通性的需求

连通性是确保单个湿地组分物种的集合种群与其各自的物种总种群在基因方面始终保持关联的重要特征。如果没有这种连通性，单独的种群将出现基因近亲繁殖，在发生随机性或大规模灾害时将无法在湿地再次大量繁殖。

许多物种需要在不同的湿地或不同的生境类型之间进行季节性迁移。小的迁移活动如青蛙从周边的林地移动至某个池塘交配产卵；大的迁移活动如鲑鱼等海洋鱼类的迁移，它们必须到达冰凉的淡水区，在河流上游产卵。

迁徙层级的最顶端是许多迁徙水鸟，这些水鸟每年从其位于北半球北极的繁殖区飞越千里，到达位于亚热带地区甚至远至南半球的澳大利亚和新西兰等越冬觅食地。这些水鸟在迁徙途中需要停歇地来重新补充能量或短暂休息。

如果这些迁徙停歇地丧失，将对迁徙物种带来巨大压力，迫使它们过度集中在现存的部分湿地，那些日益稀缺和遥远的停歇地之间的距离也越来越远。

对于那些必须沿河道上下洄游的鱼类来说，水坝、堰和其他屏障可能对它们构成严重的阻碍。

当原生物种群落相继在本地灭绝，无法通过再次大量繁殖而重建群落时，孤立的小型栖息地斑块将无法保护那些具有存活力的种群，物种丰度也将不断降低。

2.8　实现可持续利用

2.8.1　“源”和“汇”

渔业资源增长的速度快于鱼类的自然死亡和人类捕捞采集速度的地区可划归为“源”区。“源”区可产出不断增长的鱼类种群，这些种群可分散在各个地方产卵、繁衍，或者促进很远地方的周边种群的发展。相反，渔业资源的死亡率和人类捕捞采集的速度超过其自然增长率的地区称之为枯竭或“汇”区。“汇”区要保持其自身的生产力，必须依赖于周边“源”区渔业资源的定期不断补充。

由于绝大多数的湿地现已成为渔业资源的“汇”区，因此确定“源”区（如重要的鱼类繁殖区）并对之开展更好的保护非常重要。我们可以在溪流、湖泊、红树林、珊瑚礁和其他海洋地区确定关键的渔业“源”区。

2.8.2　最佳/最大可持续渔获量

根据水质、食物可获取性和各种死亡率因素（捕食、疾病等）的不同，天然水体拥有不同的最大渔业资源承载力。人们可以通过捕捞来利用渔业资源。如果渔获量一直较低，渔业资源的捕捞量就不会超过其自然承载力。但是，如果渔获量太高（过度捕捞），捕捞量超过其自然增长速度，鱼类种群就会崩溃。最大可持续渔获量（MSY）是指渔业资源仍可保持其最大增长率的最高捕捞水平。由于在这个点上，捕捞率等于增长率，因此可实现可持续发展。最大捕捞率通常为渔业资源承载力的30%-50%。世界各地的许多渔场都利用这一方法来确定可最大限度获得捕捞效益的适宜限额。最大可持续捕获量$(H)=\dfrac{Kr}{4}$，其中，K为最大种群或承载力，r为内在增长率。

不过，许多生态学家和自然保护主义者批评该方法的使用，并认为这种方法导致了鱼类种群的崩溃。因为该公式并未考虑到各种关键要素，如鱼类的年龄结构、繁殖能力、副产品的间接后果或整个大的环境等。此外，还存在一种风险，即一旦市场对渔获量存在一定的需求量，要使实际的渔获量不超过既定的渔获量就将非常困难。在一个情况糟糕的年份，这可能导致对鱼类种群的长期影响。因此，现在全社会更加认同的理念是力争实现“最佳可持续渔获量”（OSY）的目标。最佳可持续渔获量可最大化总收入与总成本之间的差额，即边际收益等于边际成本时。最佳可持续渔获量可最大化所利用资源的经济利润或租金，并通常小于最大可持续渔获量。在环境科学中，最佳可持续渔获量是指在不削弱鱼类种群或其环境支持最大经济渔获量的能力的前提下，长期可实现的对可再生渔业资源的最大经济渔获量。

计算出可维持最大经济渔获量的水平是一回事，而要确保渔民团体将其捕捞量控制在这一水平上则是另一回事。这就是一个典型的“公地悲剧”案例。人们已尝试利用多种方法来解决这一问题。

2.8.3 议定限额

如果通过对照监测渔业资源的种群数量与渔获量之间的情况，可以测算或推算出最佳可持续渔获量，那么，所有利益相关方可以就总的可允许渔获量限额达成一致意见，并以某种可以接受的方式共享这一限额。不过，只有在所有利益相关方就这种限额达成一致意见（利益相关方本身是一个权利有限的群体），并且建立了某种机制来制约每个利益相关方所承担的限额，这种议定限额才能真正发挥作用。要实现这一目标，必须拥有高水平的合作和监管。

逆向限额制度可能更加易于操作。逆向限额制度包括就适当的存活量达成一致意见，允许一定数量的成年鱼类个体持续繁殖，或者允许每年有一定数量的鱼卵和幼鱼得到抚育。只要这种存活量得到保证，就无须进一步监测其余的捕捞状况。例如，海龟孵化和补充育苗计划中可能存在这种逆向限额制度。

2.8.4 禁捕季节 / 禁渔区

禁捕季节是实施禁捕逆向限额制度的一种方法。严禁在某个关键的时期，通常是鱼类的繁殖季节捕捞特定的鱼类资源，以便为河道提供充足的时间来补充鱼苗。

2.8.5 鱼类大小限制

实行鱼类大小限制，是确保一定比例的鱼类种群在每个季节得到捕捞的另外一种方法，并确保足够数量的幼鱼个体能够存活，成为未来的繁殖者。这种方法通常用于管理休闲渔业。在休闲渔业中，即便必须将所有体型偏小的鱼类丢回到水域中，人们仍然可以享受到捕捞小鱼的乐趣。

2.8.6 网眼大小

有关特定渔场可以使用的渔网类型，特别是网眼大小的规定，在确保幼鱼或体型偏小的鱼类或非目标物种的低死亡率方面极其有效。不过，某些渔网的确对一些珍稀受保护物种，如潜鸟、海洋哺乳动物和海龟等非常危险。在确定在保护区可以允许哪些捕鱼方法和渔具时，湿地管理者必须非常慎重。

2.8.7 严禁使用的捕捞方法

中国法律严禁采用毒药、电捕和炸药等方法捕鱼，但许多地方执法力度相对较弱，导致上述捕鱼方法仍然长期存在。

在现实生活中，主要存在两种类型的毒药捕鱼方法：

1）潜水员将有经济价值的食用鱼、章鱼或观赏鱼驱赶到珊瑚礁的洞里，然后用塑料瓶和喷嘴向洞里喷射氰化钾。这样可将要捕捉的鱼类弄晕，从而可以很容易地将它们从藏身之地打捞至捕鱼袋或篓子中。

2）渔民搅打或削下部分有毒植物，如毛鱼藤（*Derris elliptica*）或滨玉蕊（*Barringtonia*

asiatica）的根、叶或果实，然后将其圆滑的纤维丢入水池、池塘或其他流动缓慢的水体中。毒物通过水和鱼鳃缓慢扩散，导致鱼类处于瘫痪或呼吸困难的状态。然后，鱼会漂浮到水面，从而可以很容易地捕捉到，或者漂流到下游，被其他渔民或渔网捕捉到。这是一种非常有效的捕鱼方法，不仅可用于科学研究、用来评估鱼类种群数量，也可用于渔业管理活动，用来消除过多的或不想要的捕食性鱼类或病鱼等。不过，这是一种不可持续的，也是一种极度自私的捕捞方式，在中国和绝大多数的国家作为一种商业捕捞方法已遭到禁止。

此外，其他一些不可持续的捕捞方式也导致鱼类资源量的快速减少，对渔业环境造成长期破坏，或杀死许多非捕捞对象的物种，这些捕捞方式也在当地法规中或特定地方遭到禁止。此类方法包括使用细筛网、刺网、拖网捕鱼等。

2.8.8　亚洲的部分传统捕捞方法

东亚地区使用的部分本地特有的传统捕捞方法包括：冰下捕鱼，使用鸬鹚、植物毒剂、科钦捕捉器、海上捕捉器捕鱼及驾船驱赶海豚等（图 2.12）。这些捕捞方法都可以可持续的方式进行，但也可能强度过大，导致过度捕捞。

图 2.12　东亚地区使用的部分传统捕捞方法

如捕蟹网（左上）、科钦捕捉器（右上）、冰下捕鱼（左中）、驾船驱赶海豚（左下）及简单的竹子捕鱼工具（右下）等鲸鱼和海豚属于温血哺乳动物，受到公众的喜爱。是否杀死它们在动物权利及濒危状况保护方面仍存在很大的争议。各种国际协定，如国际捕鲸委员会（IWC）制定的条约对鲸鱼和海豚的利用状况做出了规定。

2.8.9　过度捕捞的迹象

过度捕捞的典型迹象包括：总渔获量的下降、单位捕捞作业的渔获量下降、被捕捞鱼类的平均尺寸减小、高价值鱼类在总渔获量中的比例下降、关键鱼种的丧失及鱼类指示物种（食鱼鸟类等）的减少等。过度捕捞可能导致恶性循环不断加剧。随着鱼类资源的减少，渔民会极力通过加大捕捞强度来弥补其收入减少的状况。这与矫正过度捕捞所需要采取的行动完全相反，将导致鱼类资源更快地减少。

2.9　保护区设计中的生物地理原则

对生活在不同面积海岛中的鸟类和生活在不同面积珊瑚礁中的鱼类开展研究，使得生物学家们意识到，许多地理原则将决定某个特定土地单元或水体中的生物多样性的丰度。鉴于单个的自然保护区在丧失或保存物种方面的方式与其他岛屿大致相同，这些岛屿生物地理原则可用于自然保护区和自然保护区体系的设计。

通常，面积较大的岛屿或土地单元包含和保留的物种数量多于面积较小的土地单元，而保存完好的栖息地单元保留的物种数量超过破碎化的栖息地或保护地。

一个基本的原则是：当某个特定保护地本地物种的灭绝速度与新物种在该保护地定殖或重新定殖的速度相当时，该保护地就可保持物种丰度的平衡。

可使物种灭绝率低的因素包括：保护地面积大、栖息地多样、邻近其他保护地或与其他保护地相连接、气候温和、水量充足等。

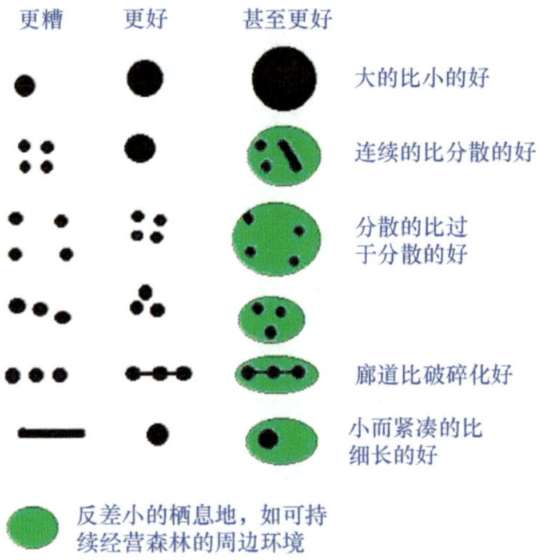

图 2.13　自然保护区设计中的生物地理原则（MacKinnon et al.，1987）

可使外来物种定殖率低的因素包括：保护地面积小、地理位置偏远、栖息地多样性低、气候恶劣或周期性缺水等。

面积较小的孤立岛屿或自然保护区往往更容易丧失大型动物、营养级高的捕食者和陆地动物，而非分布更广的小型动物、飞禽或游泳动物。保护好某个保护地营养级最高的食肉动物，是本地生态健康状况良好的一个指标。

图 2.13 是展示如何将生物地理原则应用于保护地设计中的一个范例。湿地管理者在增加保护地面积方面的对策通常有限，但在增强水道和其他栖息地（如森林或草地）廊道的连接性方面，往往可以有许多选项。

2.10　减缓气候变化

当前，人们日益将气候变化视为对湿地的威胁因素（参见2.4.12节）。人类持续燃烧利用化石燃料已大大改变了我们的大气状况，使得大气中的二氧化碳含量增长了一倍，同时增加了其他温室气体，如甲烷、一氧化二氮的含量，使得保护性臭氧层减少。所有这些变化已对全球气温和气候造成了显著影响。湿地也正受到以下几种方式的不利影响。

2.10.1　极端天气事件

全球总体变暖导致植被区和动物分布的重新调整。湿地管理者必须预先考虑到新物种的入侵及其他物种的消失状况。例如，许多鱼类必须到冷水中繁殖。但是，此类栖息地可能在特定的流域中消失。在气候总体趋势转向全球变暖的同时，气候变化变得日益极端，因此将出现更多的寒潮、热浪、水灾和旱灾。沿海地区将经历数量更多、强度更大的暴风雨、飓风和潮汐风暴潮。

2.10.2　永冻土层

中国最北端、青藏高原高地和部分山地在被称为永冻土层的土壤表面下拥有永久性冻结水。地下永冻土层受到的影响，将对中国最北部地区的植被、湿地的分布和性质产生重大影响。

气候变化与永冻土层之间的相互关系错综复杂，有时甚至相互对立。森林砍伐、森林面积的减少、土壤和植被覆盖的减少，都将**削弱**大气层的隔热性能，导致地球上的更多热量流失到大气空间中，导致地下冰块总质量的**增加**，降低非冰冻土壤的季节性深度。永冻土层的增加，将阻碍树木（甚至包括根系浅的落叶松）的生长，妨碍地面排水，最终导致沼泽湿地面积的**增加**。

另外，气候变化正导致地下冰块的快速融化。这种状况通常会导致以下局面：

- 部分湿地草甸上树木更快地定殖（能够在非冻土上扎根）
- 排水状况更好，因此导致部分湿地水位的**下降**
- 土地剖面的塌陷（冰块比可流走的水拥有的水容量更大）
- 冻结甲烷气体的释放。甲烷是一种温室气体，可导致气候变化的进一步加剧

不过，在部分地带中，尽管融化后的永冻土层释放更多的水量，但其排水状况仍然较差，这可能导致水位增加及湿地面积的**增加**，或者导致**新**（或称新生）湿地的出现。长期冰封的天然泉可能突然之间再度开始流动。

监测永冻土层、湿地草甸和沼泽及柳树（*Salix* spp.）、桦树（*Betula* spp.）和落叶松（*Larix gmelinii*）再定殖的变化状况非常重要。为保持草地、森林与湿地之间的最佳平衡，可能必须采取部分干预措施，如加快或减缓排水的速度。

随着海洋的变暖，海洋的面积也将扩大，同时其能量也将增加。这将导致海平面

的上升及海潮力度的加大，从而侵蚀和淹没沿海地区和红树林。随着气温的继续增长，极地和格陵兰岛的大冰原开始融化，导致海平面的进一步上升。尽管人们对长期海平面上升速度的估算存在差异，但预计至少将上升几米，许多沿海城市（如上海等）可能被海水淹没。在 21 世纪内，尽管我们努力将全球变暖水平控制在比工业革命前的温度不超过 2℃的范围内，但海平面上升的幅度仍可能超过 1m。

2.10.3 珊瑚白化

随着更多的二氧化碳释放到大气中，许多二氧化碳也被海洋捕获，导致海洋酸化，阻碍了珊瑚的正常生长。珊瑚礁由于酸化而濒临死亡。在气候变化条件下，其他生物快速生长，水母和有毒浮游生物面临的瘟疫也将不断增加。

2.11 辨认主要的水鸟

湿地鸟类科目概况见图 2.14。

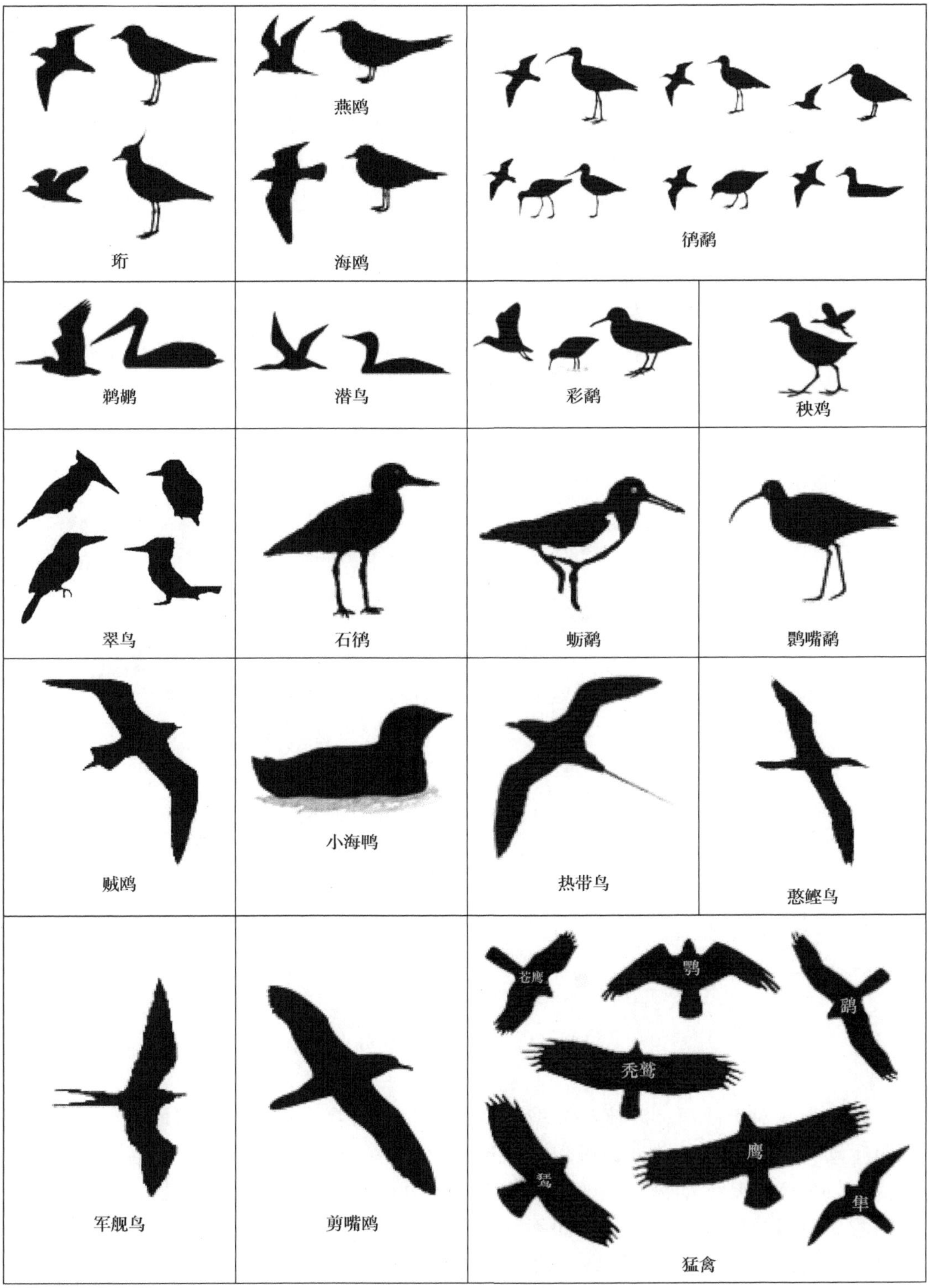

图 2.14　湿地鸟类科目概况

2.12　辨认主要的软体动物

中国的珊瑚礁和潮间带水域分布有所有 7 个纲的软体动物（图 2.15）。头足纲软体动物包括鱿鱼、章鱼和鹦鹉螺等，可以很容易地得到识别，同时鱿鱼和章鱼也是重要的经济鱼类。许多海洋双壳纲软体动物如蛤蜊和牡蛎等也是重要的商业渔业资源，部分物种，如生活在珊瑚礁中的大蛤蜊因为被过度捕捞而处于高度濒危状况。许多腹足纲软体动物如海螺可以食用，其他一些则因其漂亮的贝壳而很有价值。另外 4 个纲的软体动物很稀少，也更不为人所知，不过在中国海鲜餐馆中可以发现有大石鳖这种多板纲软体动物。仅有两个纲的软体动物，即双壳纲和腹足纲软体动物在淡水和陆地栖息地中定殖。

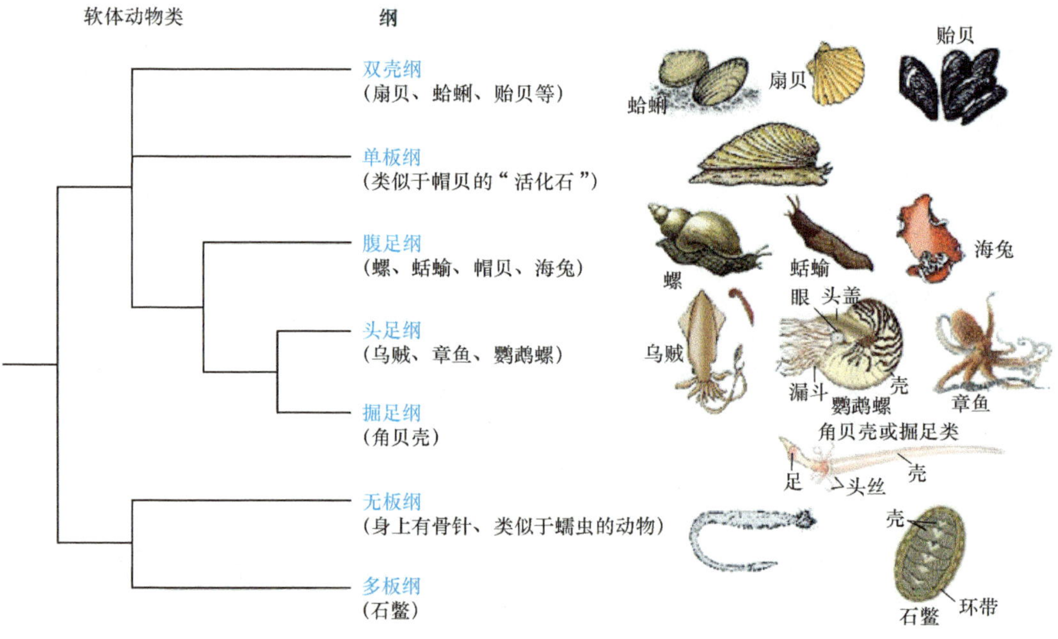

图 2.15　湿地软体动物分类

贻贝和螺对于江河、溪流和池塘保持健康非常重要。它们可为鳄鱼、水獭和鸟类等野生动物提供食物。贻贝通过鳃可以过滤出水中的小型颗粒物，并将这些颗粒物转化为可供鱼类和其他动物食用的食物。由于贻贝可以过滤水中的颗粒物，它们在进食的过程中可以帮助净化水源。然而不幸的是，贻贝的这种进食习惯使之很容易受到水污染的影响，并经常死于水污染。贻贝与人类一样需要洁净的水源才能生存。

许多贻贝种类的保护状况令人担忧。鉴于绝大多数的贻贝终生只生活在一个地方，它们需要稳定的生存环境。河上筑坝是贻贝面临的最大威胁因素。这是因为河上筑坝会减少对许多贻贝生存必不可少的水流，甚至导致这些水流完全消失。在河床上开采砂石也将对贻贝的栖息地——河流渠道和河床造成巨大干扰。此外，除草剂、农药、化肥、采矿废物及家庭垃圾和牲畜粪便对水的污染，也会导致贻贝和其他水生动物的死亡。

2.13　辨认主要的鱼类

中国是世界上淡水鱼类最为丰富的国家之一，共有 920 种鱼类，超过美国（800 种）和欧洲（233 种）。中国淡水鱼类共 33 科，不过鲤科（Cyprinidae）的种数占鱼类种数的一半以上（共 473 种）。另外，爬鳅科（Balitoridae）、鳅科（Cobitidae）和鮡科（Sisoridae）的种数占 25.11%，其中主要是鳅科和鮡科。

中国绝大部分的食用鱼为鲤科。此外，鳇鱼、鳟鱼 / 鲑鱼、鲶鱼和鳝鱼也是在中国常见的食用鱼。

只有专家才能准确辨认特定湿地的所有本地鱼种。不过，如果湿地管理者掌握主要鱼科的基本知识，也是很有帮助的。

2.14　辨认两栖动物科

图 2.16 显示的是中国两栖动物的科及其相互之间的关系。

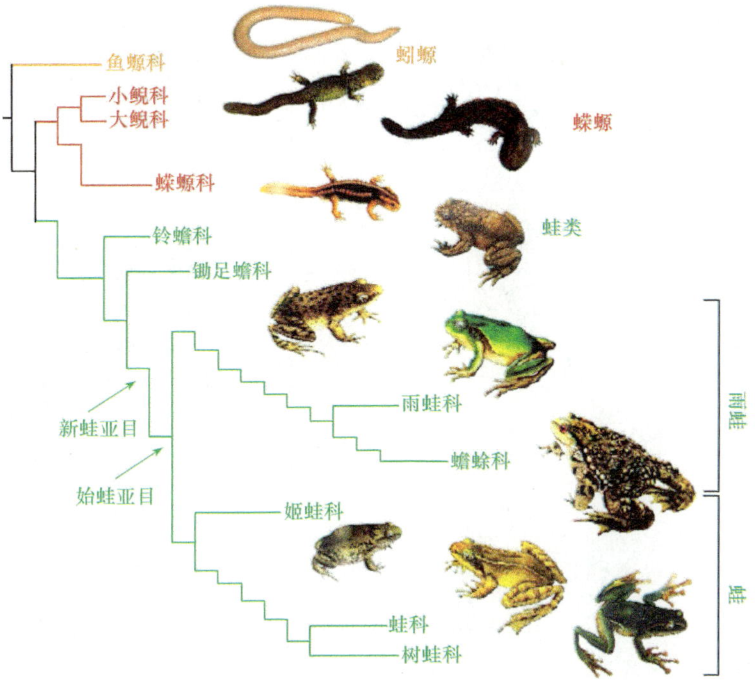

图 2.16　中国主要两栖动物科目

2.15　辨认主要的水虫

图 2.17 为有关中国主要水生昆虫群组的示意图。

昆虫群组	幼虫	成虫
水甲虫		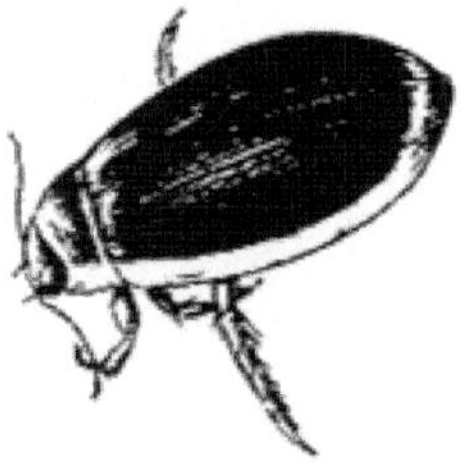
池黾	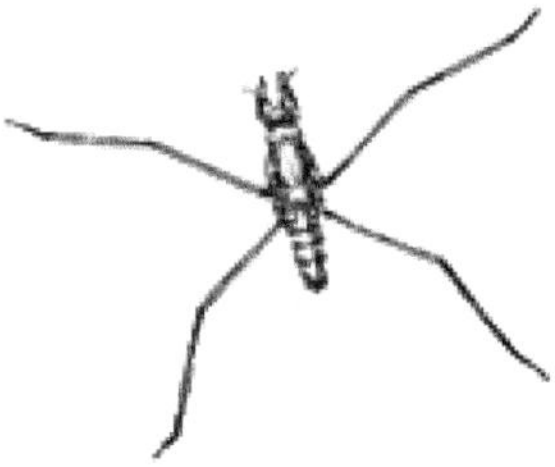	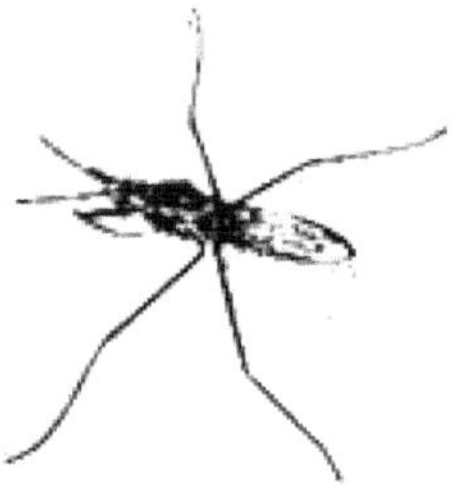
豉甲虫	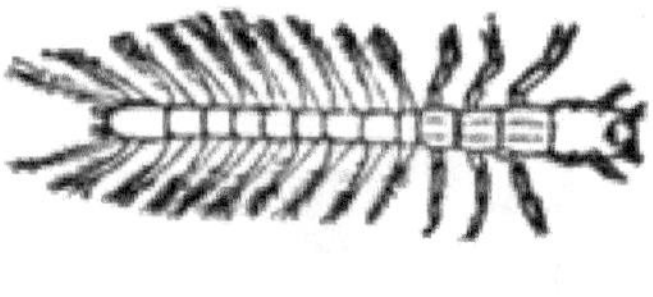	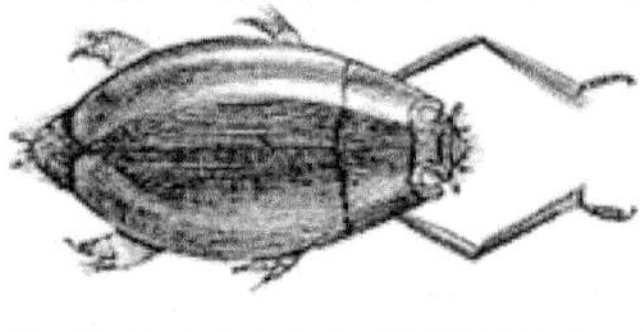
水蝎		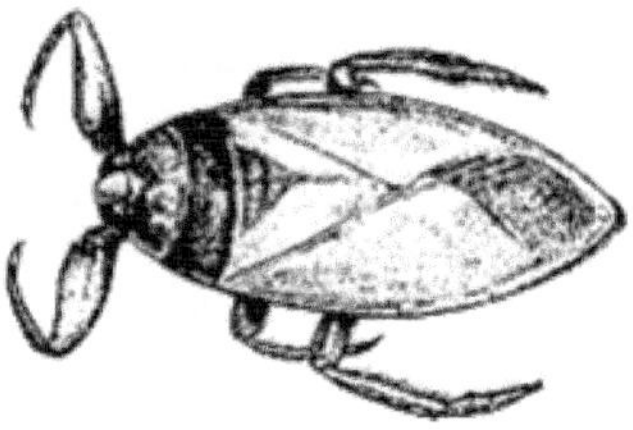
蜻蜓	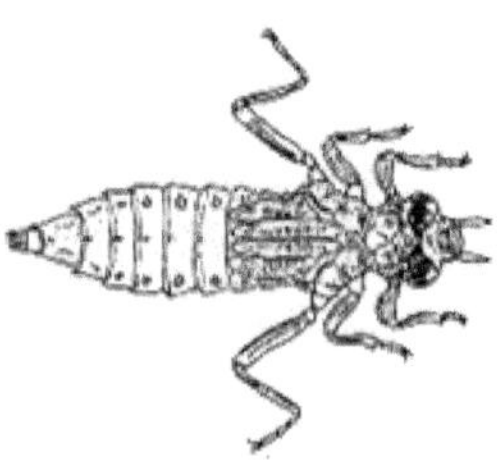	

昆虫群组	幼虫	成虫
豆娘	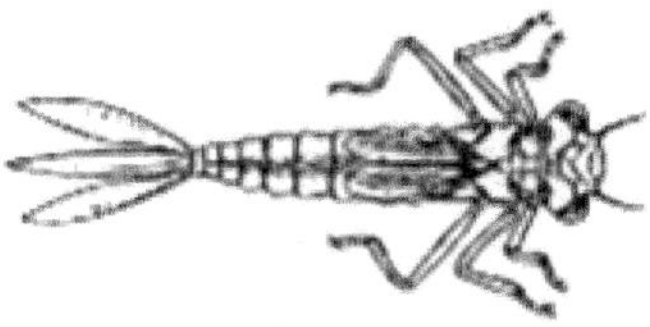	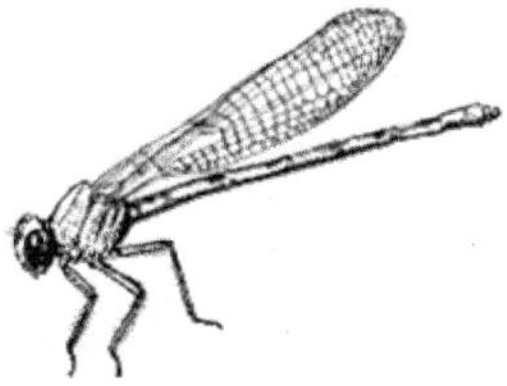
石蝇	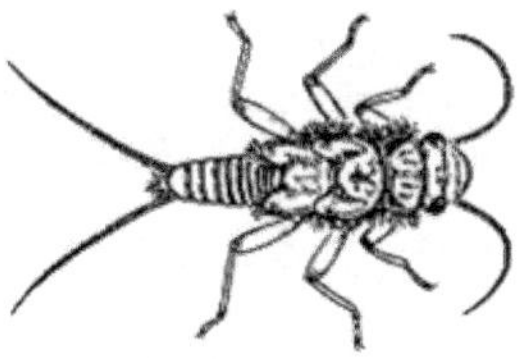	
石蛾	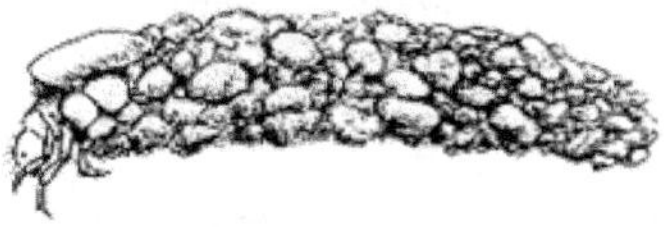	
蜉蝣	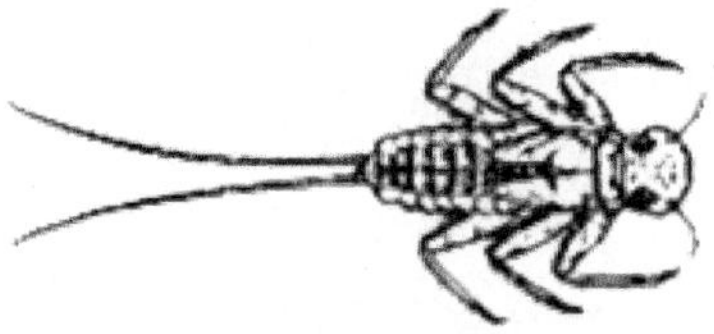	
蚊子	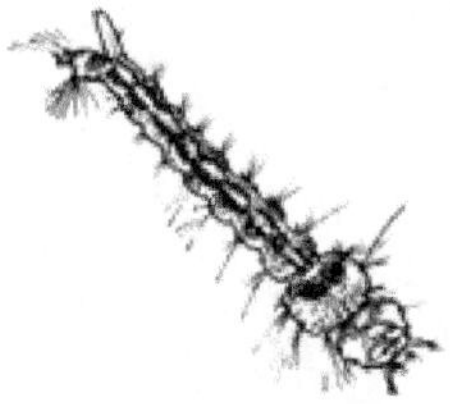	
蠓虫	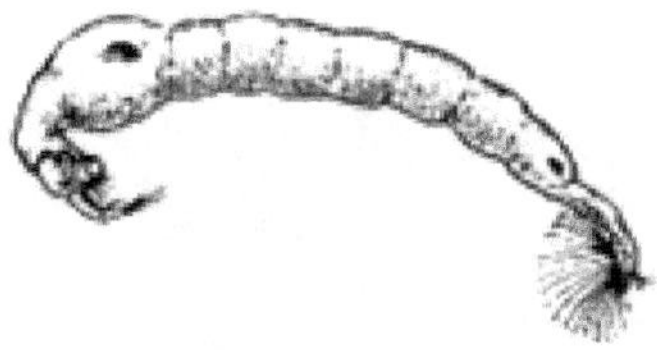	
划蝽		

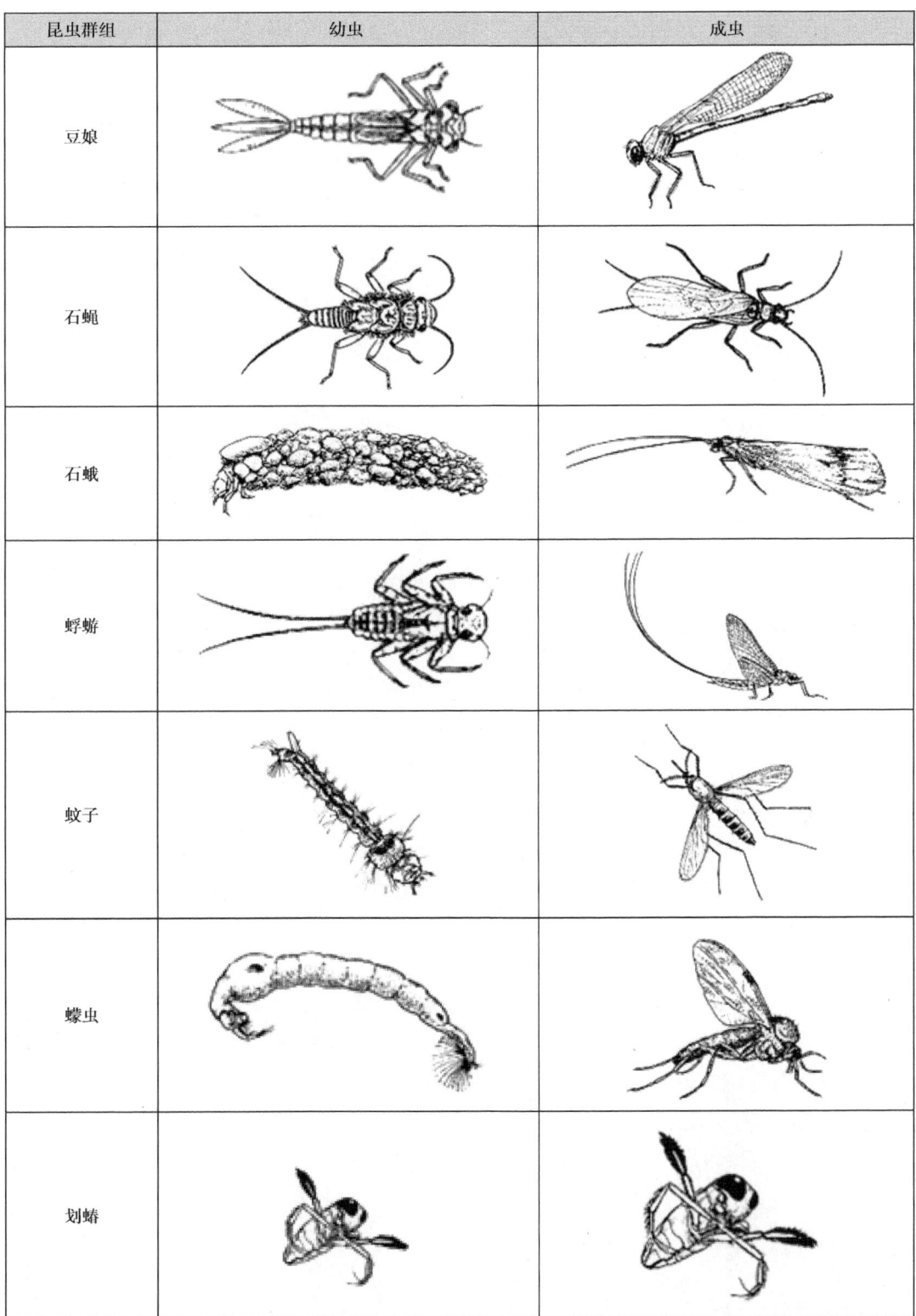

图 2.17　主要水生昆虫群组

2.16 辨认主要的湿地植被

主要湿地植被详见图 2.18。

莎草

苔草

香蒲

浮萍

芦苇

芦竹

睡莲

鸢尾花

荷花

稗草

画眉草

荩草

芒草　马唐　牛筋草

雀麦　早熟禾　籽粒苋

图 2.18　主要湿地植被

2.17　红树林生态环境

红树林主要生长在海南、华南其他省份热带海岸线的潮间带。如今，80% 以上的红树林遭到破坏或改变为其他土地利用方式，特别是盐田、稻田和鱼塘。这种土地利用方式的改变，导致全球重要生物多样性的丧失，滨海环境遭到破坏，并给依赖健康红树林的经济行业带来许多直接和间接损失。

随着人们对红树林重要性意识的提高，红树林对人类价值的估算值也不断增长。红树林是许多海洋鱼类的哺育场，对维持近海渔业至关重要。同时，红树林在采集淤泥，形成新的土地方面发挥着重要作用，同时可以保护周边的珊瑚不会被淤泥堵塞。此外，红树林在支持重要的生物多样性、为生态旅游业提供潜力、提供可持续产品、减缓沿海风暴、保护人类的生命财产等方面也发挥着重要作用。2004 年发生在印度尼西亚的海啸事件清楚地表明，那些将红树林保护得很好的沿海村落遭到海啸破坏的程度很小，而那些将红树林砍伐或改为其他土地利用方式的村落则遭到了海啸的巨大破坏，数百座沿海村落遭到严重破坏。中国处于地震带的沿海地区发生海啸的可能性很大。据 Costanza 等 1997 年估算，全球红树林的平均生态系统服务价值约为 3713 美元 /

(hm^2 · a)。在 2014 年的论文中，Costanza 等将这一数字进行了修改，大幅提升至 786 193 美元 / (hm^2 · a)。

健康的红树林生物群系不仅包括红树林，还包括其他植物、滩涂和浅潮水。只有红树林生态系统具有丰富的多样性，才可以抵御海啸的破坏，并且可以适应泥沙、海岸和气候状况的变化，红树林生态系统的健康才能得到保证（图 2.19）。

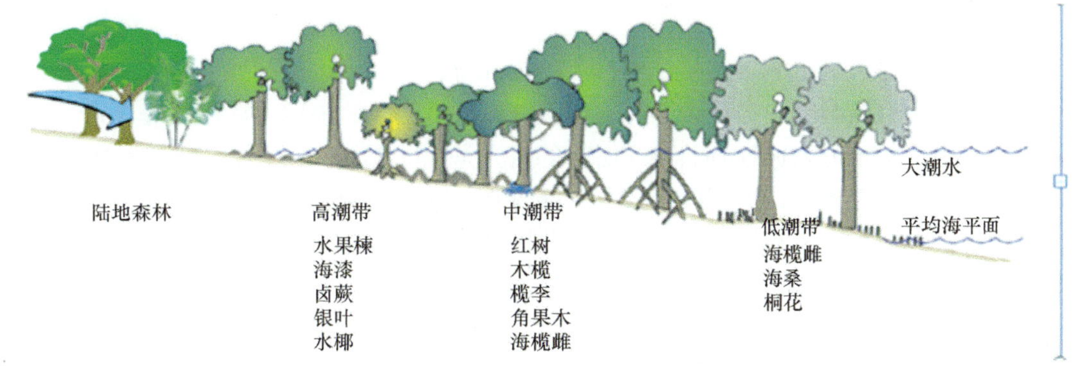

图 2.19　红树林潮汐画面

红树林是一个复杂的生态系统，包括植物、鸟类、鱼类、许多无脊椎动物，以及潮间带水流环境及沉淀下来的泥沙和其他冲积物（图 2.20）。

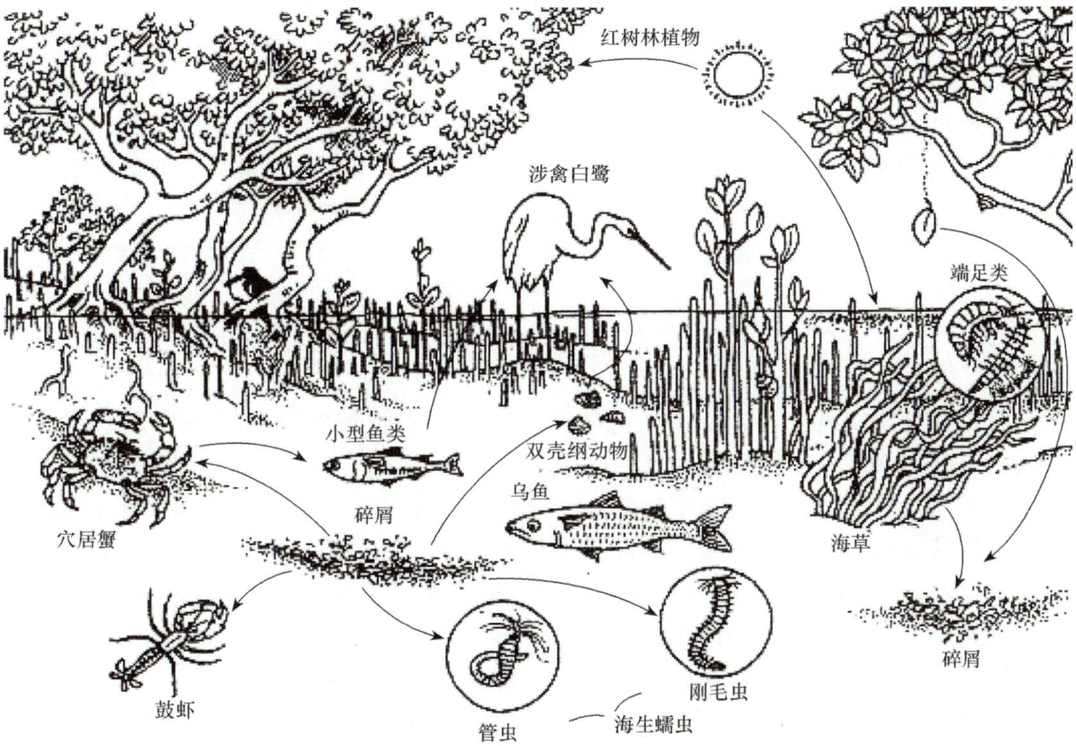

图 2.20　红树林复合生态系统

- 红树林生态系统由森林、草本植物区、池塘、溪流、沙滩、滩涂、开放水域和潮间海滩等组成。不同的物种需要不同的栖息地要素。不过，红树林生态系统要保持其多样性和恢复力，所有这些要素均必不可少。

红树林沿高潮带向低潮带的剖面生长，根据物种应对洪水淹没的能力和盐度的不同，不同的物种生长在这一剖面的不同地方。

在与海岸线和河道保持平行的区域，生长着条纹状的植被。不同的植物种在不同的地区占据主导地位。

红树林是更大的滨海生态系统的一个组成部分，并与周边的其他生态系统紧密地结合在一起。在红树林的前方是水更深的滩涂和其他浅海。这些地区往往是海草床的重要栖息地。更远的地方是珊瑚礁。在红树林的后方是内陆森林。在土壤和水相对较少的沙地上，通常生长着粗草这类海岸植被或海滩森林。陆生鸟类和无脊椎动物在不同的栖息地之间移动，因此，从某种程度上讲，这些栖息地相互依存。一个栖息地遭到破坏或与其他栖息地分开，将直接影响其周边其他生态系统的状况，并导致这些生态系统的退化。

蜜蜂、飞蛾和其他昆虫和部分鸟类在红树林中的植物和树木授粉方面发挥了重要作用。织叶蚁、食虫动物、蜘蛛和蜻蜓及许多鸟类在控制那些可能破坏红树林及其他植物的害虫数量方面具有重要作用。其他一些鸟类作为传播种子的媒介，将无花果和小型红树林及红树林伴生树种的种子传播到各个地方。蟹和龙虾在泥沙里钻出很深的洞，有助于排干土壤基质和使基质透气，去除污染物，使红树林的根系能够吸收到二氧化碳。不过，蝙蝠的作用往往被忽视。事实上，蝙蝠对于许多毛刷状的花卉，如红树科（Rhizophoraceae）及番樱桃属（*Eugenia*）和玉蕊属（*Barringtonia*）的授粉可以起到重要的作用。

再造林应考虑到红树林伴生植物种，这些植物种对于支持和吸引那些红树林为维持一个健康的生态系统所必需的部分动物具有非常重要的作用。此外，毗邻红树林的红树林伴生植物区或森林非常适宜于大量的经济植物和其他作物生长。这些经济作物包括芒果、椰子、木瓜、荔枝、香蕉、腰果等。此外，这些地方还可养蜂酿蜜。

2.18　中国东北地区的林火生态状况

尽管森林本身并非湿地，但它们对于维持下游湿地的健康状况往往不可或缺，可作为雨水的集水区、土壤形成的源头，并为上游集水系统提供枯枝落叶和其他营养物，这些枯枝落叶和营养物再流入下游的江河和湿地。事实上，许多森林都含有小型的湖泊、池塘或湿地。如果森林遭到破坏，可能对湿地生态系统造成严重的负面影响。因此，必须保护好森林，使其避免受到不可持续砍伐及自然或人为火灾的影响。

东北地区的落叶松林地区是世界上为数不多的、雷电经常导致自然林火（春季尤其如此）的地区之一。树龄较长的树木往往都有以往遭受过雷击的印迹。不过，此类火灾通常规模较小，并不会带来很大的热量。它们会周期性地清除林下灌丛，烧毁草地中的青草和牧草，并且仅仅烧死幼龄小树。通过周期性地烧死地处草地和草甸边缘地带的入侵幼龄树，林火可帮助确定森林斑块和草地之间的明确界线。大兴安岭湿地的许多令人赞不绝口的景观就是这一过程导致的。此外，林火还具有丰富生境类型的作用。通过不断创造新的定殖物种可以定居的空间，并在不同的演替阶段创建出不同的森林斑块，可以为各种动植物提供更为多样的微生境。

由此我们可以得出结论：自然林火的这种背景模式是可取的，可能对于维护该地区最初的丰富的多样性必不可少，应允许自然林火继续合理地发展下去。

第 3 部分　湿地保护地的规划

由于湿地管理者并非土地所有者，因此无法按照自己的意愿来管理湿地。湿地管理者仅仅是一个管理人，代表国家和人民接受委托来管理湿地。如同博物馆馆长不能出售或租赁珍贵文物一样，湿地管理者的任务也是负责保护湿地。为保护地管理局寻求更大的经济保障或者利润，可能是帮助其实现湿地保护目标的一种方式，但绝非真正的职责所在。而其真正的职责应该如下。

- 确定威胁因素及其驱动力
- 确定管理中的阻碍因素
- 设计一系列可以减少这些阻碍因素的行动 / 计划
- 监督这些拟定行动的实施情况
- 监测这些行动对保护地的影响
- 对管理对策进行相应的调整
- 宁可过于谨慎，也不可掉以轻心——预防原则

最安全的管理对策如下。

- 最大限度地扩大面积和增强连接性
- 遏制或控制偷猎和非法采集现象
- 最大限度地提高栖息地的多样性
- 最大限度地减少外来入侵物种
- 将敏感地带划为特殊区域，避免受到可能的危害
- 在土地利用或开发中采用预防为先的原则
- 对所有的决策进行存档
- 聘用专家小组

3.1　系统规划和网络

必须在不同的时空尺度上制定对湿地保护地系统的规划。

湿地保护地系统规划应该在流域、地区和全国等更大尺度上开展，有关保护地发展的政策应面向景观尺度进行制定。保护地系统是一个由多个保护地组成的网络，这些保护地可向各种类型的栖息地、物种提供各种程度的保护，并提供预期的重要生态服务。

系统规划以地区性或国家政策为指导，以需求评估为依据。该规划基于差距分析、空间覆盖、连接性问题和保护地未覆盖的范围，确定保护地系统的代表性。系统规划

的时间跨度可以是中长期规划。

湿地保护地系统规划旨在确定哪些区域必须纳入保护地系统，哪些保护地的边界或分区必须进行调整，哪些地点存在连接性的问题，确定污染源或实际的威胁因素，以及保护地与周边土地利用方式之间是否存在冲突等。由单一部门制定的规划很难得到其他各方的认同。湿地保护地系统规划必须与整个地区其他部门和管理机构制定的开发计划保持统一。

此类系统规划主要由更高一级政府部门的专家团队负责制定。不过，湿地管理者应参与规划工作，确保规划考虑到保护地在面积、连接性或其他方面的需求。

与所有其他规划一样，湿地保护地系统规划也并非一成不变。相反，它只是提供一个总的路线图。随着各种新问题不断出现，规划中的部分目标可能无法得以实现，或者不再成为优先领域。因此，所有规划必须不断完善：规划—实施—监测影响—评估现状或定期修订（图 3.1）。由于中国政府的社会经济发展规划以五年为一个周期，如果湿地保护地系统也改为以五年为一个周期，则是最为理想的做法。

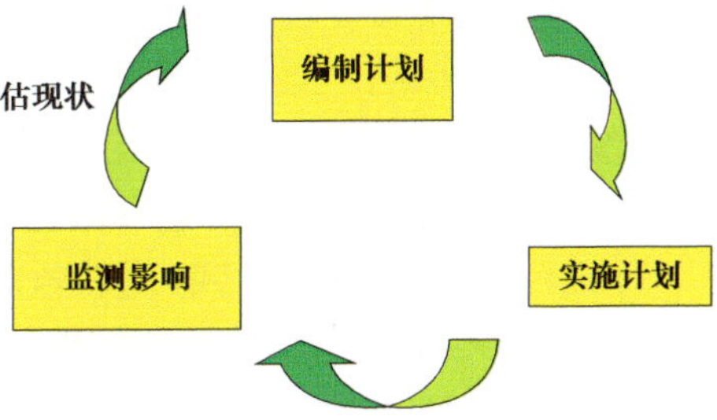

图 3.1　评估周期

3.2　保护地层级的管理规划

此处所指的规划主要针对单个保护地的管理规划（图 3.2）。保护地层级的管理规划应评估保护地的目标、边界和分区的适宜性，确定应解决的主要威胁因素和应保护的主要特征，并简要列出为实现这些目标所必须采取的行动，这些行动计划涉及执法、

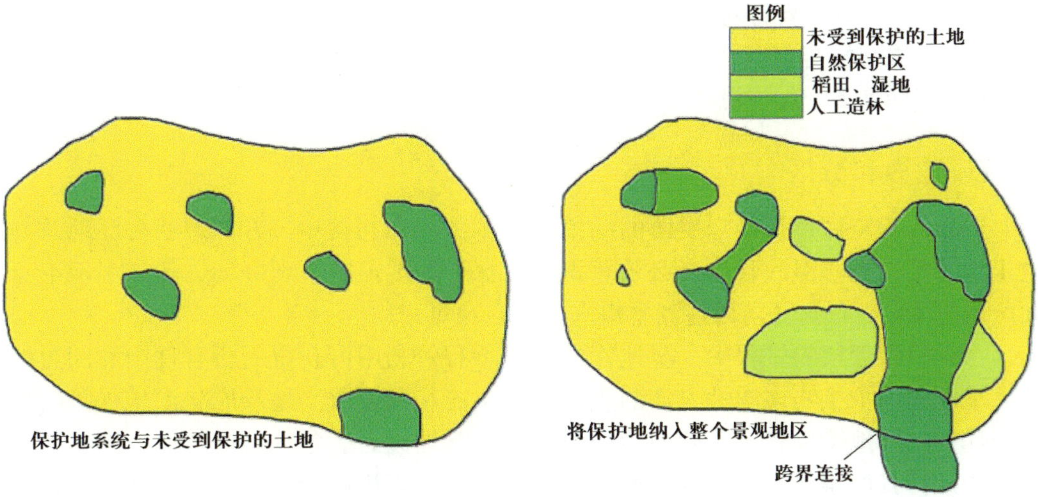

图 3.2　保护地系统规划的景观途径

物种和栖息地管理、研究需求、参观者的出入和教育功能、与当地社区之间的关系、基础设施需求、保护区人员的能力需求及最终的预算要求等。

管理计划不仅是有关预算要求或申请的电子表格，也不是新建大楼的施工总体规划。相反，管理计划应作为一个指导工具，减轻项目管理者在今后决策过程中的负担。此外，管理计划还应作为一个沟通交流工具，让其他机构、发起人和利益相关方了解保护地管理活动希望实现的目标。

管理计划最好由一个独立的专家小组来进行审阅。专家们可能要求保护区管理方对管理计划进行修订，对计划做出的创新举措或申请预算进行阐述或进一步说明其理由。管理计划申请的资金应由专家小组负责审批。

管理层还应接受独立的监督和绩效评估，确保管理计划始终得到良好的落实。

目前，中国制定保护区管理计划的水平仍然相对薄弱。保护区的管理者往往缺乏有关湿地保护的技术背景。尚未建立相应的专家评估和审批制度，缺乏后续监督或绩效评估。不过，政府已意识到了这一问题，并计划开展有关保护区管理和投资方法的改革。

政府的“十三五”规划政策涵盖许多的“深入改革”。改革意味着仍有不完善之处，我们必须改进这些不完善之处。以下是保护区管理中常见的问题列表。

- 为保护区管理设置专门的道路和围栏
- 核心区处于中心地位
- 办公地点远离保护区
- 目标不明确
- 缺乏对生物多样性影响状况的基准数据和监测
- 生态补偿体系不恰当
- 种植的树木地点和树种不适宜
- 相当于救护中心 / 繁殖中心和博物馆
- 孤立——缺乏协调和主流化
- 总体规划主要针对开发而非保护
- 忽视气候变化
- 网捕鸟类，如在机场等地

3.2.1 边界和分区

边界的确定往往是谈判妥协的结果。许多部门和利益相关方都想获得湿地及其周边土地的使用权。湿地管理者必须意识到，保护湿地需要管理好远远超过湿地本身水面区域的活动，最远甚至可追溯至集水区所在地。

在竞争管理权的过程中，湿地管理者必须有效利用可获得的所有理由和同盟的支持。湿地在保护生物多样性方面的重要性并不能足以说服当地政府将湿地保护的权力授予他们。湿地管理者还必须向当地政府展示湿地生态服务的经济价值，阐明如果湿地无法维持其健康状况必须采取修复措施或替代工程措施所带来的巨大代价，以及湿地在保护人类健康、提供休闲娱乐、防洪、应对气候变化等方面的额外价值。

在谈判建立湿地保护地的过程中，湿地管理者必须对急需得到保护的关键湿地及为维持湿地的生态健康状况所必需的连接性有清楚的认识。

要申请建立新的保护地，必须拥有一个强有力的正当理由，并提供良好的地图、基准数据、可行性研究和成本效益分析。

必须尽快通过边界划分，以及通过设立边界布告牌，公开宣布某个保护区的保护地位及其禁止开展的活动，加强土地利用的安全性。湿地管理者应组织相应的活动，增强当地社区对新建保护区重要性的意识，获得他们的支持，确保这一保护地位在各种不同的规划部门，特别是国土资源部门的土地清册中得到充分的记载。

在确定保护区的边界后，保护区管理者应在保护区内建立一个不同的管理区系统。

现有的《中华人民共和国自然保护区条例》(1994)仅按照联合国教科文组织“人与生物圈”(MAB)计划倡导的方法将自然保护区分为三个区：核心区、缓冲区和实验区。从以往来看，中国的首批自然保护区主要是人与生物圈计划保护区，其成立的目的是研究人类活动与自然环境的相互关系。不过，后来新建的许多保护区的目标各不相同，包括保护关键物种、关键生态系统、景区或生态功能等。旧的分区系统不再适用，正因如此，许多新建保护地选择不加入“自然保护区”名录，而是采纳加入不同的保护地系统，其地位也不明确，如森林公园、湿地公园、国家公园等。

确定具有高自然保护价值 / 重要性 / 理由的保护地要素如下。

- 关键物种（濒危、特有、野生亲缘种）
- 有代表性的生态系统
- 生态系统服务
- 文化景观 / 无形知识
- 休闲娱乐 / 生态旅游
- 可再生资源的可持续利用

分区：建议在每个登记注册的自然保护地设立核心区，以便保护最为敏感的物种。不过，至少 80% 的分区应该与保护地总体分类保持一致，另外 20% 可以用于其他分区。例如，

- 荒野区——基本为荒野地区，但可在获得特殊许可的条件下用于研究、摄影、监测等
- 管理区——用于开展栖息地或物种管理的区域
- 游客出入区——游客可有限出入的区域
- 限制利用区——按照管理计划的目标可部分利用的区域
- 行政区——用于建设行政办公楼、仓储设施、公共设施、道路等的区域
- 外部缓冲区——在保护地边界以外，可以开展部分受限制活动的区域

核心区不应仅仅位于保护地的中心区域。这些中心区域通常位于山顶或湖泊最深的部分，而这些区域受到的威胁最小，生物多样性也最少。因此，核心区应用于保护保护地当初成立时所保护的关键特征，应扩展至保护地边缘地带，并与邻近保护地的邻近核心区连为一体。

核心区应包括：
生物多样性最丰富的区域
最脆弱的区域
情况危急的区域
囊括最多样的生境类型的区域
不可替代的关键繁殖地、栖息地等

3.2.2　更大范围的缓冲区和红线

除了内部的缓冲区外，所有湿地保护地都要求在保护地以外的区域，特别是上游地区建立更广泛的缓冲区，确保外部活动不会因不当的干扰、污染或沉淀对湿地的功能和健康状况造成损害。此类缓冲区将扩大至保护地管理权限以外的区域。不过，湿地管理者必须能够对更大尺度的土地利用规划进行审定、提出意见和发挥影响，确保将保护地充分纳入整个景观和流域尺度的主流化管理之中。

来自中央政府的新政策向湿地管理者完成上述任务提供了有力的支持。党的十八大做出了实施“生态红线”管控制度，保护关键生态服务的重大决定。针对那些被迫放弃可能威胁此类生态服务提供的规划或其他潜在活动的个人，生态补偿资金将用来为他们提供补偿。计划将超过30%的陆地面积划定为“生态红线”，在现有15%作为正式保护地的基础上大幅增加。

管理计划的提纲

- 保护地简介
- 指出保护地的自然保护价值、关键生态服务、主要威胁因素，并确定自然保护的优先行动
- 说明被列入某种自然保护地类型的理由
- 按照优先顺序阐明保护管理目标
- 确定边界和分区，并提出修改意见
- 列出针对以下内容的子计划：保护和监测，面向管理的研究需求，栖息地改造、控制、恢复或物种管理，适宜的控制措施（如外来入侵物种的控制）、传播和宣教责任；游客利用设施、信息；消防应急措施、防治荒漠化、应对自然或人为灾难
- 当地社区的参与或利益共享计划
- 基础设施建设投资或设备采购计划及其理由
- 人员发展需求的详细情况，包括培训和其他能力需求
- 投资预算、人力成本、运营预算申请的详细情况
- 针对通过用户交费、补偿机制等方式所募集资金的使用建议

3.3 活动规划

所有运营活动和其他主要活动都需要制定自身的计划，这有助于确保提高这些活动的效率。与此同时，这些计划必须得到上级管理部门的批准，以便得到实施，或必要的预算支出得到批准。以下运营计划的核查提纲（表 3.1）可能有助于此类计划的编制。

表 3.1　运营计划

内容提纲	备注
活动简称	加入唯一的活动编号
日期、具体时间和地点	复杂的活动可能有多个举办的日期和地点
目标	如将非法渔网从湖泊北端全部清除
活动简介	如“爱鸟日”公众庆祝活动，邀请 500 名儿童参观保护地
预定时间表	活动或活动前准备的时间安排
单独活动列表	列出每项细分活动的时间表、任务和费用
所需材料	列出所需的实际材料及获取方式（采购、借用、租用等）及其费用
职责	确定由谁负责哪项具体的活动或采购
推广需求	如何宣传活动或行动，是否必须告知媒体等
报告	如何汇报行动或活动
待批准的总体预算	所需的总预算及来自哪个项目或保护地的预算项

第 4 部分　栖息地管理

全国湿地面积占国土总面积的 5.58%。中国的湿地类型众多，包括沼泽、泥炭地、湿草甸、湖泊、河流、泛滥平原、三角洲、滩涂、红树林、水库、池塘和潮间带等（参见 2.1 节）。青藏高原内陆湖的含盐度各不相同，尽管降雨的补给很少，但经常得到来自冰川融化的雪水补给。

中国的湿地提供大量的经济、生态和社会效益，因此保护这些湿地非常重要（参见 1.2 节）。这些服务包括防洪减灾、蓄水、气候调节、水净化、侵蚀控制、土地形成和创建风景区和休闲环境等。此外，湿地还为大量动植物提供栖息地，包括鸟类、鱼类、两栖动物、甲壳动物及可作为人类食物的许多动植物。湿地也是非常重要的虾类栖息地。

一个鲜为人知的事实是，中国的泥炭地所储存的碳超过中国所有森林所储存的碳。因此，保护和防止这些泥炭地变干或燃烧（泥炭变干或燃烧后都将把碳释放到大气中），将对全球减缓气候变化的工作做出重大贡献。

中国国土内分布着数千个湖泊。这些湖泊对于鱼类和许多其他水生物种至关重要，同时也是天然水库，在暴雨期间可以存储大量的洪水。

4.1　湿地栖息地管理所面临的挑战

湿地面临严重的威胁（参见 2.4 节），是一种需要得到特殊保护管理的栖息地类型。湿地对于许多重要物种的生存至关重要，同时提供许多关键生态系统服务，如作为天然水库、碳库及水净化和防洪等。在国务院颁布的政策中特别强调了这一点：湿地保护需求必须例行纳入所有政府规划之中。

湿地管理错综复杂，湿地生态系统面临的重大挑战在于其开放性（水和物种的流入、流出）。湿地很容易受到湿地保护地管理者所无法控制的许多外部因素的影响，并通常被各种利益相关方用于多种不同的目的。另外一个难题在于湿地的动态属性，其特征和位置时常处于变化之中。洞庭湖在不同日期的历史分布图体现出的湿地分布状况不断变化的特征（图 4.1），就很好地阐释了这一点。

4.1.1　明确职责以维护湿地生态系统功能

维护湿地生态系统功能的“秘诀”包括以下内容：

- 确保湿地生态系统得到保护，确保湿地资源得到可持续的采集
- 防止水污染，检查淤积水平，避免富营养化

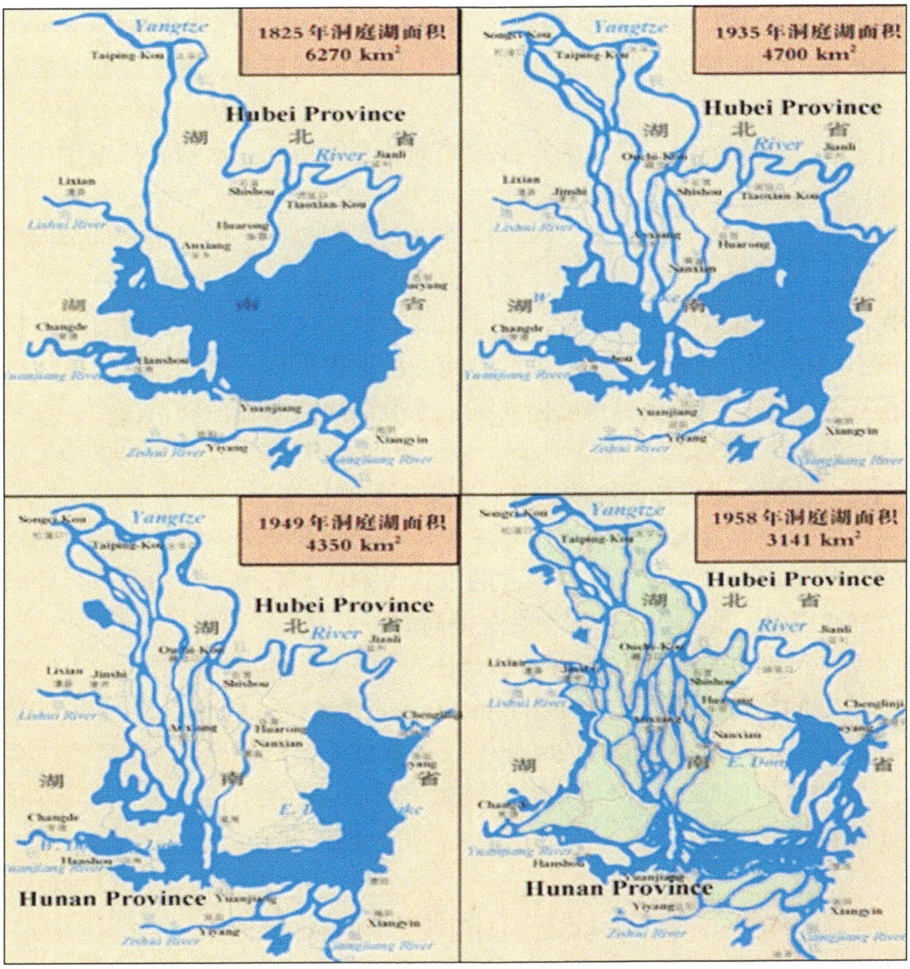

图 4.1　湖南洞庭湖的动态属性

- 通过修筑梯田和治理侵蚀工程减少侵蚀，维持湿地周边土壤的深度及其健康
- 通过修筑梯田和利用拉沙坝，减缓水流速度
- 通过维护好灌溉水道、沟渠、水管和竹沟等，减少水浪费
- 让农业废弃物腐烂到土壤中形成绿色肥料，避免火灾（火灾会杀死许多土壤分解物）
- 维护高生物多样性，减少病害
- 维持邻近湿地自然保护区健康的自然和半自然生态系统（林地、湿地、草地）
- 不开垦坡度大于 25° 的斜坡，将陡坡耕地退还为草地和植被
- 禁止砍伐、禁止焚烧和禁止放牧，保护周边森林

4.1.2　需要现场巡护员

无论计划多么周全，或者监测设备多么先进，只有在现场部署护林人和巡护员，

保护性管理工作才最为有效。现场护林人和巡护员相当于管理的“耳目”，他们主要负责查看保护区的实际情况，他们是保护区引人注目的保护者，理应得到当地人的尊重和合作。

护林人和巡护员如果只是闲坐在保护站，就毫无用处。他们必须走出去，在保护区周边地区进行巡护。他们必须全天候地时刻保持警觉。

良好保护的“秘诀”，就是以不可预见的方式定期巡护，并且巡护的范围涵盖保护区所有区域。

如果没有一个有效的保护体系，管理层就不能仅仅基于巡护员所看到的状况采取行动。所有巡护员应在每次巡护结束后提交一份巡护报告。这些报告应包括对重点关注物种的观测情况，以及栖息地总体状况、人类活动或家畜活动情况等。可在基本的巡护报告后附上更具体详细的监测职责。巡护报告表的最简单形式见 9.8 节，不过，每个保护地可以编制一份更加详细的表格，收集更多的信息。

如果巡护的内容要求过细，将影响巡护员的士气和动力。因此，尽可能减轻巡护员填写报告的负担。当巡护员看到自己的报告得到认真对待，作为采取行动的依据并且影响管理决定时，他们的巡护积极性是最高的。

如果保护区面临严重的偷猎现状，巡护员可利用由世界自然基金会（WWF）开发和推广的空间监测报告工具（spatial monitoring and report tool，SMART）软件系统（http://www.smartconservationsoftware.org/）。该软件系统利用全球定位系统（GPS）记录方法来监测巡护路线、观测动物或其迹象、偷猎证据等，也可用于加强巡护工作，确定更加需要巡护的地方，并重点放在最急需巡护的区域等。

4.2 保护不同类型的栖息地

4.2.1 湖泊

湖泊是在低洼地带形成的面积较大的水体。湖泊通常由某些江河或溪流供水，再通过其他江河或溪流排水。

中国的湖泊分布在从高海拔地区到沿海低洼地带的各个地区，其特征千差万别。部分高原湖泊的盐度也各不相同。

湖泊拥有大量的水生动植物，不同的特有物种分布在不同的流域中。许多珍稀特有物种在中国的分布范围非常狭窄。因此，如果我们希望保护绝大多数的水生物种，就必须保护好大量不同的湖泊。

4.2.2 水库

水库是出于各种水文目的，如蓄水、防洪、灌溉等而人工建造的湖泊。不过，许多水库也是极其重要的生物多样性保护地，也常常用于多种目的。

鉴于水库是非天然湖泊，管理者可以有更多的方式来合法“改善”其生物多样性功能，这些方法包括部分工程方法、种植特殊植被和引入各种本地野生动植物等。

4.2.3　池塘和沟渠

湿地是一种动态性很强的生态系统。人工池塘和沟渠常常会受到植被的堵塞，因此需要定期维护和清理，才能维持其最初的功能。即便是天然水池，也会随着时间的推移而发生变化，甚至会逐渐被沉积物和杂草填满。因此，湿地管理者必须确定如何才能最佳地实现湿地保护地的管理目标，是任凭这些自然和演替过程继续改变保护地，还是将这些变化控制在合理的状态，控制植被和水流以维持这种状况。

4.2.4　滨海栖息地

中国漫长的海岸线组成了一个狭长的，但对于许多生物多样性非常重要的栖息地。这些栖息地类型包括滩涂、沙滩、悬崖、砂石滩、滨海沼泽和边缘红树林，它们拥有不同种类但都很丰富的软体动物、甲壳动物、鱼类和植物。它们是涉禽和其他鸟类重要的觅食地，也是那些每年在北方繁殖地和热带越冬地之间迁徙的鸟类的重要停歇地。许多涉禽在这些海岸线沿线筑巢，海豹等海洋哺乳动物甚至在这些沿海地区繁殖。

滨海栖息地可以为人类提供多种食物，同时也是养殖鱼类和甲壳动物的鱼池所在地。在加固大江大河所冲刷的来自中国内陆的冲积泥沙方面，沿海植被具有极其重要的作用。这些冲积层被植物捕获，并积聚形成新的肥沃的农业用地，甚至形成上海和厦门等沿海城市的新土地。

4.2.5　海洋栖息地

中国的海洋分布范围广泛，从北部的寒温带海洋、半封闭式渤海、东部沿海，一直到南部的西沙和南沙群岛热带珊瑚礁。

北部海域拥有海豹等海洋哺乳动物及大量的海鸥和其他鸟类，而在水下则生活着金枪鱼、鲭鱼、鱿鱼和鲈鱼等各种鱼类。

南部海域拥有色彩斑斓的珊瑚礁，各种热带鱼及奇形怪状的软体动物、海豚、鲨鱼和海龟等出没其中。

深海有大鲸鱼，它们不时发出低沉的“悲鸣”声，并且每年在极地水域间迁移。

中国的海洋动物种类十分丰富，包括鱼类、软体动物、蟹、虾和海参等。许多珍稀中式菜肴就依赖于这些海洋动物的持续供应。

4.2.6　与海洋相关的其他问题

与陆地保护区相比，中国海洋保护区的发展相对滞后。其部分原因在于海洋保护区难以划定或巡护。海域的主权归属存在分歧也是一个问题。

该海域面临的其他问题包括：世界海运业及捕捞船队对该海域的大量利用；污染和淤积的威胁已超出负责中国海洋保护区管理的国家海洋局的控制范畴。

4.2.7　保护河流集水区

湿地依赖于各种水源，如降雨、溪流、河流、地下水和潮汐等。保护这些水源可能涵盖数百公里的上游地区，并延伸至河流的上游集水区，同时还需要让足量的水通过大坝、分水渠、水井等人工屏障。中国政府已制定了多项旨在保护和增强集水区功能的政策。从宏观尺度上看，环境保护部已划定了多个关键生态区，其选择的主要指标是它们在保护水源方面发挥的作用。此外，中国政府正实施一项建立生态红线区域的计划。按照该项计划，允许开发地区的范围和类型将限定在那些对维持生态系统服务（主要是集水区）至关重要的区域。与此同时，中国大部分地区现已禁止采伐，国家林业局正积极参与天然林的保护和植树造林工作，以便改善集水区的状况。

森林在改善集水区的功能方面发挥着重要作用。首先，森林有助于土壤的形成。土壤和深入渗透到土壤中的树木根系，使得地表更加易于渗透雨水，以便集水区海绵体可以存储更多的水，减少直接流入河流，导致河流泛滥的水量。其次，森林中的枯枝落叶和下层林木有助于保护脆弱的土壤，避免土壤遭到雨水的破坏和冲刷。因此，上游森林可以从多个方面对湿地有利。

通过增强土壤的渗透性，森林可以减少洪水径流，延长一年中水流从地表排出到河道中的时间。

通过形成和保护土壤，森林可以改善供水的水质和清澈度，从而使水源得到过滤，并且更加洁净和健康。

落叶和其他森林凋落物的分解，可以丰富溪流和湖泊中的水分养分，为许多食物链提供食物来源。

森林可以使上游河道避免受到阳光的直射，使水温变凉，并且可以存储更多的氧气。

湿地管理者必须了解湿地的供水情况，呼吁在上游集水区开展良好的森林保护活动，并考虑到建坝、引水或将水抽出河流系统及受到污染或富含泥沙的水流入河流系统中对供水所造成的威胁。

许多水源区都存在供应短缺的状况。即便在拥有许多常绿湿润森林生态系统的海南地区，人类的用水速度也超过了供水速度，同时水库水位和地下水位不断下降。除湿地管理者以外，其他许多水资源使用者都迫切地希望获取这些有限的水资源。当地政府必须利用常识和公平原则来决定哪些人应获得哪种水资源。不过，湿地管理者必须提出强有力的理由，表明自身有必要获得充足的水流。要确保这一点，必须将国土资源部制定的土地利用规划与当地发改委部门制定的土地利用规划进行统筹协调。

4.2.8　草地和火灾管理

火烧是控制植被生长的一种良好工具。火灾的破坏能力已得到公认，因此，人们对火灾的一个固有看法就是：火灾的影响通常是负面的，森林管理者应尽力防止野火的发生，扑灭那些确实已发生的火灾，并且尽快恢复被焚烧的植被区。不过，对生物多样性而言，火烧的影响并非总是负面的。事实上，火烧形成了许多的自然景观。火烧可以抑制森林和灌木的生长，促进草地的生长。

如果湿地管理者必须保留或增加草地面积，就可能需要将火烧作为一种管理工具，同时通过砍伐、放牧或洪水淹没等手段控制树木的生长。

芦苇滩和各种不同的草地拥有特定的动物群。不过，许多其他物种也需要矮草或开阔空间，而非密集的大片高草作为栖息地。在自然界，本来拥有许多大型食草动物，如鹿、野牛甚至大象等，它们可以维持那些保存完好的高芦苇或高草与那些已被开发或用于放牧的斑块之间的平衡（图 4.2、图 4.3）。在没有这些原始食草动物的情况下，湿地管理者可以利用有节制地收割芦苇或除草方法来开发部分湿地，以便为动植物提供更广阔的栖息地，维持更多的物种数量。

图 4.2　农民正在收割芦苇

图 4.3　农民用牛来控制草的生长

4.3　栖息地修复

4.3.1　修复退化生态系统

过去数十年来，许多重要生态系统的退化，导致中国许多生态系统服务的丧失。

此前，许多森林遭到砍伐、焚烧、开垦为耕地、分成许多小型斑块。受薪柴采集、家畜放养和林火的影响，这些森林的自然恢复速度更加缓慢。通常，森林可以保护湿地及其水源。因此，维护森林的健康，将直接关系到湿地生产力的保护状况。

以下章节将详细阐述部分方法，利用这些方法，湿地管理者、护林人员和农民可为国家做出贡献，并通过管理好和帮助恢复其景观周边的自然生态系统，改善其自身生态环境。

拯救野生稻

云南金河村的村民王兴生（音译）向当地政府部门报告，在原定改为橡胶种植园的森林中发现生长有药用稻（*Oryza officinalis*）。目前，当地村民已与农业部达成协议，村民负责保护这一地区，不在周边区域施用化学除草剂，并取消了原定的茶叶和橡胶种植计划。作为回报，农业部的《野生植物保护办法》负责为该村提供修建养猪场的技术知识，并提供其他补偿。

4.3.2 坡地退耕还林

坡地耕种会导致中国最为珍贵的资源——沃土的大量流失。每年，中国的土壤流失总量达到数十亿吨，由此可以想象流失情况的严重性。随着具有吸水性的土壤海绵体的流失及作为高降雨量时排水区的河床被淤泥堵塞，坡地耕种已证明使得每年由洪水导致的土壤破坏雪上加霜（图 4.4）。

图 4.4　四川急需还林的坡地耕种

因此，政府实施了一项特殊的计划，鼓励农民将坡地退耕还林。政府向农民提供植树的幼苗并且对退耕还林的坡地提供补偿（有时以提供粮食的形式）。

农民必须退耕还林还存在其他原因。树木茂盛的地区可以增加村庄的吸引力，改善生活环境，提供良好的遮阴挡风之地，改善土壤状况及其保水能力（降低向土壤加水的必要性），并且促进当地可防治病虫害的野生动物的生长。

在决定种植何种树木时，应避免过多种植同一树种。这是因为单一种植在风景营造方面令人感到单调乏味，生物价值更低，对野生动物的吸引力降低，提供环境服务的效率降低，并且更易于受到病虫害、疾病和野火的侵袭。

同样，应避免种植大片同龄的树木。这是因为，种植大片同龄的树木会导致树冠浓密，遮蔽下层林木的光线，使得林下层寸草不生，外形不美观，并且难以维持生存。应尽量建立一个拥有各种树林分布、可以并存的本地物种组合，以便在即使没有更多人工管理的条件下，该地块也可自我维系生存。应种植各种适宜的林下灌木丛、幼树苗和幼龄树等。

减少坡地土壤遭受侵蚀的方法

- 修建梯田
- 修建实体的或生物性（如成排的竹子）田基，减少流下山的水流量
- 在易于遭受侵蚀的沟渠地带修建拉沙坝
- 尽快利用植被覆盖遭受侵蚀的土地
- 通过植树造林增强土壤的渗水性能（农林复合系统）
- 鼓励在休耕期施用大量的绿肥
- 种植防风林

在落叶阔叶林地区，如温带林地或热带季雨林地区，应与本地的多种常绿树种（如温带的针叶树和冬青，以及热带的针叶树）混种。

如果确保林地中的中空树干、树枝或巢箱可以为野生动物提供庇护所，林地可以提供对那些以花蜜为食的鸟类和昆虫具有吸引力的观花树木（如热带的刺桐和肉桂或温带的丁香和酸橙），特别是果树（如热带的无花果或温带的山楂和蒙古栎），那么林地将对野生动物具有更大的吸引力。

要获得建立此类林地的植物资源，可能还需开展更多的工作。比如，可能必须建立一个小型的苗圃，根据从本地自然生境中获取的种子或树苗来培育树木和灌木。此外还可能必须获得特殊的许可，从公园、森林或自然保护区中采集此类种子。

4.3.3 建立苗圃

在农场或村庄的角落确定一处可用于建立苗圃的地方。苗圃应部分背阴，以便使栽种的植物避免受到阳光直射。与大型苗圃相比，小型苗圃更加便于管理维护。

确定该地区的本土植物，从这些本土植物中选出可在苗圃中培植的植物。部分植物可以直接从其原生地移植到景观花圃（图 4.5）。

仅培植那些在景观花圃中必须定期更换的植物。否则，苗圃将很快充满各种生长过快的植物。

在将幼苗或小树从苗圃移至开阔景观地时，必须精心平整好土地。如果在这一关键时期因为投机取巧导致培育树苗的努力付诸东流，将是一大悲剧。

移植树苗时提高其存活率的几点建议

- 为每颗树木或树苗准备单独的树坑，增加适宜的土壤结构，在种植后充分浇水
- 在一年中植物面临来自干燥（高温和风）或严寒压力最低的时候移植树木。在温带地区，树叶掉落的冬季是移植树木的良好时期
- 保护幼龄树，避免它们遭到鹿、猪或其他动物的食用或干扰
- 在移植期间，利用种植袋或减掉根系系统，同时附着土壤（可以采用特殊的工具来开展这项工作），最大限度减小对根系的干扰
- 在移植后的前几周大量浇水，直至新的根系长出来，在植物底部周围建立土壤剖面，以便捕获雨水和蓄水

图 4.5　准备将野生树苗用于移植

4.3.4　建立林地的步骤

恢复森林覆盖的各种工具和步骤见图 4.6。

4.4　恢复和建立湿地和池塘

过去几十年间，受排水、蓄水及地下水过度利用（利用井眼进行农业灌溉导致）的影响，中国的湿地面积不断下降。如今，我们正受到地下水位下降、湿地侵蚀加剧及湿地蓄水能力丧失的影响，导致洪水频率增加及气候变化的速度不断加快。政府制定了遏制和扭转湿地丧失的政策，并要求农民为实现这一艰巨的任务做出贡献。

部分大湖（如湖南的洞庭湖）已实施了退耕还湖的工程。这些地方的农民可以在夏季将其圩田放水淹没，用于养殖蟹、龙虾、鱼、鸭等，作为种植农作物的替代生计方式。

不过，对于中国各地的众多小型湖泊和湿地而言，只需通过阻断排水渠，让变干的湿地重新为水淹没，即可使之得到恢复。利用沙袋将排水渠堵住（图 4.7），往往就足以对小型湿地进行恢复，并使许多重要的湿地生物群及湿地生态系统功能得以恢复。

此外，利用黏土或塑料衬管堵住出水口，并注满地表水，就可建立新的湿地或池塘。通过改变水深和基质（泥浆、砾石、岩石等），可以建立多种栖息地环境，从而支持更多的生物多样性种类。这项工作既可以在单个的农田中开展，也可以全村共同建立一个村庄池塘。

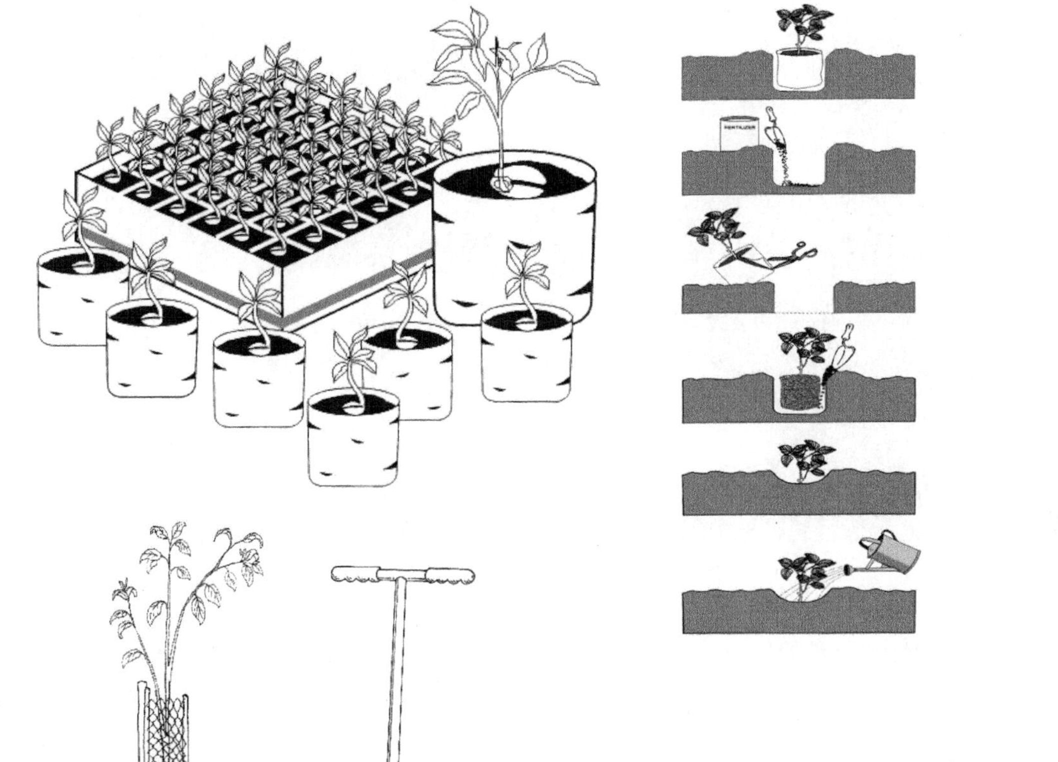

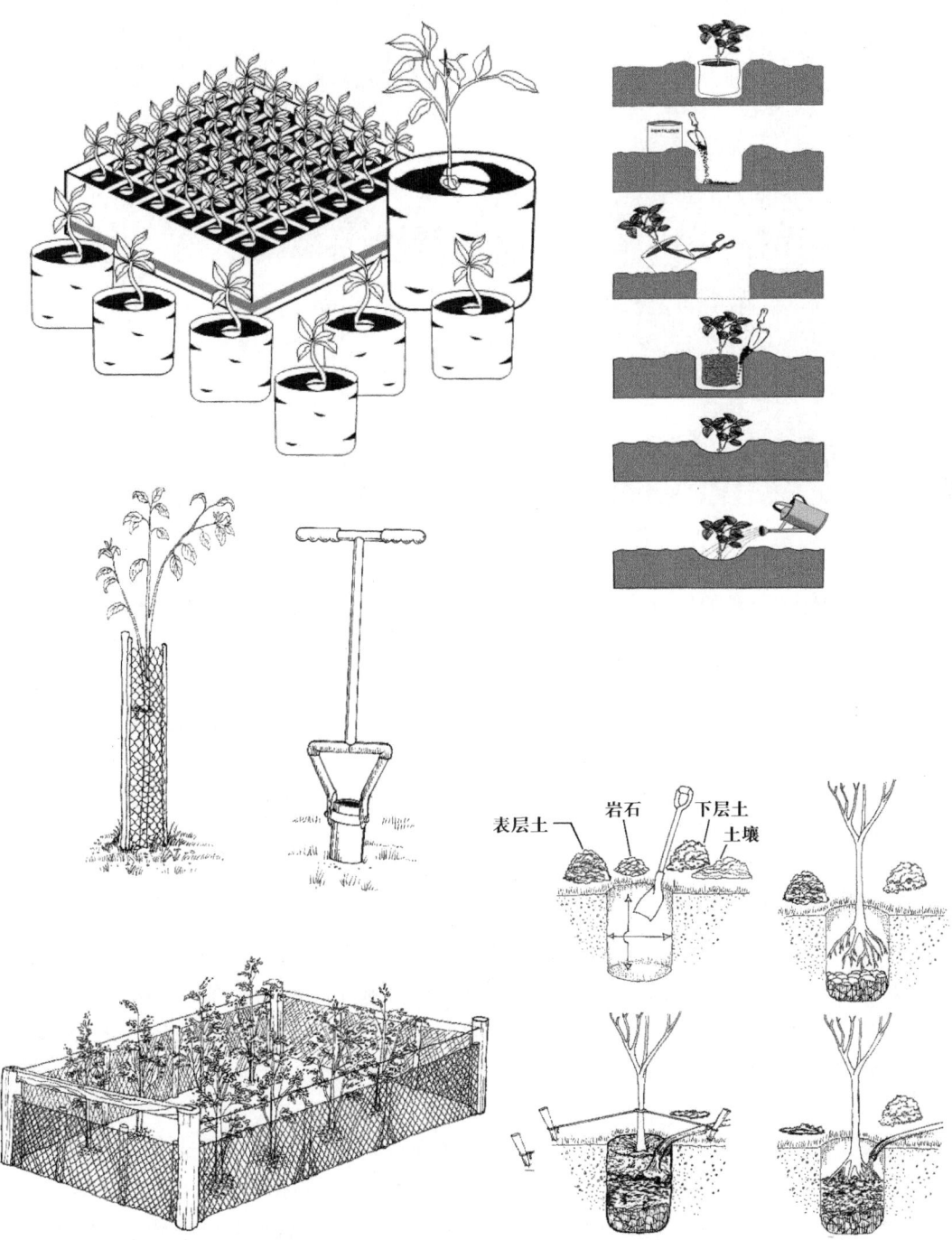

图4.6　恢复森林覆盖的各种工具和步骤

图 4.7　村民利用沙袋堵住湿地排水口

池塘给农民带来的好处在于，池塘可以承担多项功能。例如，作为给农作物浇水的蓄水池；家畜的饮用水源；可用于养鱼；改善当地的小气候；为各种食用昆虫的动物，如青蛙、蜻蜓和鸟类提供栖息地等。如果池塘中建有小岛，可以吸引很多鸟类前来。在小岛上，这些鸟类可以远离食肉动物的攻击，并将小岛作为它们的栖息和繁殖地。

将小型湖泊作为拉沙坝，可以减少沟壑的侵蚀程度，从而为当地的水土保持工作做出贡献。

4.4.1　确保水道的健康

洁净和富氧的水源看上去更加令人赏心悦目，可以容纳更多的生物共存。相反，受到污染、满是泥泞或者富营养的水源会杀死许多水生生物，形成难闻的气味，导致浮垢丛生，并且成为蚊虫和有害细菌滋生的温床。此外，这些受到破坏的水源还会泄漏到河道以外的整个生态系统之中，导致更大范围的环境破坏，并对人类健康带来威胁。

过多的水槽会堵塞水道，带来在清理和处理过程中的大量管理成本，同时还会减少水中的氧气含量，导致急需氧气的物种（如部分鱼类）的相继死亡。

要实现良好的水道和湿地管理，必须确保良好的自然或人工过滤和清理，避免它们受到污染，保持良好的水流，以利于增氧通气。生活在河道水生环境中的生物数量越多，河道系统将更加稳定，更不易于发生各种物种突然死亡、出现藻华或水草阻塞的现象等。

4.4.2　创建人工池塘

让水流经砾石基质河床及利用滤食性生物（如双壳类软体动物），可以过滤水中的泥沙和污染物。滤食性生物需要软质的泥沙或泥浆基质生存。泥浆基质只能用于水湍流很小，并且泥浆不会被搅入水中的地方。利用种植适宜的水草，可以将泥浆固定（图 4.8）。

图 4.8　带有塑料衬管的人工池塘可以拥有丰富的鱼类和其他生物体

某些最初看上去非常好看的水生植物可能成为危害严重的杂草，如凤眼蓝（*Eichhornia crassipes*）或大薸（*Pistia stratiotes*）等。应尽量根除这些植物，避免它们再度大量繁殖。

在农业景观的角落里保留部分永久性或季节性沼泽地通常是可行的。这些沼泽地本身可以成为极其重要的微型自然保护区。从全球范围来看，湿地面临人类开发的巨大威胁，仍处于自然状况的湿地已所剩无几。这些湿地可能繁殖有许多濒危沼泽植物（包括野生稻、荸荠和其他具有很高种质资源价值的物种）和野生动物（如两栖动物、龟、蜻蜓和鸟类等）。记住，蜻蜓是农田中控制病虫害的益虫。

水鸟和龟类可能需要安全的栖息地或捕食地。龟通常聚集在突出水面的岩石或生长在湖泊中的树干上。鹭类和鸬鹚则通常聚集在突出水面或长在小岛中的树桩上。这些栖息地可能成为许多动物夜间的栖息地或繁殖地。其他一些鸟类（如秧鸡、小鹭和林莺等）则需要浓密的芦苇滩或灌木丛作为栖息地。

在某些国家，人工湿地中也可用于保护鳄鱼。例如，在中国的华东地区，人工湿地可用于保护扬子鳄。

4.5　管 理 水 位

湿地管理主要涉及水位的控制问题，以便维持合理的湿地生境系统的持续运行。在湿地自然水文情势保持完好的情况下，最好的自然保护措施就是了解和维持水文状况的自然运行。

不过，在很多情况下，受引水、施工、地下水位下降及建坝或其他活动的影响，原生或自然水文情势已经受到人为改变，湿地管理者必须利用工程或机械手段来提升或降低所需的水位。

要认识到何时在某个特定的湿地需要更多或更少的水资源，必须深入了解当地生态状况，并且制定一项有关期望成果的明确计划。水位调节的时机将取决于迁徙水鸟何时到访该湿地，湿地管理者希望支持的物种类型，以及对不同物种需求的了解。涉禽可能需要泥质和沙质河岸，水深在 5-10cm，而雁鸭类需要的水深可能在 20-50cm。

放水淹没或排水的时机将影响哪些类型的植物能够茁壮成长或衰败，因此，湿地管理者必须了解湿地的野生动物主要依赖于哪些物种，同时可能必须在那些迁徙水鸟到达前为这些植物的生长提供适宜的水资源。

要获得上述知识和技能，通常需要数年的时间，而保护区管理者普遍缺乏这些知识和技能。这些知识不可能通过短期培训班快速获取到。正因如此，发达国家保护区的管理者通常是生态学者而非纯粹的行政管理者。他们管理保护地的目的是出于对野生动物的热爱和欣赏，并将之视为一项神圣的使命，而不仅仅是将其作为一项工作。

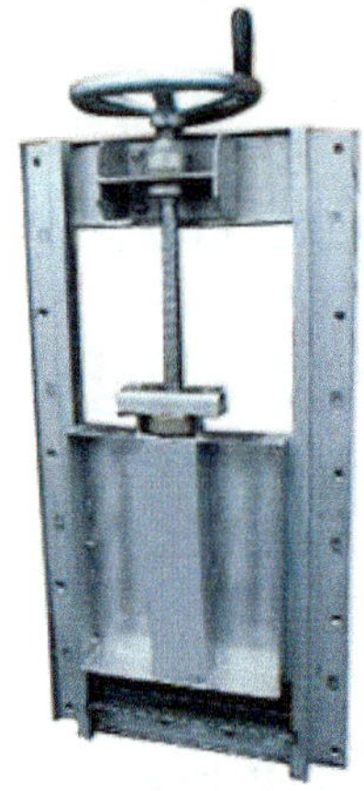

图 4.9　用于控制小型溪流的简易水闸

通常，我们可以将水位操控方法分为两大类型：①通过加快某个地区的排水速度达到降低水位目的的方法；②通过阻断或堵塞排水渠道达到提升水位目的的方法。

在极其罕见的情况下，湿地管理者可能必须利用机械装置，如水车、风车或水泵将水引至高处。

一套如图 4.9 所示的小型简易水闸系统可以满足绝大多数的水位控制需求。由水利部门管理的大型水闸可以用来控制大型湖泊的水位。在必须将排水口永久堵塞时，可以利用沙袋来封住排水口。可供湿地管理者使用的结构体很多，英国皇家鸟类保护协会（RSPB）就出版了有关这些结构体的应用技术指南（https://www.rspb.org.uk/Images/Water_management_structures_tcm9-214636.pdf）。

4.6　防火报告和消防

火灾可以很快毁掉一个生态系统，几乎所有的火灾都是由意外或故意的人为活动所导致的。闪电过后通常会伴随暴雨，使得野火不可能发生。不过，在某些情况下，雨水是在高空形成的，在到达地面之前就已停止。在出现此类暴风雨的情况下，可能出现“干雷暴”（即不会伴随暴雨的雷电），从而导致野火的爆发。

高温、干燥和多风天气是引起野火的最危险因素。不过，在中国，几乎所有地区都容易发生火灾，有时消防服务可能都无法控制，火灾会焚烧掉大面积的森林、灌木、农田，甚至毁掉房屋，造成人员死亡。

所有农村人口都有责任和义务注意防火，随时警惕、提前发现野火，并在野火发生时协助专业消防人员灭火。

在出现干燥的天气时，当地媒体应发布警示，提醒这是有可能发生野火的危险时期。在此期间，更需要保持警惕。在中国，由于男性吸烟者众多，并且有人会随意丢弃烟头，就很容易引发火灾。

中国的许多林区现已建立了防火带网络、火灾瞭望台和发生火灾时的灭火规程。

如果你正好生活在野火频发的地区，必须确保拥有适宜的防火服装和基本的消防工具。穿着质量好的靴子，对于在被大火烧毁的地面上行走时防止脚部被烫伤非常重要。应穿着颜色鲜艳的服装，以便当你受困或需要救援时他人能够迅速地发现你。穿着的服装应尽可能遮盖身体大部分的皮肤，但不能过于厚重造成行动不便。面罩、帽子和手套都极其有用。手边准备一把斧子或切灌刀，一把用于灭火的笤帚或平刃消防专用工具。铁锹可用于清理沟渠、清除干燥的地面覆盖物，也可用于扑灭火焰。

以下是在扑灭野火时有用的基本指导原则。

森林大火通常不可预测，由于它无法得到控制，也很危险。扑灭森林大火首先要把火势控制住，继而扑灭火焰。必须分两步走：第一步是将森林大火**控制**在某个范围内，阻止火情继续扩散；第二步是**扑灭**所有的热点地区，将大火完全扑灭，并继续对**边缘地区进行巡查**，确保不漏掉任何角落（图 4.10）。

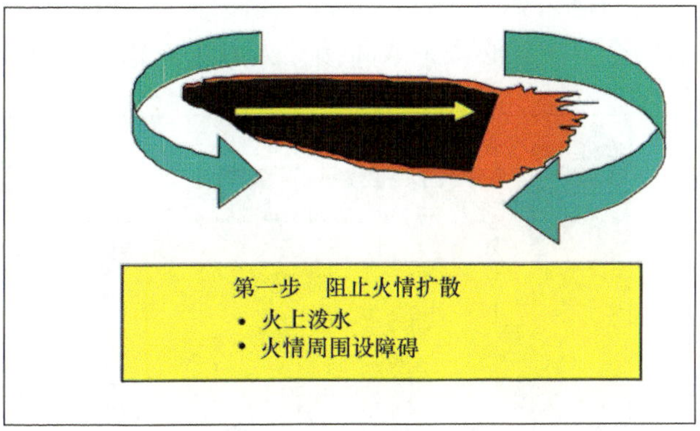

图 4.10　防火——阻止火情扩散

如果安全的话，可以采取**正面**灭火的方法。这通常是在火灾扩散速度很快，同时热气、火焰和烟雾非常严重的情况下。如果正面灭火不安全，可以采取一个替代的灭火方法，即**侧面**灭火的方法，首先从后面开始，再转向正面（图 4.11）。

侧面灭火方法是一个更为安全的灭火策略，消防员可以随时撤退至被大火烧毁的地面。这是正面灭火的替代方法。

当火焰和烟雾很大，正面灭火非常危险时，消防队的领队可能要求消防队退回至大火边缘地区最远 20m 的地方。然后，建立一条与大火边缘地区相并行的火情控制线（图 4.12），并确保能**随时看到火焰**。在建立火情控制线时，火情控制线与大火边缘地区之间的燃料必须全部燃尽。

在采用并行灭火方法时的安全注意事项如下。

如果大火边缘地区与正在建立的火情控制线之间的范围过大，大火可能突然爆发，并袭击消防员。许多重大事故都是这样发生的。正因如此，在建立火情控制线时必须

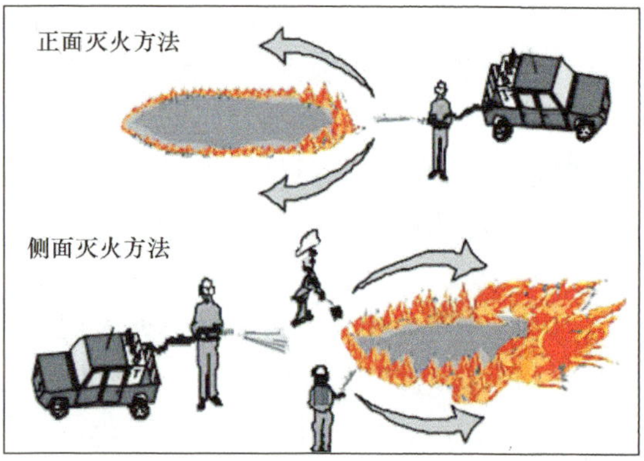

图 4.11　防火——正、侧面灭火法

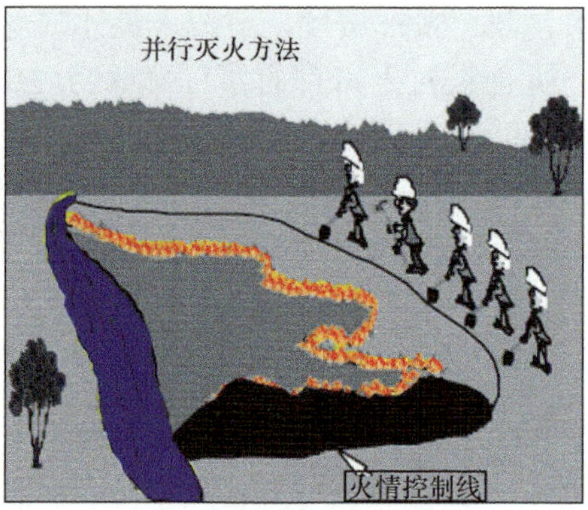

图 4.12　防火——并行灭火法

确保剩余地带的燃料全部烧尽，同时必须随时关注大火边缘地区的情况。

在采用间接灭火方法时的安全注意事项如下。

“迎面用火”是一种间接灭火方法（图 4.13）。必须选择一个适宜的防火边界线，可能距离森林大火很远。在适宜的条件下，沿防火边界点火，并朝着步步逼近的森林大火回烧。如果条件不允许，新火只会逃逸，导致火情更为严重，并危及消防员的生命安全。如果必须采取间接灭火方法，指挥官将给予明确的指示。

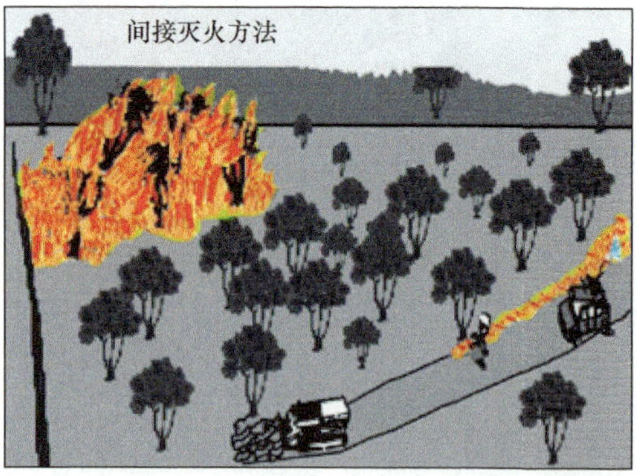

图 4.13　以火攻火——迎面用火（backburning）
当森林大火太强，无法用直接灭火法扑灭时，有时可采用“迎面用火”法。不过，这可能是一种危险的策略，必须随时由经验丰富的消防负责人进行管理

图 4.14　安全控制森林大火的方法

消防手段通常包括以上描述的所有方法。具体消防策略的选择取决于天气、燃料、坡度、火情、安全和可用资源。

第二步包括认真巡查大火边缘地带的每个地方，扑灭所有的明火和阴燃。这是消防活动中最艰难同时也是最重要的一步。因为任何一个火星逃逸，就有可能导致整个消防工作前功尽弃，被迫重新开始。

处理森林大火的方法取决于森林大火的火情（图 4.14）。辐射热量是一个重要的安全考虑因素，同时还必须考虑大火的移动速度、正在燃烧的燃料类型及大火的总体火势。如果火灾规模小、移动速度慢，并且天气或坡度不会对消防员带来明显的危险，那么火灾就可以在火焰边缘地带得到扑灭。这被称为“直接灭火”方法（图 4.15）。

这种方法直接针对火焰，使林火变小，但消防员仍必须在热浪和烟雾中作业。如果由于处于陡坡地带或燃料分布密集，使得消防员无法靠近林火边缘地区或者这样做不安全，就不能采用直接灭火方法。

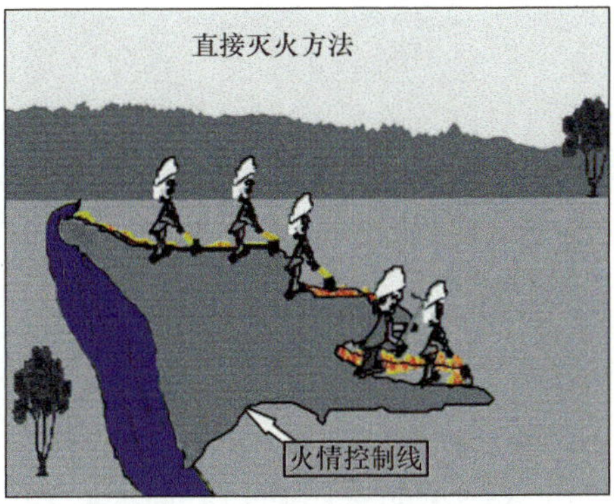

图 4.15　防火——直接灭火法

第5部分　物 种 管 理

5.1　辨认有价值的物种

我们未能认识到其价值的许多物种实际对生态系统和湿地环境具有很大的好处。然而，如果我们不重视这些物种的价值，并且不采取保护措施，我们就可能很快丧失掉这些物种及其免费提供给我们的各种惠益。有价值的物种可分为以下几大类：

- 控制病虫害的物种——猛禽、捕食昆虫的鸟类（图 5.1）、蛇、两栖动物、鱼
- 传粉物种（图 5.1）——蜜蜂、蝴蝶、飞蛾、甲壳虫、蝙蝠

图 5.1　湿地中部分有价值物种——传粉动物和食用昆虫的鸟类

- 防风林——林带、林地、周边森林
- 有助于土壤形成的生物体——蚯蚓、分解者、真菌等
- 野生食物——蕨类植物、蘑菇、水果、可食用的树叶
- 药用物种
- 特殊用途植物材料——纤维、树干、竹子、燃料、鱼毒、用作包装和做饭的树叶、茅草等
- 林地物种——薪柴、蘑菇、药用植物的来源
- 改良土壤的物种——固碳豆科植物等
- 家用物种的野生亲缘种——小麦、水稻、水果、鸭、原鸡

5.1.1 净化水源的服务

湿地最具价值的生态功能之一是其在净化水源方面发挥的作用。净化水的媒介主要是泥中的许多细菌和其他微生物。此外，许多以过滤物为食的生物，如珊瑚虫、软体动物、蠕虫和其他生物也有助于水源净化。其他一些软体动物和某些鱼类食用大量的水藻，避免了湿地遭到水藻的全部覆盖和淤塞。

5.1.2 传粉媒介

太阳鸟、捕蛛鸟、蝙蝠、飞蛾、蝴蝶、苍蝇、甲壳虫，特别是蜜蜂等均是许多湿地草本花卉的重要传粉者。植物的多样性增加了湿地的自恢复力和生态效率，因此，保护和维护充足的传粉者是湿地管理者一项重要的管理职能。

定期的监测应确保传粉物种的种群数量得到足够维持。但是，如果这些物种的种群数量开始减少和消失，必须确定其原因，并采取措施扭转此类发展趋势。

事实上，人们认为，杀虫剂在农业活动中的广泛使用，是导致世界各地许多传粉物种种群数量下降的主要原因。

在欧洲，尽管一个良好的自然保护区系统可以为绝大多数的传粉物种提供良好的栖息地和健康的食用植物，蜜蜂、飞蛾、蝴蝶和许多其他物种的种群数量仍然下降，这种状况令人担心。洞穴遭到破坏及住房屋顶建设标准提高，是蝙蝠大量减少的主要原因之一。

中国也不可避免地受到了这些变化的影响，某些地区的蜜蜂数量很少，无法充分对果蔬作物进行授粉，野生湿地草甸中的情况也非常类似。

有利于传粉者的方法包括：

- 允许在湿地内及其周边区域养蜂蜜
- 种植特别有利于花蜜流淌的灌木丛，促进蜜蜂野生种群数量的增长
- 与当地农民达成协议，减少他们对杀虫剂的依赖
- 为某些蝴蝶和飞蛾（参见 5.8.1 节）种植新的食用植物
- 留下中空的树干、洞穴等，供蝙蝠栖居；放置蝙蝠箱，促进蝙蝠野生种群数量的增长

5.1.3　种子传播者

所有植被类型都必须利用各种种子传播机制来散布种子，在新的可用栖息地进行定殖，并确保生态系统的丰富性。某些植物（包括一些红树林物种）的种子会掉落到海里，潮汐作用会将这些种子移动至潮间带周围。某些树木（如柳属、杨属）利用风来散布种子。许多其他植物拥有钩状的种子，这些种子会附在路过的动物毛皮上，甚至人的裤子上，在不知不觉中被带到其他地方。可结出食用坚果的树木如橡树（栎属、石栎属）或栗子（栗属）会利用松鼠或松鸡来带走其果实，埋藏、储存或丢下部分果实，从而促进物种的传播。不过，果实甜的树木和灌木需要吸引那些食用果实的鸟类或哺乳动物的注意力，这些动物要么直接把果实带到其觅食的栖息地，要么在吃完果实后通过粪便将种子排泄出来。

果树和食用果实动物之间的相互合作，是自然界自然更新和填补空缺的一种手段，同时也大大增加了湿地的多样性和吸引力，将更多的植物和食用果实的动物种类引入湿地。

榕、山楂、花楸、柿子、野梅、马缨丹和许多定殖灌木（如悬钩子、接骨木等）主要以上述方式生活在湿地中，但是，如果湿地没有食用果实的鸽子、椋鸟、巨嘴鸟、八哥、暗绿绣眼鸟和松鼠，这些植物也将很快消失。

一种值得关注的鸟是啄花鸟。这种色彩鲜艳的小鸟以桑寄生的黏果为食。一些种子会粘在啄花鸟的鸟喙上，这些鸟经常将黏性的种子从鸟喙上拍落至树枝上，通过这种方式将种子传播到新的生根发芽之地。桑寄生的花卉还可吸引漂亮的太阳鸟前来。当人们看到这些太阳鸟在林间飞快地飞来飞去，心情也会更加愉悦。

由于蝙蝠主要在夜间出没，它们在维持植物多样性方面的作用也往往被忽视。然而事实上，蝙蝠在对红树林、玉蕊和其他树木轻软柔和的花卉进行传粉方面具有重要的作用。大的蝙蝠也食用果实。它们会采摘大的果实，将之带到附近的栖居地，再慢慢享用这些果实。芒果、榄仁、某些大无花果、番木瓜等的种子都几乎全部是由蝙蝠进行传播的。其种子在很大程度上依靠蝙蝠进行传播的植物往往拥有下垂的果实，这样，那些倒着吃食的动物就可以更加方便地获取到这些果实。

某些倒悬在河道上的果树（如野生无花果）依赖鲶鱼吃掉落到水里的果实，鲶鱼再将这些种子带回到河流上游的新河岸地带。

5.2　对食草动物的控制

食草动物的多样性和类型将决定牧草和青草的高度和密度。应尽量维持最佳的野生食草动物数量。食草动物过多将导致草地的退化，湿地管理者必须找到减小放牧压力的方法。通常，当野生食草动物遭到狩猎后其密度降低时，灌木丛和乔木将入侵草地，湿地管理者可以允许部分家畜吃草，增加食草动物的数量。

5.3 救助和康复

5.3.1 通常无须建立救助中心

中国的许多湿地自然保护区投资建设了一些特殊的中心，用来容纳和照顾那些被救助或者受伤的鸟类或其他动物。不过，此类中心的价值不大。它们的修建、配备设备和人员、维护需要花费大量的费用。与那些需要得到更好保护的野生健康鸟类的数量相比，需要救助的鸟类数量毕竟很少。

鉴于中国每年有数百万只鸟死于鸟网，而许多湿地自然保护区所保护的鸟类数量只有数千只，显而易见，建立一个特殊的救助中心，救助几只或者数十只受伤的鸟类或哺乳动物是一件低优先级的事项。

因此，绝大部分的保护区无须建立救助中心，而应将其有限的资源集中用于更好地对野生种群开展就地保护，从而更好地开展自然保护工作。仅针对一些极度珍稀的物种，才有必要为维护救助中心而投入资金修建房屋、配备人员、开展兽医护理、喂食和动物护理工作。不过与此同时，救助中心也会分散保护区的注意力，并将人员从主要的自然保护任务中抽调出来。

5.3.2 对油污鸟类的救助和治疗

在油轮或炼油厂出现重大的原油泄漏时，浮油将对水鸟造成巨大的危害，那些覆满油污的水鸟如果不能得到及时的救助和治疗，就可能在痛苦中死去。鸬鹚、鹈鹕、剪嘴鸥和其他水禽最容易受到油污的影响。这些水鸟在拼命想将自身羽毛上的油污清除时，就会吞食一些毒性很大的原油。

绝大多数受到原油污染的鸟类都会死亡。那些不会很快死去的鸟类也会存在长期的健康问题，同时寿命和繁殖成功率都会降低。污染物也可以传递给鸟蛋和幼鸟，甚至传递给那些可能吞食油污鸟类的捕食动物。

被冲上海滩的油污鸟类的数量仅仅是那些在海上死去的鸟类数量的很小一部分。某些油污水鸟，如剪嘴鸥、鸬鹚、鹈鹕和潜鸭等即便采用最佳的救助和治疗技术，其生存率也极低。不过，另外一些油污鸟类，如天鹅、雁类和其他水禽可能对这些技术反应良好，几天之后就可释放。美国国家湿地研究中心（NWRC）就发表了一篇有关如何治疗油污鸟类的论文：http://www.nwrc.usgs.gov/wdb/pub/wmh/13_2_8.pdf。油污鸟的成功康复需要以下6个基本过程：

- 对油污鸟及时干预和找回
- 让油污鸟保持安定
- 清除油污鸟羽毛上的原油
- 清除油污鸟羽毛上的清洗剂
- 恢复鸟羽毛的防水功能

- 让鸟适应新的环境，准备放归野外

通常，预防更多的野鸟进入溢油区，比拯救那些已经受到溢油污染的鸟类更为重要。可以采用不同的措施来做到这一点，如借助巨大的爆炸物或者其他装置来恐吓水鸟，让它们远离危险的溢油区。与此同时，必须努力彻底清除这些溢油。溢油可能导致湿地恢复过程缓慢，并可能长达数十年的时间。

5.4 人工繁殖和再引入

目前，中国部分重要湿地物种仍在丧失之中。近年来，白鳖豚（*Lipotes vexillifer*）和此前的斑鳖（*Rafetus swinhoei*）一样已宣布功能性灭绝。长江鲟也可能在野外灭绝，其他一些极危物种包括江豚、几种鲟鱼等。

尽管大部分的湿地物种和所有的湿地生态系统服务功能通过就地保护和恢复栖息地可以得到最佳保护，但仍有部分物种已在野外灭绝或即将灭绝，这些物种只有通过人工繁殖和再引入计划才能得到拯救。不过，也有一些过度投资和过度依赖于维持人工繁殖种群的例子。

目前，部分鱼类在其成鱼摄食区和传统的河流上游繁殖地已不复存在，导致它们在当地的灭绝。湿地管理者在人工繁殖和再引入方面的技能必须保持和提高。

人工繁殖和再引入——最后的补救办法。中国的三个范例

朱鹮（图 5.2）体形漂亮，曾经广泛分布于中国、日本和韩国，现已在绝大部分地区灭绝。1981 年，有关专家在陕西省发现了 7 只朱鹮，这是朱鹮最后的一个小型种群。1986 年，6 只幼鸟从野外被带回，构成了北京动物园人工繁殖群的基础。此后，北京动物园一直在开展成功的人工繁殖和再引入计划。部分鸟蛋进行自然孵化，但更多的鸟蛋采用孵化器进行孵化。这也是世界上最为成功的朱鹮人工繁殖和再引入项目之一。目前数百只朱鹮仍生活在各种人工繁殖站，另外有几百只朱鹮被放归野外，包括重新引入日本。朱鹮的总种群数量现已超过 500 只。

图 5.2 朱鹮（*Nipponia nippon*）

扬子鳄（图 5.3）此前曾广泛分布于长江下游的众多湖泊和河道中，不过，受湿地围垦成稻田、污染、扬子鳄食用有毒老鼠及其他人为因素的影响，尽管扬子鳄体型不大，并且从未主动攻击过人，但仍面临几乎灭绝的命运。自 1979 年以来，安徽和浙江先后建立了人工繁殖扬子鳄保护区，并取得了极大的成功。目前这两个保护区人工饲养的扬子鳄数量已超过一万只。这些人工饲养的扬子鳄应作为将之重新引入其分布范围内许多受保护湖泊湿地计划的基础。不过，由于人工养殖业中商业利益的介入，对自然保护的关注变得并不明显。

图 5.3 扬子鳄（*Alligator sinensis*）

麋鹿（图 5.4）此前曾分布在华东的绝大部分地区，但在19世纪，由于遭到大量的猎杀而濒临灭绝，仅在北京南海子（南苑）皇家猎苑拥有一个人工繁殖种群。20世纪初，麋鹿在中国完全灭绝，仅有几只麋鹿被法国传教士大卫神父（Pere Armand David）带回欧洲，继续进行人工饲养，此后数量不断增加。1985年，22只麋鹿被带回中国，其中20只在南海子放归。1987年，另外18只麋鹿在江苏大丰沼泽湿地放归。在精心保护和良好的兽医护理下，这两个种群的数量均不断增长，现已有数百只。此后，多个麋鹿群被放归湖北和其他长江流域的新建自然保护区。同时，部分麋鹿个体偶尔逃脱自然保护区，形成了部分自由活动的野生种群。中国现有的麋鹿种群数量估计超过2000只。

图 5.4 麋鹿（*Elaphurus davidianus*）的再引入

5.4.1 采取人工繁殖的若干原则

1. 仅在野外拯救某个物种的尝试看起来可能失败，并且野生种群数量仅剩下最后很少几个个体时，方可将人工繁殖作为最后的补救方法。

2. 针对极危物种的人工繁殖计划应完全以自然保护为导向，不得以炫耀珍稀动物或商业目的为出发点。

3. 取消部分野生个体，建立繁殖群的决定不得由湿地管理者或者国家主管部门单方面做出，而应在世界自然保护联盟（IUCN）物种生存委员会的建议并与其取得一致意见的基础上做出，同时最好与获得批准的物种生存计划保持一致。

4. 此类计划应在兽医和生物学家的监督下，在安全、健康的救助中心开展。

5. 应按照被救助动物个体的历史、来源和亲代背景维护本计划中所有个体的种畜登记簿，并应对所有动物安装追踪器，用于个体识别。

6. 应精心选择亲代，确保各个亚种之间不会混交，并确保允许多个雄性成年动物为下一代做出贡献，以保持基因多样性。

7. 应精心选择人工繁殖动物的野外放归保护区，确保这些地区条件适宜，在正确的地域范围内，并不会受到直接狩猎的影响。在全部放归之前，将被放归的动物可能需要接受放归训练或适应新环境。应密切监控被放归动物的状况，评估是否成功放归，并汲取经验教训。这可能包括对部分个体安装无线追踪器。

5.5 物种的再引入

湿地是一种动态生态系统，常常面临本地物种的快速流失。当湿地的性质发生变化，湿地遭受极端天气事件或引水时，就会出现物种丧失的状况。在自然界中，物种可以快速迁至其他适宜的栖息地，或者在新的或经过改变的栖息地重新定殖。不过，由于中国绝大多数的湿地已破碎化，并且与整个水系统之间的连接性很差，上述自然界中的状况已不再可靠。这就要求湿地管理者通过实施监测、控制和再引入（必要时）政策，协助湿地保持或增加其极其多样和适宜的生物群。

再引入鱼类、植物和部分昆虫相对容易，但要再引入鸟类和哺乳动物的部分脆弱物种就可能极其困难。

世界自然保护联盟物种生存委员会是一个专门致力于物种再引入这一主题的专家小组，并定期发表有关物种再引入的方法和安全规程的最新指南（http://www.issg.org/pdf/publications/RSG_ISSG-Reintroduction-Guidelines-2013.pdf）。

物种再引入已成为一个被大量滥用的借口，被用于寻求预算资金来修建更多的建筑和设施及增加人员。然而，物种再引入通常并不必要，也不是湿地保护地应有的职责。在有必要开展物种再引入的时候，该项工作应由专业的物种再引入中心来开展，而非自然保护区或湿地公园的活动范围。

两种形式的物种再引入具有重大意义。第一种形式是移位，包括在一个湿地捕获物种个体，再在另一个湿地野放。这种活动可能非常必要，可恢复此前因人为原因被相互隔离的种群之间的连接性，或者恢复此前灭绝的当地种群。

此前保持连接的湿地现被大坝或堰隔离，或者成为不适宜许多物种生存的受污染水域。

在移位的情况下，必须让鸟类或哺乳动物存留的时间尽可能短。它们只需要极少量的检查，确保它们身体健康，并且能够在“新家”中照顾好自己。将动物养在箱子里，减少对它们造成的压力，或者放松布袋，以便动物在挣扎时不会伤及自身，也不会让人或者狗感到惊恐。

物种再引入的第二种主要形式是将圈养的动物直接放归，或者将此前圈养但因受伤而被救助的野生动物恢复健康后放归野外。除了按照一个既定的计划将物种再引入某个新的地点外，被救助的鸟类或其他动物更多的是被放归到离它们当初被发现时尽可能近的地方。

5.5.1 鱼类的增殖放流

通过人工放养鱼苗可以提升许多湿地的休闲价值和商业价值。通过释放人工繁殖场中的鱼苗，可以增加商用鱼类或游钓鱼类的数量。那些数量已经非常罕见或在当地灭绝的天然鱼类可以通过上述这种方式重新引入。如果成鱼到达上游繁殖地的天然通道被下游的堤坝或其他人为开发项目阻挡，上述方式就可能必不可少。

必须注意不要引入那些可能对湿地本地生物群造成危害的物种。相反，应尽可能从本地物种中选择适宜繁殖和放养的物种。

5.6　控制野生动物

5.6.1　控制顶级捕食者

当食物链中的捕食者抑制其被捕食者的丰度时，就可能出现营养级联效应，从而使下一个低营养级免遭捕食（如果中间的营养级是食草动物，那下一个低营养级就是草类）。例如，如果某个湖泊中大型捕食性鱼类的丰度增长，那么其被捕食者——浮游动物食性的鱼类丰度就会下降，而大型浮游动物的丰度就会增长，浮游植物的生物量随之就会相应减少。该理论已推动了生态学许多领域的全新研究。营养级联效应对于认识将顶级捕食者从食物链中去除的影响也非常重要，目前人类在许多地方已通过狩猎和捕鱼活动在食物链中去除了顶级捕食者。

通常，在湖泊中，大型捕食性（食鱼性）鱼类可能大大减少小型浮游动物食性的鱼类种群数量，而浮游动物食性的鱼类将大大改变淡水浮游动物群落的状况，浮游动物的摄食反过来也将对浮游植物群落的状况造成重大影响。去除捕食性鱼类可促使浮游植物繁茂生长，使湖水从清变绿。同样，去除大型和中型捕食性鱼类也可能导致藻类和蠓虫的数量过多。

5.6.2　减少外来或有害捕食性鱼类的数量

由于几乎所有的河道都受到很大程度的干扰，它们的性质发生了很大变化，其相互间的连接性被中断，加之各种地方许多外来入侵物种的引入，使得许多湖泊和其他类型湿地的鱼类区系的自然状况可能严重失衡。人类可能必须控制鱼类的管理，而这取决于湿地管理者希望实现何种目标。如果湿地管理者希望湖泊中有更多可供捕钓的鱼（如鲑鱼），可能就必须清除部分捕食性鱼类，如吃小鱼的梭鱼。一种可使用的方法就是电捕。操作员将一个小型发电机背在身上或者放在船上，将正负两极投入水中，距离约半米。这种装置可将鱼类吸引至负极端子，并形成一个钢丝网，可以将鱼捞出水面。然后，操作员可以选出那些需要从水中清除的鱼类，并将那些不需要清除的鱼类再放回水中。

5.7　吸引鸟类前来

通过提供适宜的栖息地及种植那些可提供花果、栖息处、掩蔽处、巢穴等的树木，可以吸引鸟类前来栖居。如果栖息地受到保护，鸟类就将做出良好的响应。因此，让某个地区避免受到狩猎、网捕或噪声的影响或干扰，可吸引鸟类更多地利用这一地区。在有湖泊的湿地，鸟类会觉得岛屿更为安全，并将小岛作为其栖息地和筑巢地。

5.7.1　利用巢箱

不同鸟类对巢箱的喜好情况见表 5.1。

表 5.1　不同鸟类对巢箱的喜好情况

带孔的小型巢箱	带孔的大型巢箱	正面敞开的巢箱	超大型巢箱
山雀	雨燕	知更鸟	红隼
麻雀	八哥	鹡鸰	灰林鸮
五子雀	啄木鸟	斑鹟	野鸽
红尾鸲	小鸮	鸳鸯	寒鸦

建造

在开始建造巢箱前，请仔细研读图 5.5 有关巢箱的插图、尺寸、使用材料的建议及孔径尺寸。确保建成后即可搭建好巢箱，巢箱可用的时间越长，吸引鸟类前来栖居的概率就更大。

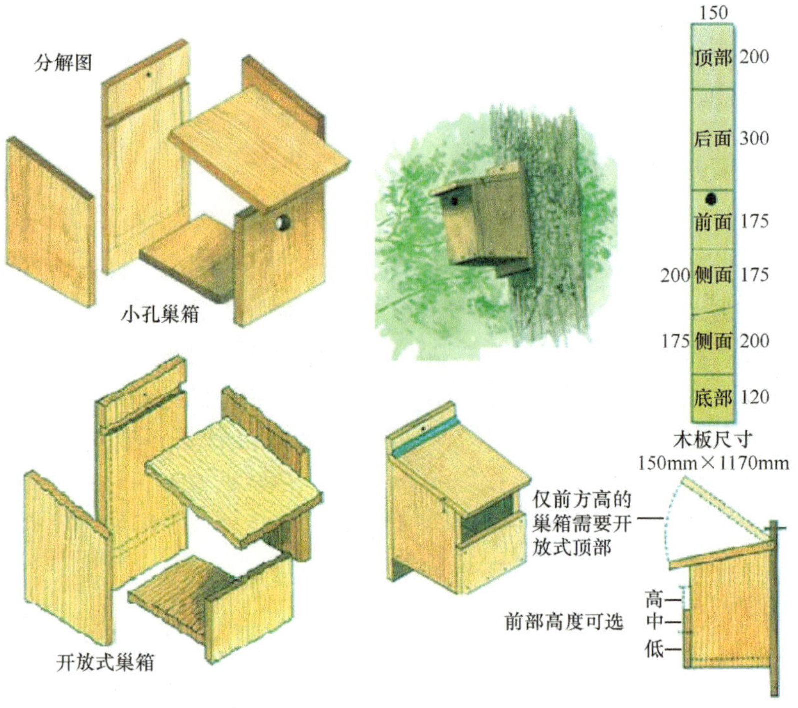

图 5.5　两种巢箱的结构示意图

材料

必须注意的是，巢箱内部不能太冷或太热，并且必须确保巢箱持久耐用。

- 巢箱必须采用木头制作。金属和塑料不宜作为巢箱的材料，这是因为金属和塑料材料可能导致巢箱内部温度过高，或者导致巢箱内部水珠冷凝，使得鸟蛋和小鸟受潮。
- 所使用的木材类型并不重要，不过硬木（如橡树和山毛榉等）的使用寿命超过软木（如松树）。
- 相反，木材的厚度更为重要，木材厚度至少需要达到15mm，才能提供足够的隔热性能，防止变形。
- 将巢箱用钉子钉牢，而非用胶水粘在一起（应使用镀锌或不锈钢钉，防止生锈），便于排水。
- 在巢箱的底部钻几个孔，确保进入巢箱的雨水能够很快流出。
- 不要在巢箱前方留栖息处，这是因为设置这种栖息处没必要，并且可能招来捕食动物！

孔径尺寸

不同尺寸的孔径适宜于不同的鸟类。

- 25mm 或以上的孔径，针对蓝冠山雀、煤山雀和沼泽山雀；
- 28mm 或以上的孔径，针对大山雀和树雀；
- 32mm 的孔径，针对家雀。

便于检查和清扫

巢箱应便于检查和清扫。

在巢箱顶部安装一个防水的铰链，以便可以方便地吊起，不会掉落。汽车轮胎内胎或丁基橡胶是最为理想的防水材料。首先按照巢箱的宽度切割橡胶，再将橡胶钉在巢箱背面和顶部。

通过为鸟类提供筑巢平台，也可吸引一些大型鸟类前来栖居。在中国，可以为生活在高原草地上、捕食啮齿动物的猛禽建造筑巢平台，或者在东北地区为白鹳的繁殖提供木制平台。

5.7.2 鸟类喂食器

在青黄不接的干旱季节为鸟类和部分哺乳动物提供额外的食物，可以帮助它们渡过难关，同时也能让游客或研究人员更好地对其进行观察。为此，如果将鸟类喂食器放在靠近游客中心或餐厅等的地方最好。

不同的鸟类需要不同类型的食物，因此有必要利用和建造各种不同的喂食器，可以通过网上购买或从专门的供应商那里订购。

通常，最好是将食物和水放在远离地面的地方，否则前来觅食的鸟类很容易遭到猫科动物或其他天敌的捕食。

如果松鼠可能干扰鸟类喂食器，最好将喂食器放置在松鼠无法进入的笼子里。

图 5.6 展示了部分标准鸟类喂食器。

图 5.6　鸟类喂食器和水浴器

5.8　吸引昆虫前来

5.8.1　吸引蝴蝶前来

蝴蝶是农田和湿地中的益虫。它们不仅是果树和豆类植物良好的传粉者，同时也可为宁静和美丽的乡村环境“增光添彩”。蝴蝶保持健康状态，是环境状况良好的表征之一。如果某个乡村地区缺少蝴蝶，则表明这里的农田中可能施用了过多的杀虫剂。在湿地和空旷的原野，蝴蝶通常在野生植物上产卵。不过，通过确保为珍稀或外形特别漂亮的蝴蝶提供适宜的食用植物，可以吸引这些蝴蝶前来这些地方。同时，可吸引成年蝴蝶在某些外形特别漂亮的花朵上觅食。要养好蝴蝶，主要需要为希望吸引的蝴蝶种植适宜的食用植物。许多蝴蝶选择产卵的地方相对比较特别，而且此类信息通常广为人知。在搜索到蝴蝶的食用植物后，许多详细信息可从网上下载。

下面列出了主要的食用植物科属及食用这些植物的主要蝴蝶种类。

热带地区：

柑橘——几种燕尾蝶

马兜铃藤——许多鸟翼凤蝶（图 5.7、图 5.8），包括裳凤蝶和暖曙凤蝶
肉桂——西番莲
夹竹桃——几种天蛾
温带地区：
荨麻——许多蛱蝶科
蓟——部分蛱蝶科
野生胡萝卜——常见燕尾蝶
柳——闪蛱蝶
野豌豆——许多灰蝶科
蔷薇——许多种蝴蝶
女贞、蓬子菜——天蛾

图 5.7　正在交配的鸟翼凤蝶

图 5.8　照料计划用于饲养鸟翼凤蝶的藤蔓植物

吸引成年蝴蝶食用花卉和其他诱惑性食物

通过种植食用植物，可以吸引蝴蝶幼虫。同样，也可吸引成年蝴蝶觅食花卉和其他诱惑性食物。在吸引外形艳丽的蝴蝶飞往风景观赏区及人们经常前去的地区方面，某些花卉相对之下更有吸引力。

在温带和热带地区，对蝴蝶很有吸引力的植物分别有大叶醉鱼草（图 5.9）及木槿和马缨丹等。不过，马缨丹必须得到很好的控制，否则可能变成外来入侵野草（参见 5.9 节）。

图 5.9　醉鱼草对蝴蝶很有吸引力

其他类型的诱惑性食物也可吸引一些外形非常艳丽的蝴蝶前来。例如，动物（灵猫）粪便或腐肉对闪蛱蝶（*Apatura*）和螯蛱蝶（*Charaxes*）很有吸

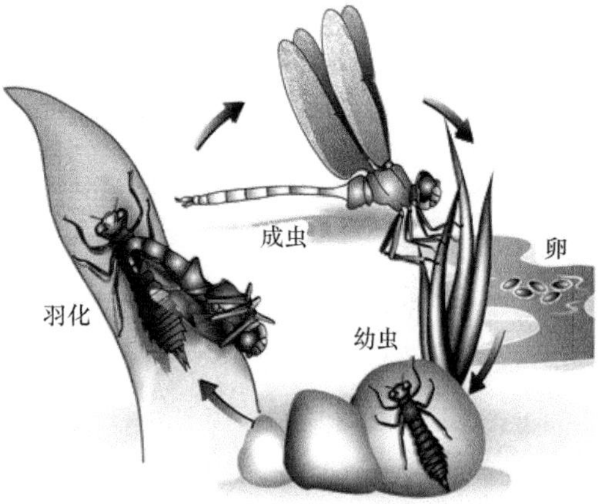

图 5.10 蜻蜓的生命周期

引力。而动物的尿液或者腐烂的水果（如香蕉）对螯蛱蝶、蛱蝶和燕尾蝶具有吸引力。水坑或富含矿物质水源的地面则对许多热带蝴蝶具有吸引力。粉蝶、剑尾蝶、青凤蝶（*Graphium*），甚至外形非常艳丽的裳凤蝶（*Troides*）都经常光临这些地方。

5.8.2 吸引蜻蜓前来

蜻蜓通常在生长在水上的植被中产卵，其生命周期中的绝大部分是作为水生食虫动物。蜻蜓的幼虫最后从水中爬出，使成虫得以羽化，并能在湿地上空的周边地区飞来飞去，捕食在半空中捕捉到的小飞虫。蜻蜓的整个生命周期（图 5.10）取决于水源的健康状况及适宜的被捕食昆虫的丰度。蜻蜓是反映湿地健康状况的良好指标性生物，通过保持各种类型的湿地栖息地（如湍急的流水、流动缓慢的水源、深水、浅水）和各种类型的水生植被，可以确保蜻蜓的多样性。

5.9 控制外来入侵物种（AIS）

当地湿地管理者可采取以下措施控制外来入侵物种：

- 密切关注新物种的引入和扩散
- 向当地管理部门报告潜在有害的外来物种
- 切勿开展有关引入外来植物种子、作物或观赏性植物的试验，除非已得到当地管理部门的许可
- 如果已发现有外来入侵物种，与其他土地管理方合作，采取早期控制措施
- 寻求当地科研机构或高校的技术援助，控制已导致严重问题的外来入侵物种

辨认对中国湿地造成最大威胁的外来入侵物种（图 5.11-图 5.20）。

必须采取以下应对措施来控制外来入侵物种：

- 制定旨在控制外来物种引入和放归的全国和省级法律法规
- 建立有关外来物种分布状况及如何辨认这些外来物种的全国数据库
- 提高人们对外来物种问题及已在各个地区得到确认的外来入侵物种的意识
- 在保护地层级对外来入侵物种问题开展监测
- 对造成严重问题的外来入侵物种开展控制或根除项目（计划）

警告：与使用毒素相比，使用生物控制方法对于非目标物种的破坏性更小。不过，生物控制方法也可能导致另外的外来入侵物种的引入！

这方面一个有价值的资源是“全球入侵物种计划（GISP）全球入侵物种数据库（Global Invasive Species Database of the Global Invasive Species Programme）”，该数据库

图 5.11 福寿螺

入侵稻田。密切关注粉红色的卵块。福寿螺可以食用

图 5.12 褐云玛瑙螺

入侵南方的果园和农田。玛瑙螺会吞食许多植物

图 5.13 黑耳蟾蜍

一种很好的控制害虫的动物，但也可能导致当地濒危两栖动物的消失！黑耳蟾蜍有毒。原生长在中国最南端地区，在其他地区则为外来物种

图 5.14 马缨丹

对蝴蝶很有吸引力，可形成一道绿篱，但不适合动物食用，并会入侵草场，很难根除

图 5.15 路易斯安那小龙虾

从美国引入。很美味，但会取代本地物种，并在堤坝内打洞，从而削弱堤坝的防洪功能，影响灌溉系统

图 5.16 罗非鱼

从非洲引入。很美味，但会取代本地鱼类

图 5.17　凤眼蓝

花朵非常漂亮，但会堵塞水道、阻塞船道，减少水中的氧气含量，导致鱼类死亡。凤眼蓝很难完全清除

图 5.18　大薸（水浮莲）

扩散速度很快，会堵塞水道，减少水中的氧气含量，导致鱼类和甲壳动物的死亡

图 5.19　泽兰

部分物种现已入侵中国南方 2/3 的农田和空地。其随风传播的小种子扩散范围很广。一些农民很愿意将这一物种作为绿肥犁回地中

图 5.20　互花米草

从美国引入，已入侵中国沿海地区，并快速扩散，导致当地的海滩植被和红树林消亡。互花米草很难得到控制，并且成本高昂

（http://www.issg.org/database/species/ecology.asp?si=1890&fr=1&sts=&lang=EN）应翻译为中文。

在上面的数据库中，按照“国别：中国”（country/China）及“栖息地类型：湿地”（habitat/wetlands）进行搜索，可以生成一个有关中国主要外来入侵物种的列表，包括每个外来入侵物种的简介及有关其详细的生物学状况、威胁及控制方法。

5.10 鼓励开展自然害虫控制措施

国际水稻研究所（IRRI）积极关注维护稻田景观的生物多样性，特别是恢复捕食动物和拟寄生物的多样性，以增强与害虫入侵和管理相关的生态系统服务。

水稻是一种短期作物，在水稻栖息地中生存的食草动物通常是*r*对策生物或物种，它们的寿命较短、体型小、经常移动、产卵多。因此，入侵的雌性食草动物在进入一处新的稻田后，会产下大量的卵。如果没有捕食动物和拟寄生物的多样性，这些卵存活的可能性会达到90%-100%，最终对稻田造成巨大的破坏。但是，如果稻田拥有丰富的生物多样性，入侵害虫的产卵绝大部分将无法存活，通常其存活率不到5%。因此，这些害虫对水稻作物造成的破坏较小，不会带来任何损失。破坏生物多样性的因素包括杀虫剂、极单一种植方法和很小的遗传变异。

国际水稻研究所开展的稻飞虱项目通过鼓励农民增加植物多样性，增加提供给捕食动物的资源，如花蜜和遮蔽处，来恢复当地的生物多样性。

一对有5只雏鸟的仓鸮在一个繁殖季至少可以吃掉3000只啮齿动物。仓鸮并非地盘性种类，可以为它们提供许多巢箱，10多只或20多只仓鸮能够集中在一个地区，构成一个完整的群落。在一个研究区域内，在48个巢箱中生活的仓鸮在8周内吃掉了至少17 000只啮齿动物。如果没有仓鸮加以控制的话，8周内就可能繁殖数千只啮齿动物。这样，有效避免了啮齿动物可能带来的灾害。

有助于维持自然害虫控制的部分实用方法：

- 为关键的猛禽类物种保留天然植被覆盖，并为其安装巢箱
- 保护洞穴和屋顶中的蝙蝠窝
- 接受燕子和雨燕在粮仓和屋顶筑巢
- 将蛇视为对农业生产有益而非有害的动物（绝大多数的蛇是无毒的，在没有受到刺激的情况下，蛇不会主动攻击人）
- 保护蜻蜓、青蛙等的湿地栖息地
- 为伯劳、猛禽、八哥、卷尾等鸟类设置方便的栖息处
- 限制农田中杀虫剂的使用

5.11 水产养殖业中的生物多样性问题

与专业的鱼池相比，半野生水产养殖系统中的生物多样性通常更高，因此受到病虫害影响的可能性要小得多。同时，由于半野生水产养殖系统需要额外添加的食物较少，因此造成的富营养化和污染程度也更小。不过，其缺点是存在引入外来入侵物种、与野生物种杂交或将新的疾病传染给野生生物群的巨大风险。

淡水养殖物种

中国养殖的淡水经济水产物种大约有 50 种。最为常见的养殖水产物种包括鲤鱼、鳊鱼和武昌鱼等。自 20 世纪 80 年代以来，随着国内外市场需求的日益增加，中国开发了许多新的水产物种或从国外引入中国进行商业养殖，这些水产物种如日本鳗鲡、鳜鱼、鲟鱼、甲鱼、河蟹、泥鳅、黑鱼、小龙虾、罗氏沼虾、罗非鱼、虹鳟鱼、白鲟、鲶鱼、蛙和欧洲鳗鲡等。2003 年，中国的淡水养殖产品的总产量达到 17 782 734t。各种水产养殖物种的产量在整个淡水养殖业产量中的大致比重如下：

物种名称	所占比重
鲢鱼和花鲢	30.10%
草鱼	20.20%
鲤鱼	13.20%
鲫鱼	10.00%
黑鲢	1.30%
鳊鱼和武昌鱼	3.30%
罗非鱼	4.20%
河蟹、大虾和甲鱼	3.40%
鳝鱼	1.0%
其他	13.30%

海洋和咸水养殖水产物种

中国养殖的海洋和咸水物种大约有 40 种。传统的海洋养殖水产物种主要限于四大组群的软体动物：牡蛎、蚌蛤、血蚶和花蛤。扇贝和鲍鱼是在 20 世纪 80 年代开发的品种，海藻是在 20 世纪 50 年代开发的品种，而虾养殖业是 20 世纪 80 年代的一项主要经济产业。中国养殖的海洋和咸水虾主要包括：明虾、日本对虾、草虾、南美白对虾、墨吉对虾、长毛对虾和刀额新对虾等，其中南美白对虾现已成为产量最大的虾类。海洋养殖鱼类从 20 世纪 90 年代开始大规模养殖，主要的养殖鱼类包括：海鲷、遮目鱼、海鲈鱼、日本比目鱼、乌鱼、黄花鱼、石斑鱼和河豚等。从国外引入中国，并在中国得到成功养殖的鱼类包括海鲈鱼、大嘴鲈鱼、多宝鱼、拟棘鲷等。2003 年，中国海洋养殖产品的总产量为 12 533 061t，各种海洋养殖物种所占的比重如下：

物种名称	所占比重
海洋鱼类	4.10%
甲壳动物	5.30%
软体动物	78.60%
海藻	11%

图5.21展示的是根据联合国粮农组织统计的有关中国水产养殖总产量的数据。

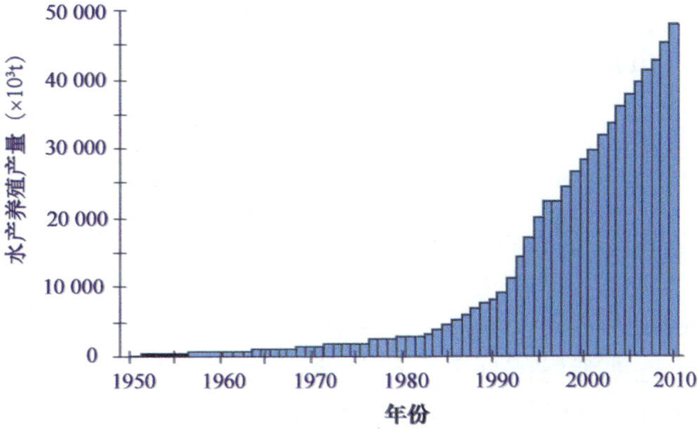

图5.21　1950年后中国报告的水产养殖产量

联合国粮农组织渔业统计数据

第 6 部分　主流化与传播

6.1　开展跨部门主流化的必要性

尽管国务院指定国家林业局为负责中国湿地保护工作的牵头单位，但国家林业局本身不可能单独完成这项艰巨的任务。除林业部门以外的其他许多部门也利用湿地，同时湿地也为这些部门提供服务。因此，国家林业局必须协调其自身活动与其他利益相关方需求之间的关系，并呼吁其他利益相关方管理其在国家林业局控制范围以外的各种类型的活动。我们将这种合作称为“主流化”（mainstreaming）。

要实现如此之多部门和机构的规划、活动和需求之间的协调一致性，必须建立多个跨部门的协调委员会，并将湿地管理计划和策略纳入总体发展规划之中。

6.2　当地社区的参与

生物多样性保护工作不能简单地由政府下达命令或委托某个机构来执行。该项工作必须赢得当地社区的认可和支持，他们的活动将最终决定当地的生物多样性是得到破坏还是得到保护。当地社区必须提高对生物多样性问题的意识，更多地参与到与那些旨在保护生物多样性和修复当地生态系统的活动相关的规划和决策过程之中。同时，必须向当地社区提供恰当的培训和激励措施，以便他们可以更充分地参与到自然保育、生态系统保护和监测活动之中。

激发当地社区更好地参与湿地保护的激励措施可能包括生态补偿计划、对开展自然保育活动的付费，鼓励报告有关生物多样性的最新消息和信息，确保土地利用安全，促进惠益共享，保护本地知识产权等。在对更为有效的生物多样性立法进行审查和修订的过程中，必须谨记以上各个方面。

此类活动还将有助于缓解当地农村的贫困状况。中国的贫困人口绝大部分生活在农村地区，他们的生计和福祉直接依赖于当地的生态系统服务，这些生态系统服务包括食物生产、淡水供给及避免遭受各种危害等。因此，投资于这些生态系统服务的维护和恢复，将有助于改善农村人口的生计，成为帮助他们脱贫的基石。

阿波礁的故事

与世界上许多其他珊瑚礁一样，在菲律宾阿波礁（Apo Reef）上生活的渔民其生计也主要依赖于他们从海上捕捞的渔获。然而，过度捕捞一度导致海上渔获量不断下降。菲律宾一所大学的海洋生物学家安吉尔·阿尔卡拉（Angel Alcala）

教授成功地说服当地村民开展了一项试验。他们将 20% 的捕捞区划定为“禁渔”保护区。试验实施后两年内，保护区内的鱼类数量恢复到足以产生“过剩”幼鱼的程度，这些幼鱼开始迁移到周边仍在被捕捞的珊瑚礁中。25 年来，该地区总的渔获量一直不断增长！阿波礁的成功使阿尔卡拉教授说服菲律宾更多的渔民建立自己的“禁渔区”。

6.2.1　负责任的渔民

渔民是指其生计依赖于或享有利用渔业资源权利的个人。负责任的渔民是指以公平、合法并且可持续的方式利用渔业资源，不会对渔业环境和渔业资源种群造成不可逆转损害的渔民。负责任的渔民尊重他人对同一资源所拥有的权利，不会浪费或破坏对他人可能同样必需的资源。负责任的渔民尊重被捕获的动物权利，仅捕获他所需要的鱼类资源，并且以人道主义的方式杀死这些鱼类资源，尽量不对被捕获的生物造成不必要的痛苦或残害。

6.2.2　要求农民减少农业化学物的用量

目前，中国面临日益严重的污染问题，很多污染问题是由水污染导致的。

化肥和农药（杀虫剂、除草剂、杀真菌剂）是农田主要施用的化学物质。所有这些化学物质对环境都是有害的，可能对人类、鸟类和其他野生动植物的健康状况构成威胁。受这些化学物质的污染，从农田中排放到溪流中的水质也会下降。最大限度减少污水处理的必要性，将有利于环境保护和经济发展。

化肥造成的代价非常昂贵，它们将增加除草的成本，如果未能得到控制，它们将会导致河道出现藻华，或者池塘和湖泊出现富营养化。施用新的化肥，常常会导致鱼类的突然死亡。

减少施用化肥必要性的方法：

- 选择适宜本地的强健作物种类
- 在一年中的自然生长期种植作物
- 让杂草在地上腐烂而非将其堆在一起烧毁
- 施用天然肥料（泥炭、堆肥、绿肥）而非化肥
- 将作物生长区用栅栏隔开，减少来自行人、车辆和动物的干扰
- 使用适当的栽培方法，改善土壤结构和健康状况，促进草皮的生长
- 开展土壤实验，确定植物所需的准确养分

限制杀虫剂的用量或影响的方法：

- 人工除草
- 仅施用生物可降解化学制剂
- 施用针对具体目标对象的制剂

- 避免在邻近河道的农田或在雨天使用化学制剂
- 在景观中使用护根覆盖物
- 在科研的支持下使用生物控制措施，如引入可作为食用植物的野草
- 确保具有丰富的生物群，其中有针对目标害虫的天敌
- 安装适宜的栅栏
- 为害虫设置阈值，仅在必要时处理某些区域
- 了解化学制剂对健康的危害，辨别害虫，并对化学制剂进行正确的处理

6.2.3 施用绿肥

尽管绿肥制作方便，价格便宜，并拥有众多好处，但它们通常未得到充分的利用。这些好处包括：

- 首先，绿肥是保持土壤养分的好方法。在冬季，土壤中的养分很容易被冲刷掉。在春季，可能从土壤中淋洗掉的养分被保持住，用于春季作物的施肥。轻质沙性土壤对冬季松土不利，在这种情况下施用绿肥最为理想。
- 其次，绿肥可抑制杂草的生长。两种植物不能在同一地方生长，施用绿肥可防止杂草在一个地方生根。
- 最后，绿肥可改善土壤结构，增加腐殖质。植物根系会使土壤变得脆弱，并吸收可能以其他方式被浪费掉的矿物质和养分。

除了保持土壤养分，部分绿肥还可固定空气中的氮素。绿肥的固氮能力意味着植物实际是在为人们提供肥料。

尽管自然保护纯粹主义者可能将各种不同的泽兰作为外来入侵物种，但泽兰毫无疑问是良好的绿肥物种，中国南方的农民一直从有利的方面来看待它。泽兰生长浓密、高大，可遮挡更难去除的杂草；当它们被修剪至贴近地面的高度时，就会很快死去；同时当它们被丢弃腐烂或在下次播种前犁入土中时，会很快地分解掉。

满江红属（*Azolla*）蕨类植物是水田中一种有价值的固氮绿肥，其他半水生的豆科植物，如田菁属（*Sesbania*）、黄芪属（*Astragalus*）和合萌属（*Aeschynomene*）也是如此。

从生物学角度看，豆科作物，如豆类、三叶草和紫花苜蓿等也能够固定来自大气中的氮素。这不仅有利于豆科作物本身，同时也有利于间种作物或后种作物，从而减少或消除施用氮肥的必要性。对豆科作物生物固氮（BNF）进行更好的量化，将为农民管控氮素，以便最大限度地提高生产力和减少对环境的危害提供更好的指导。许多此类植物可以作为喂养家畜的良好饲料型植物，作为养蜂蜜的花卉，也可以作为休耕地在犁地栽种前的良好绿肥。

6.3 共　管

共管是指政府部门和非政府部门共同分担管理职责，是更深层次的合作方式，

而不仅仅是与当地社区的合作，包括在规划、监测、报告和工作量方面分担部分管理职责。

共管不是指职能的简单重复，而是组建一个有效的团队。在这个团队中，每个成员的最大优势可以得到最佳利用。表 6.1 列出了政府机构和当地社区可能最适宜于履行的职责。

表 6.1　政府与社区各自的职责

政府机构	公众 / 社区
规划 / 分析	调查 / 咨询
巡逻 / 执法	监测
报告（正式）	报告（非正式）
区域控制	辅助
游客中心	文化娱乐
保护管理	其他劳力
监管 / 控制	度假村
公路、码头维护	交通（船只、马匹等）
管理条例	缓冲活动
游客中心商店	手工艺品
汇总 / 分配收入	股东
承包	租借

6.3.1　森林管理责任制度

对集水区森林开展适宜的保护，往往是维护湿地健康和丰度的一个基本要求。

按照最近颁布的法规规定，政府将此前集体林的林区土地权属重新分给当地的村庄或农户进行保护管理。土地权属的期限从几年到最高 70 年不等，如果管理效果令人满意，有时可能续约。其结果参差不齐。当然，有很多成功的案例。不过，也有许多失败的案例。在部分地区，当地管理机关对承租人提供的管理水平感到失望，从而被迫终止土地权属协议，将土地收归政府管理。

在归属国家天然林保护工程管理的地区，承租人在森林的管理、砍伐或再种植方面很少有自主选择权。不过，在绝大多数此前是集体林的森林中，承租人在管理方法方面确实拥有自主选择权。

产量低和生态状况不佳的原因如下：

- 土地权属的持续时间不够长
- 由于农户森林归属权持有不够集中，导致管理低效
- 缺乏林业管理技能和能力
- 无法获得最好的种子库
- 对生态服务提供的补偿不够

基于迄今为止的经验，政府正通过延长土地权属期限，提供续签合约的保证，以及允许土地权属的自由出售和转让，来解决前两个问题。此外，政府通过开展各种生态补偿机制的示范活动，来改善生态系统服务的补偿状况。在这些生态补偿机制中，流域下游的受益者将为上游良好的流域保护工作所提供的生态系统服务支付费用。

6.3.2 家庭旅馆（homestead）生态旅游

目前，户外旅游业在中国蓬勃发展，可以为经济贫困地区创造大量的经济收入。不过，游客也给当地的自然保护地造成了很多的负面影响，可能轻易地扼杀旅游业“这只会下金蛋的鹅”，同时当地贫困社区往往并未享受到旅游业创造的经济效益，这些经济效益通常流向“外部”投资者。按照定义，生态旅游是一种破坏性较小的旅游形式，可以在环境保护方面带来净效益，同时有利于促进当地社区的发展。从这个意义上讲，推动基于社区的生态旅游是一项很好的政策——目前基于社区的生态旅游在中国并不发达，拥有巨大的发展潜力。

图 6.1 是两种不同的生态旅游发展模式。在传统的生态旅游发展模式（图 6.1a）中，大量的旅游活动是通过大型的私营企业来开展的；而在优选模式（图 6.1b）中，大量的旅游活动是通过当地社区来开展的。图中红色箭头代表旅游收入的流动方向，绿色箭头代表必须提供生态补偿。

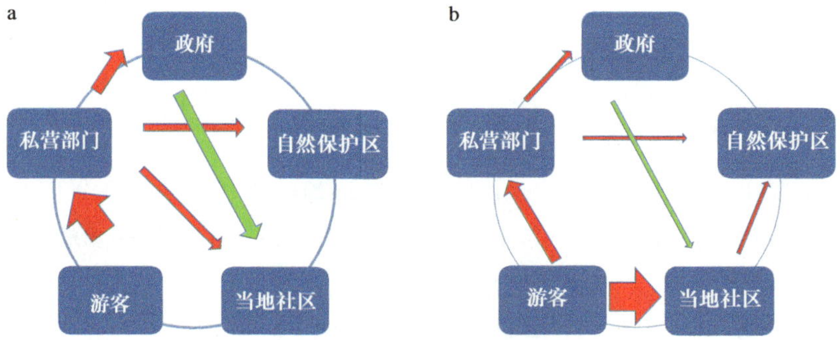

图 6.1 两种生态旅游开发模式

优选模式中两个尤其成功的国际范例分别是泰国北部地区和马来西亚沙巴州（Sabah）。

中国当地社区可以提供的基于社区的旅游活动类型如下：

- 家庭旅馆体验
- 当地正宗菜肴
- 划船、骑马、有导游的爬山
- 个人野生动物摄影观兽旅行 / 飞钓
- 冬季越野滑雪
- 皮划艇 / 漂流 / 攀岩

- 骑山地自行车 / 悬挂式滑翔
- 少数民族文娱活动（音乐、猎鹰等）
- 手工艺品

几户家庭可以相互协调，建立一个小型的协会，共同经营一个网站，宣传其旅馆设施和娱乐项目，接受网上预订，并且多个不同的家庭就各自接受哪些预订达成一致意见。此类旅游方式可以为游客带来更加个性化的本地体验，同时有助于改善当地人口的生计状况，从而使得他们更愿意帮助保护当地的特色、风景和野生动物。

注意：如果没有建立相应的设施来足以接待大量的游客，生态旅游可能面临诸多风险。

6.4　公安与执法

中国湿地的执法力度仍显薄弱，其原因包括：

- 法律法规略显薄弱
- 湿地的保护权限与利用权限之间有时存在冲突
- 湿地保护地人员通常没有管制权，而森林公安在帮助湿地的执法方面略显动力不足
- 法院对湿地的重要性所知较少，通常总是同情当地贫困人口
- 巡护、报告和执法的力度不够
- 通常缺乏确凿的证据
- 管理人员不熟悉刑事诉讼程序

必须有效应对以上所有因素。目前，国家和各省的湿地管理机构正在加强和修改相关法规。

保护地管理者必须确保当地社区及当地公安部门和法院对湿地相关法规的内容及其重要性有充足的认识，同时应与当地公安部门和法院的负责人建立良好的工作关系。

保护地管理者必须建立一个值得信赖的公开信息提供人网络。

必须给每个保护站配备至少一名相当于警察身份的巡护员，拥有逮捕和没收非法物品的权力。

必须加强巡护和报告体系，并填写所有巡护工作的报告表，涵盖所有有关非法人为活动的信息。非法人为活动的报告应提供手机照片证据。9.8 节中附有一个简单的报告表模版。

巡护线路和时间必须保持变化，确保不会错过发现任何非法人为活动的机会。同时，由于许多非法活动发生在夜间和节假日期间，在此期间也必须开展巡护活动。如果发现有严重或持续的非法活动，可能必须启动特殊的执法程序。在此情况下，必须尽可能长时间地对执法程序的具体细节和时间进行保密，避免巡护人员将这些情况透露给他们的亲戚朋友。重大的执法活动必须有公安人员陪同参加。

6.5　非政府组织（NGO）的作用

自20世纪80年代以来，许多国际自然保护组织进入中国，与不同层级的政府部门或保护地合作，协助和资助了许多自然保护项目的开展，这些国际组织包括世界自然基金会（WWF）、大自然保护协会（TNC）、保护国际基金会（CI）、野生动植物保护国际（FFI）、湿地国际（WI）、世界自然保护联盟（IUCN）等。20世纪90年代以来，一些本土非政府组织开始在中国出现。不过，迄今为止，只有几百家国内环保非政府组织与政府部门、媒体和当地社区开展合作。这些国内非政府组织可以分为以下几种类型：注册的非政府组织、未注册的志愿团体、非营利性企业、学生环保协会、观鸟协会、网络团体等。所有这些非政府组织都可给湿地管理者带来新的专业知识、人力、动力和支持，推动重要湿地保护地及其物种的保护。非政府组织可在公众、媒体和政府之间架起一座重要的桥梁，并可长期在基层开展工作——许多政府机构也可以如此。

有必要修订现有的相关法律法规，如1994年颁布的《自然保护区条例》。同时，新拟定的有关国家公园和其他类型保护地的法律，也有助于推动非政府组织参与保护地的共管工作。

6.6　了解“护鸟队”！

如果你在某个周末来到大连周边的山上闲逛，可能会遇到一批身着迷彩服、神情严肃、身体强健的年轻人。你可能会以为他们是一个特殊的武装部队。不过，你再仔细看看。一些年轻人的笑脸娇小可爱，表明他们不可能是解放军战士！相反，他们是一支“护鸟队”（Bird Force）（图6.2）。

正如志愿者Grace Xiao所说：“我们都是志愿者。没有资金资助我们，所以我们全部自费来阻止非法捕猎活动。我们的负责人是一位军事迷。他现在是一家装饰公司的经理，他创建我们这个团队的目的是保护大连老铁山的鸟类。他向我们提供货车、锯子、小刀、双筒镜和他收集到的许多军事装备。每个周末，我们的团队成员都要来到老铁山。巡护的时间很长，但也很有趣。我们常常要在山上行走6个小时或者更长的时间。如果我们发现有必须销毁的猎网，就会记录下其GPS的位置。此外，如果有人告诉我们哪儿有鸟儿受伤，我们通常也会在我们的微信群中公布这一消息。谁有空就可以前去把受伤的鸟儿转移到大连市的鸟类救助中心对其进行治疗。”

这些年轻志愿者的专业性也毫不逊色。事实上，他们如同军队一样接受训练，爬山、悬挂、绕绳下降、举重物、在崎岖不平的山上跑上跑下。不过，事故时而发生。一次，一个女孩不幸摔断了腿，被救护车紧急送往医院。

“护鸟队”负责在山上和森林中四处搜索有无非法的捕鸟网和捕鸟器。有时，可能为时已晚，但他们仍会将非法捕鸟网取下，并将死去的鸟儿尸体处理掉。每当

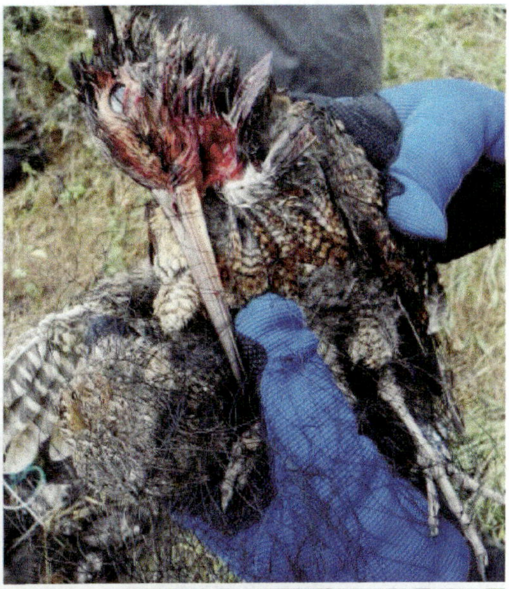

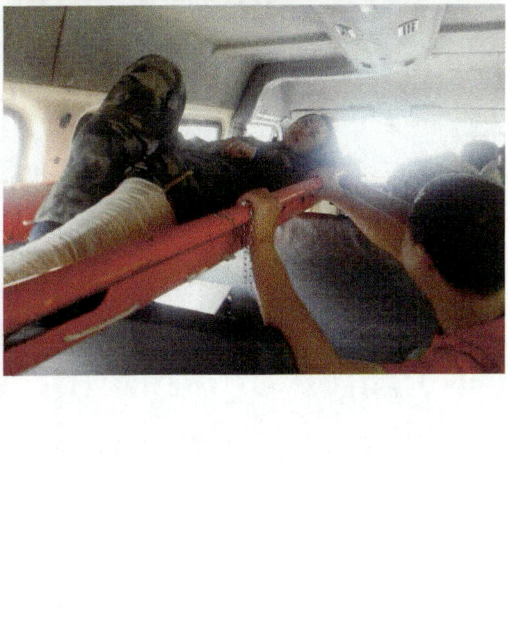

图 6.2　护鸟队在训练和工作期间的照片

他们救下被偷猎的鸟儿，并将之放归野外时，他们就会感到非常欣慰，觉得自己的付出是值得的。小型的鸟类相对容易救护。不过，由于大连地处猛禽迁飞线路上，许多鸟都是体型较大的鹰和鹗。有时，这些猛禽必须被送到大连猛禽救助中心进行体检和护理，在被认为适宜于放归野外之前进行进一步的饲养甚至施行小型的外科手术。

这是多么出色的一群志愿者！过去 30 年间，非法用网捕鸟的活动导致中国本土的鸟类种群数量大幅减少。如果中国能够有更多诸如这样的团体来表达他们对野生鸟类的爱心和重视程度，帮助减少非法的用网捕鸟的行为，该是一件多好的事情！同时，我们也对所有其他观鸟群体表示感谢，感谢他们愿意投入时间和金钱，并且投入专业知识，致力于鸟类的保护。

6.7　传　　播

6.7.1　制定传播策略

良好的传播是良好的湿地保护地管理中一个至关重要的组成部分。在实现重要湿地的良好保护的各种目标方面，一个很大的阻碍因素是公众对湿地价值的认识程度较低。在整个中国，无论是在中央层面还是地方层面，人们普遍误认为经济发展和生物多样性保护之间存在固有的冲突。所有层级的政府部门对健康的湿地和生物多样性保护与人们的生活质量、经济繁荣和人类福祉之间的关系也认识不足。

要更好地阐释湿地的价值、保护湿地的必要性，告知人们湿地面临的威胁和其他问题，并且激发人们关注和支持湿地保护地系统或某块特定湿地的保护，一个有效的途径就是制定良好的传播策略。

传播策略可以满足不同受众的传播需求。我们确定有以下3类主要的目标群体：

1．**政府决策者、当地政府机构及政府官员**，特别是湿地保护地周边的此类群体。

2．**公众**。

- **年轻人**，通过正规的教育活动和学生辩论组、观鸟协会等
- **普通大众**，通过大众媒体和网站
- 湿地保护地周边的**当地社区**

3．**中介机构和合作伙伴**，帮助将环保信息传播给前两类目标群体，但其部分能力仍需得到提高。

- **关键媒体机构**（报刊、电视）**的记者**
- **教师**，负责讲授有关湿地重要性的新材料

传播策略应明确主要参与方在特定地区的湿地及其生物多样性问题中发挥的作用，同时针对不同的目标群体设计不同的传播信息。

第一类目标群体，即政府决策者、当地政府机构及政府官员的意识增强，有望为湿地及其生物多样性保护工作划拨更多的资金，并且更好地将生物多样性保护纳入省级部门的对话、计划、规划和政策之中。中国政府官员对湿地问题的关注度加大，也将通过官方媒体报道，让更多有关湿地及其生物多样性保护的新闻见诸于世。

第二类目标群体，即公众，了解新的相关法规，与政府部门在湿地保护行动中合作甚至对政府部门开展游说，提高对湿地及其生物多样性保护的关注度和参与度，帮助降低个人和公众活动对环境的负面影响，也非常重要。

第三类目标群体，即中介结构和合作伙伴，将作为前两类目标群体之间的中间人。

为覆盖各目标群体，实现预期的特定成果，我们拟定了以下5种措施：

- 通过大众媒体、出版物、简报、网络和活动提高人们的意识
- 通过将新课程内容引入学校培训课程开展宣传教育
- 开展相关培训，提高员工、合作伙伴和中介机构的能力
- 提供相关信息，以便制定更好的决策和增强政府部门的意识
- 宣传保护地的活动和旅游景点

所有这些传播活动都非常重要，值得大力开展投资。建议每个湿地保护地聘用一名专业的全职传播官员。不过，由于预算紧张、人力有限，传播策略应确定其最为有效的方法，以便发挥最大的影响力。可以采取以下方法：

- 利用更多的合作伙伴和资源，而非单打独斗

- 利用大众媒体（电视和报纸）覆盖尽可能多的受众
- 利用明星来宣传湿地及其生物多样性保护问题
- 将湿地生物多样保护问题与品牌形象挂钩
- 投资于具有滚雪球效应或高复制潜力的方法，增强影响力，如通过中介机构传递相关信息
- 将活动时间安排在可引起高度关注的时刻
- 保持灵活性，充分利用这一不断变化社会所涌现的各种新机会
- 快速响应，提前准备好相关材料，以便快速做出回应
- 依托现有的思想体系，而不是创造全新的思想体系，如道教、儒教、社会主义、尊重人与自然保持平衡和整体发展观，部分传统信念等
- 利用政府和联合国渠道，覆盖非政府组织不易覆盖的群体，如政府高级官员、国家林业局和环保部等

通过与相关的政府机构、科研机构和非政府组织建立合作伙伴关系，来扩大影响力。可以提供强大支持的政府部门合作伙伴包括水利部门、环保部门、林业部门和农业部门下属的渔业管理部门、沿海地区的海洋管理部门等。保护地管理者应尽量将保护地的管理计划与各省的再造林计划、生物多样性保护策略和行动计划、旅游发展计划及生态红线保护计划保持一致。媒体活动的时间安排应该与媒体关注的时间段保持一致，此类活动有如年度环境日、生物多样性保护日、“爱鸟周”等。项目宣传的所有元素都应带有一个易于辨识的品牌形象，有助于提高其知名度，并监测项目对目标受众的影响。可开展“知识 - 态度 - 行为（KAP）”调查，并据此确定在信息和意识方面的需求，为最终制定传播策略奠定基础。

应基于贴近人们日常生活和经验的具体示例来编写重要的信息。向目标受众传达的重要信息必须清晰地说明生物多样性保护可以带来经济和社会效益，如①有关健康的湿地生态系统对气候变化具有更大适应力的证据；②有关湿地生物多样性对人类健康具有重大影响的证据：健康的湿地可以提供优质的水源、预防水灾等。

必须针对每类不同的目标群体采取不同的传播策略。同时，针对以下问题的答复必须包含在所有重要信息中：

- 确切地说什么是生物多样性?
- 生物多样性与我的日常生活有何关联?
- 生物多样性有何价值? 为什么必须保护生物多样性?
- 生物多样性与社会经济发展之间有何关系?
- 生物多样性与自然灾害之间有何关系?
- 哪些物种处于危险之中?
- 生物多样性丧失带来的后果是什么?
- 哪些活动对生物多样性具有破坏性?
- 为保护生物多样性，普通大众可以做些什么?
- 为保护生物多样性，政府可以做些什么?

6.8 提高人们意识的方法

在传播策略的指导下，保护地应组织一系列活动来提高人们的意识。可以采取的方法很多，而互联网可以提供一个全新的工具，更好地传播相关信息。不过，活动应满足不同受众的不同需求。嘉宾讲座可以更令成年人感兴趣，而要吸引小孩子的注意力，可能必须采取不同的形式。下面列出了保护地为提升形象、提高当地人群的意识和获得当地人群的支持，可以采取的部分方法：

- 出版物——书籍、宣传册、折页、海报、指南等
- 网络——专用信息网站、网站、博客
- 教育——教学或编写教材、课程、模块等
- 新闻制作——电视、电影、文章
- 对话——让所有相关方，特别是与自然界息息相关的当地社区参与对话
- 活动——利用精心选择的时段来引发人们的关注。通过吸引名人或当地重要人物的参与，来吸引人们的关注。

6.8.1 观鸟比赛 / 摄影比赛

观鸟比赛在湿地日益受到欢迎。观鸟比赛可以提升湿地的知名度，提高人们保护湿地和鸟类的意识，提供有价值的数据，甚至可以作为筹集资金的活动。

观鸟比赛通常为一整天，即24h。观鸟队登记参赛，每个队2-4人。比赛规则是在当天尽可能记录各种不同的鸟类。每次记录必须有该队至少两名成员作为目击证人（如果观鸟队有4名成员，则至少有3名成员作为目击证人）。团队成员必须始终在一起。活动组织者可以决定是否让通过声音辨别出的鸟作为可接受的记录。其他比赛规则通常包括禁止利用录音机作为引诱物，禁止夜间用照射灯，始终待在比赛规定的范围内，禁止进入私有土地（面向所有观鸟队开放的除外）。

裁判员会在现场随时检查观鸟队提交的记录清单。获胜团队通常会获得部分奖品。当地企业可以向活动提供赞助，并因为向自然保护活动提供支持而获得环保贡献奖。

摄影比赛也是提高公众对保护地的关注度和保护意识的一种好方法。

6.9 拍摄鸟类和野生动物

拍摄鸟类和湿地保护地其他特征的好照片，这一点非常重要。照片是用于存档的良好记录、用于辨认的证据，以及帮助记载和监测变化的良好方法。同时，照片也是制作宣传材料、展览和新闻所不可或缺的素材。公众常常要面对来自各行各业的大量精美照片。要抓住公众的眼球，我们拍摄的照片也必须质量上乘。

鸟类摄影本身是一项富有挑战，但很有意义的活动。由于鸟类通常体型小，羞怯，并且通常与我们保持适度的距离，因此我们需要一部带有功能强大的镜头的照相机，

才能捕捉到小鸟的细部特征。同时，我们需要在高速状态下的良好光线，以便对运动之中的鸟类进行定格。以下小窍门可确保您获得更佳质量的鸟类摄影作品。

6.9.1　照相机

狂热的鸟类摄影爱好者（图 6.3）带动了日本经济的发展，更能提高佳能和尼康等日本照相机生产商的经济效益。好的镜头往往也很贵重，如果掉到地上或者被水打湿很容易坏掉。因不同的管理目的，价格便宜得多又更加方便携带和拍摄的相机也可买到。全新的类单眼照相机提供从广角到长焦的各种变焦镜头。这些照相机的体积更小，价格也相对便宜。不过，照片的质量还是个问题！

图 6.3　使用长焦镜头的鸟类摄影爱好者

过去，使用长焦镜头常常必须使用坚固的三脚架或沙袋。不过，现在数字卡的速度很快，你可以使用多达数千种不同的感光度 ISO，无须携带笨重的三脚架即可拍摄快速移动的照片。

长镜头非常笨重，手持和带上山很耗费体力。此外，还必须保持光学原理和便利性之间的平衡。绝大部分人发现，一个长 400mm 的镜头无须三脚架就可以轻松对付。然而，更长的镜头就必须使用坚固的三脚架，但是要达成重量和放大倍数之间的折中，也可使用一个 1.4 倍或 2 倍的适配器。如果能靠近目标，使用长度相对短的镜头就可以获得更好的效果。

数码照相机的性能现在越来越好。现在，你可以利用手机或 iPad 就能拍摄出高质量的照片。对于拍摄风景、活动、花卉和静止不动的鸟来说，手机或 iPad 还不错。但是，要拍摄野外活动的鸟类，手机或 iPad 就无法胜任了。相反，你需要一台桥式相机（现在有的长焦镜头最高可达到 50 倍）或者一个可以利用各种长焦镜头的更大相机。该市场的两大领先竞争对手分别是佳能和尼康。

单筒望远镜。该设备对于观测大型水体中的涉禽和其他鸟类几乎必不可少。许多观鸟者十分依赖这种望远镜，在森林和其他地方也会使用它。这种望远镜的优势在于可以使用稳定的三脚架，并且放大倍数很高。通过绝大部分的单筒望远镜就可以拍摄出相当不错的照片，同时可以用适配器将单筒望远镜与尼康 Coolpix 相机等小型数码相机连接在一起。

三脚架。通常，相机的镜头越长，光量越低，所需的曝光时间越长，相机或拍摄对象需要移动的次数更多。因此，必须使用三脚架使相机保持稳定。一个笨重的长镜头必须要一个稳固的三脚架和镜头支架。

6.9.2 拍摄好片的小窍门

使用隐蔽处。许多鸟类都不怕隐蔽处，即便它们能看见里面有各种活动或听见里面有声音都是如此。正规的自然保护区可能设有供公众使用、放置在适宜地方的隐蔽处。一些特殊的地方也可修建隐蔽处，供摄影者隐藏其中，或者也可使用小型的便携式隐蔽处（图 6.4）。在线供应商也出售迷彩网，是一种快速轻便的替代隐蔽工具。小汽车可以作为移动的隐蔽处，去接近路边的鸟类。

图 6.4 小型摄影隐蔽处

摄影的良好地点。带着相机四处拍摄，可能让摄影者拍下大量质量平庸的野生动物照片，这些照片可以用于辨识或存档，但很少能达到用于出版物的水平。你必须倍加小心，找到适宜于观察野生动物的良好地点，如水坑、饲喂区、潮线、营巢区、栖息处、求偶地和该季节的最佳饲喂点（果树、粮食作物、捕捞地点及昆虫出没地等），就可以拍摄到更多质量更高的照片。

时段。俗话说："早起的鸟儿有虫吃。"必须在鸟类开始活动前准备就绪。一天的中午时分，鸟儿通常无精打采。赶紧回家去吧，大吃一顿！

诱饵和食饵。你可以放置一些有吸引力的食饵或诱饵（对水禽和鸽子等非常有效），像经验丰富的老猎人一样，吸引鸟儿到一个特定的"拍摄地"。即便是一个旧的猫头鹰标本或模型，也可能吸引许多小鸟的注意力。

声音回放。一些观鸟者利用录制歌曲或此前录制或下载的收藏声音文件的回放，来吸引本地鸟类靠近，便于观看和拍摄。不过，切勿过分使用这一技巧。这是因为声音回放可能严重干扰和搅乱正在繁殖中的鸟类活动。播放猫头鹰的叫声，可以吸引许多小型雀类靠近。

修图。在以前的胶片拍摄时代，修图被认为是整个拍摄艺术的一个组成部分。通过修图可以去除斑点和划痕，与画面无关的要素也可以得到“弱化”。可以增加或减少曝光量和对比度。采用数码照片，可以实现许多的特效。不过，正统主义者认为过多的图片处理有点类似于弄虚作假。尽量将修图的比例减少至最小。

归档。借助数码摄影，我们现在可以保存数以千张的照片，而每张都是大文件。你必须学会精挑细选，去掉那些不重要的照片和重复照片，只保留那些拍摄得很好的或你认为需要的照片。按照主题和日期对照片进行明智的归档，否则今后你可能浪费大量的时间来寻找你所需要的照片！

视频。现在，绝大多数的照相机也可摄制高清视频片段。对于保护地而言，保存质量较好的视频剪辑集，可以剪辑后用于以后其他的节目制作。所有针对摄影的小窍门也同样适用于视频拍摄。不过，视频图像必须保持流畅或稳定，因此使用优质的三脚架非常重要。应确保同时抓拍下适宜的自然声音作为背景，特别是如果视频拍摄过程中原声出现机器、音乐或人声时更应如此。

全自动照相机。现在，全自动照相机的价格越来越便宜，拍摄质量也越来越好，并被日益用于捕捉那些怕生野生动物的图像。可以将全自动照相机放置在可以拍摄野生动物经常出现的路线的地点、饲喂点或饮水点附近。全自动照相机利用红外光束来发现所拍摄场景内的运动情况，可以在夜间摄制黑白照片或在白天拍摄彩照片，也可拍摄照片或视频。翻看相机照片是一件非常激动人心的事情，可以发现所“抓拍”到的精彩镜头。如果不想让全自动相机被好奇的过路者偷走或离群的大象所破坏，切记把全自动照相机很好地隐藏起来。8.3.5节对全自动照相机的使用方法提供了更为具体详细的建议。

使用闪光灯。在有阴影的地方或者夜间拍摄鸟儿时，闪光灯就非常有用。不过，闪光灯会导致眼睛出现奇怪的反射情况，使照片看上去不那么自然。此外，闪光灯如果距离被摄制对象过近，可能严重干扰被摄制对象。请负责任地使用闪光灯。

微距摄影。一方面，拍摄鸟类可能需要大量使用长镜头；另一方面，要拍摄昆虫、花卉和广角镜头，则必须使用微距摄影技巧。景深非常重要，因此在专业的微距摄影中，经常使用闪光灯或环形闪光装置。逆光照明和小角度拍摄，可以增强特写摄影的魅力。

摄影者须知

- 切勿在拍摄照片的过程中伤害鸟儿。
- 切勿干扰或进入营巢地。
- 切勿过于靠近鸟儿，或者将其赶走。
- 切勿播放本地鸟叫声，干扰鸟儿的繁殖。
- 切勿为获得更好的视线而破坏植被！
- 切勿在危险的地方用诱饵诱惑鸟儿，使之处于危险境地。
- 保持安静！

6.10　游客管理和教育

游客到访湿地公园和其他保护地，有助于开展意识教育和公众教育，提供健康、有价值的休闲娱乐活动，为保护地获得收入，并且为当地社区带来各种创收机会。不过，游客的过度利用可能“杀死那只会生金蛋的鹅”，严重破坏湿地和干扰湿地中的野生动物。因此，必须实现两者之间的合理平衡。

保护区应欢迎和鼓励游客到访和欣赏中国这些美丽的湿地。不过，必须赢得公众对湿地维护工作的广泛支持，增强他们对湿地在促进环境质量和环境安全方面所具有的经济和功能重要性的认识程度。公众对湿地的利用程度和关注程度，证明了建立保护地的必要性，同时也鼓励当地政府对湿地保护地提供资金和其他方面的支持。

在许多情况下，湿地公园或保护地需要获得门票收入，以支付基本的运行费。危险在于，为增加游客收入的冲动，可能导致保护地的管理与自然保护的需求之间产生冲突。

确保游客对湿地保护地的利用保持在安全限度以内，有以下 3 种方法：

1. 精心开展有关游客承载力的评估，作为确定每个特定区域或季节时游客数量严格限额和控制的基础。

2. 对保护地进行分区，将游客限制在其影响相对不严重的非敏感区。应将关键的保护地划归核心区内，严禁游客经常出入。

3. 利用各种方法最大限度地减小对保护地的负面影响。这些方法包括建立影响程度小的木板人行道或小路、最大限度地减小对野生动物干扰的遮蔽物和隐蔽处，使用影响程度小的卫生间和有效的垃圾处理及收集系统，使用指导游客行为的布告牌，如保持安静、禁止乱丢垃圾、禁止烟火、禁止危险攀爬、禁止采集植物、严禁干扰动物、禁止游泳、禁止乱扔石头、切勿离开有缆绳的地方、禁止乱涂乱写等。游客参观保护地前，最好是先参观信息中心或宣教中心，提前了解到有关参观地点、穿戴设备和行为方式的建议。

6.11　游客中心的设计

保护地的游客中心（图 6.5）是主要的接待区，在这里，保护地的管理人员可以与游客进行互动，并向其提供建议和指导。

图 6.5　游客中心

游客中心可以与售票处、宣教中心甚至饭店、厕所、纪念品店和 / 或适宜的观鸟隐蔽处等连为一体。

游客中心的面积将取决于其所包含的功能及预计容纳的游客数量。

在设计或运营游客中心时，应始终牢记以下注意事项。

游客中心应体现保护地的风格及环保政策。

游客中心的外观结构应非常漂亮，但必须与周边自然环境协调一致，不能过高或过于花里胡哨，应采用当地的天然建材，如石材和木头，而非混凝土和钢材。

不同的受众所拥有的兴趣和需求各不相同。游客中心应满足各种不同的教育、增长知识和互动的目的。可以建立一个专门的儿童角，在这里，那些可能对观看信息显示屏感到枯燥乏味的儿童可以自己玩各种染色成套工具、沙坑或互动玩具。

针对一年四季和各种天气状况进行规划。例如，儿童可以在户外游玩探险游玩区玩耍，而家长可以阅读更多的宣传材料。游客中心必须拥有足够的室内空间和让游客感兴趣的东西，以便在外面下雨时为游客提供寓教于乐的机会。

法律越来越多地要求旅游地为残疾人士提供相关的设施，因此保护地必须提供轮椅坡道、宽大的门和特殊的卫生间设施等。残疾人因为身体的限制，很少能有机会出去欣赏大自然的美景。保护地应为之提供机会，让他们可以通过良好的展品间接欣赏到绝大部分的美景和野生动物。

展品必须高质量并且令人感兴趣。死气沉沉、粗制滥造的鸟类和其他动物标本应成为过去式。展品应该是色彩鲜艳、充满活力、富有魅力的照片或视频，以及有趣的说明文字或解说词。

售票区应提供相关基本信息，包括保护地地图、旅行指南、野生动物识别材料和告知游客正确行为方式的小册子等。游客商店可提供各种书籍、纪念品、观鸟设备等。

游客中心应提供使用说明和流动方向指示牌，这样游客可以非常容易地获得相关信息或快速转入下一个旅游项目，而不会朝着交叉和相反方向四处流动。

可以在游客中心设立一个教室或会议室，可以播放视频或用于召集会议、嘉宾讲座等。

游客中心如能作为自然爱好者的活动中心，或用于对学生寓教于乐的课堂，则是一件更好的事情。

6.12　提供给游客的信息

必须向游客提供基本的信息，使其可以从参观中得到更多的东西，选择最佳的参观路线、当季的活动、时间安排和关注点，就保护地内的适宜行为提供建议，并提供有关湿地及其生态环境、物种和特殊重要性的基本情况。

这些信息可以通过多种方式提供，这些方式包括训练有素的导游、保护地网站、游客中心提供的地图、宣传册、折页、博物馆、信息中心、分布在保护地各个地方的标牌和布告牌等。最好让游客随时了解到他们在保护地地图中所在的位置，向游客提出替代路线的建议，介绍在保护地内不同地点应关注的地貌、植被或物种特征，以及如何辨别这些特征。

可以在观察板或最新新闻简报中提供有用的最新信息。观鸟者也可在网络、博客和电话应用上报告最新的观察情况。

户外信息牌必须文字和图片印制清晰，并置于防水框中或雨棚之下。这些信息牌应放置在适宜的高度，并保持适宜的角度，以便于游客阅读。印刷油墨必须质量很好，否则暴露于日光下很容易褪色和模糊。

图 6.6、图 6.7 展示的是不同类型的户外信息展示牌。

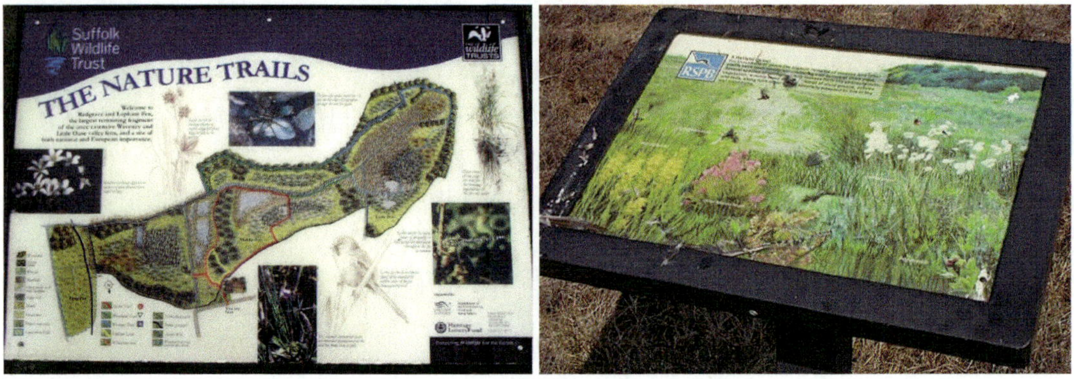

图 6.6　游客导览图

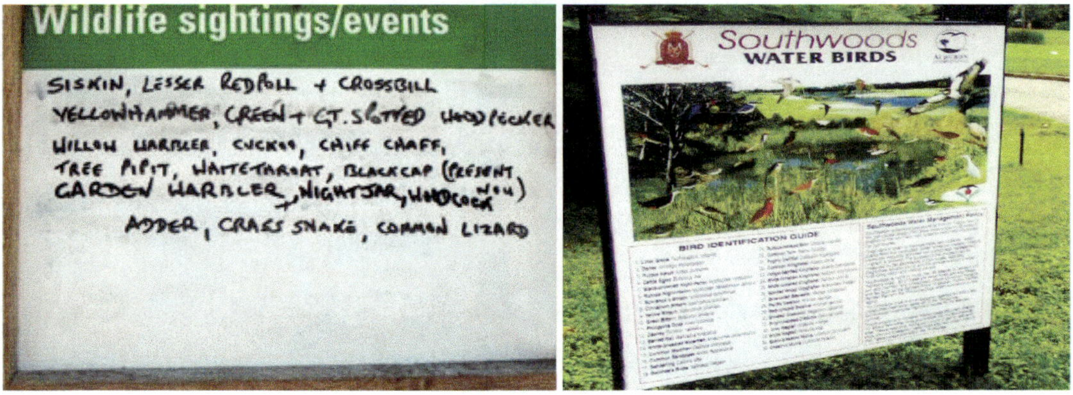

图 6.7　游客信息板

6.13　博物馆的建立和维护

许多湿地自然保护区和湿地公园都认为有必要修建和维护一座博物馆。不过，保护地的主要目的应该是保护野生动物，而非展示被屠杀的动物！同样，保护区不是展示圈养动物的动物园。请将用于无关展示的时间、金钱和人员尽量限制在最小范围内。

维护不善的小型博物馆和动物园看上去非常糟糕。这些园里的动物状况很差，十分可怜。而那里的动物标本的保护和陈列状况都很差，有的甚至开始变得斑驳破碎，并且布满灰尘和霉菌。捕获动物用于展示，无论是活着的动物还是作为标本的动物，都是野生动物的巨大损失，与保护地自然保护的角色完全不相符合。

6.14　宣 教 计 划

许多湿地自然保护区和湿地公园针对学校学生制定了成功的宣教计划，将有关湿地的知识寓教于乐。保护地可以组织实施自身的宣教计划，或者仅向当地学校或非政府组织提供其相关设施，当地学校或非政府组织再组织孩子们的探索和学习活动。

宣教计划给湿地保护地带来的好处在于，它可以有助于保护地获得更多的公众支持，培养更多的自然爱好者和未来的自然保护者。

宣教计划的具体内容取决于湿地保护地的性质和安全性、参与该计划的孩子们或学生们的数量和类型。

孩子们可以参与部分有益的活动，如帮助清理垃圾、监测各种物种状况、种植树苗等。

如果保护地管理机构能够提供带有白板的教室、各种项目设施、卫生间等，宣教计划就可能取得最佳的效果。

网上有大量的宣传材料，可以让孩子们对湿地产生兴趣。这些材料可以下载和改编，或者翻译成中文，使之适合于当地的实际情况。

各种活生生的事物最能吸引孩子们的兴趣。他们喜欢探索和发现事物，绘制各种事物，并谈论他们的体验。在孩子们的妈妈可能会觉得反感的同时，孩子们喜欢在地面进行探究，而且越脏越好！

6.15　简报还是网站？

许多湿地自然保护区、湿地公园和自然保护项目通过定期出版简报的方式来尽量提高公众的意识。一些简报富有吸引力并且包含许多有用的信息。不过总体而言，人们发现这种方法花费不菲，对保护区的管理人员来说费时费力，效果不佳，并且到达的受众数量十分有限。

另外一个费用更低、更为简便的解决方案是建立一个网站。不过，除非网站做得很好，否则通常难以吸引大量的定期点击量。以下建议可能有助于减少时间和精力的浪费及效果不佳的状况，有助于湿地管理者覆盖更广的受众群体。

第7部分　运　　行

7.1　结构、设施及其维护

修建木板路或人行道，可以限制游客对湿地产生的不良影响，确保游客不逾越游客活动区，并且还能在整个湿地或附近区域为游客提供安全便捷通道（图 7.1）。

图 7.1　小路和木板人行道典型设计方案图例

这些设施造价昂贵，并且需要定期维护。设计时必须小心，并应牢记以下事项。

路面湿滑，寒冷天气尤其如此。防滑措施包括在倾斜路面上方安装铁丝网，添加粗糙防滑板条，尽可能保持路面干燥。加快干燥速度的一种方法就是在木板人行道板材之间留出缝隙，光线可以通过这些缝隙到达地面，避免路面变得黑暗而潮湿。

建筑材料最好是木材或石头，这样与自然背景很协调，护栏没有必要营造仿木效果。道路为人工修建，护栏、安全链等呈现真实面貌即可。注意材料（如有毒油漆、防腐剂等）不能产生污染。换言之，要对修建木板人行道使用的全部材料进行筛选。

无论行人还是车辆，碎石路面都容易产生噪声。在敏感的野生动物附近，不得使用这种路面。湿地道路应悬挂减速标识，或采用减速坎，提醒过往车辆减速行驶。

为了尽量减少对野生动物的干扰，所有道路线路必须妥善规划。道路应远离繁殖区和摄食区，与隐蔽区或瞭望区应保持一定距离。必须考虑湿地观赏点与湿地保持适当距离。

7.1.1　修建桥梁和隧道

水流和连通性对于湿地健康十分重要，因此必须注意修建道路时不能妨碍水流和连通性。为了确保自然流动性，需要修建大量隧道或桥梁，并定期维护和清理。

7.1.2 修建掩体

管理方应为游客充分提供遮盖物，如避雨亭、休息场所等（图 7.2）。遮盖物应设置垃圾箱，也可提供展架，放置宣传单、地图、信息栏或观光路线图（图 7.3）。

图 7.2 简易避雨亭

图 7.3 带有遮盖物的布告牌

7.1.3 洗手间

湿地本身就是一个天然污水处理系统，但人类废弃物过量会带来富营养化问题。

如果估计大量游客会使用洗手间，洗手间应安装处理装置、化粪池、远离湿地的独立排水道或使用整体清除系统（图 7.4）。请勿使用化学处理系统，因为此类洗手间使用的化学消毒剂对湿地生物群和生态系统有害。

7.1.4 收集垃圾

湿地垃圾延伸范围每年都在扩大。人们将垃圾扔进河流、水塘和海洋。虽然不易发现，却很危险。塑料被野生动物摄入，渔网、钓鱼线、铅制钓鱼重

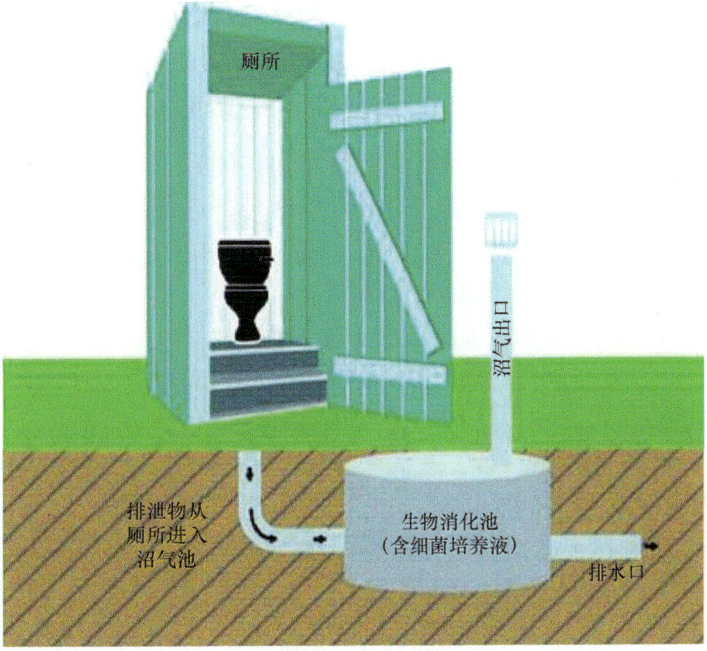

图 7.4 户外卫生间

物、尖锐金属罐和其他碎片导致大量野生动物死亡。很多垃圾无法分解，环境危害可以持续很多年（图 7.5）。

湿地管理方需要采取 3 项措施：

1. 对湿地保护地内的海滩、河岸或水体进行清理。这项工作需要投入大量工时，因此代价不菲，但是如果管理方能邀请中小学生、青年或成年志愿者参与，可以显著降低成本。

2. 如何在垃圾进入湿地前，安全正确处理垃圾，应向当地社区和农民提供相关培训，提高公众意识。

3. 充分配置垃圾处理设施，定期收集垃圾，从而将湿地保护地内垃圾产生的不良影响降至最低水平（图 7.6）。

图 7.5　到处乱丢垃圾破坏了湿地环境

图 7.6　收集垃圾的合理设施

7.1.5　隐蔽观察点、百叶窗、瞭望塔

湿地是观察和拍摄野生鸟类和其他野生动物的理想场所。但是如果游客和摄影者既不干扰鸟类，又能获得很好的视野，应使用合适隐蔽点、百叶窗和瞭望塔，因为大多数野生动物都会将人类视为威胁，所以使用这些装置不会被它们发现。

为此，很多湿地保护地专门修建瞭望塔或隐蔽点。有时会修建百叶窗，这样人们可以走进野生动物，而不会被它们发现。

摄影者可以修建摄影专用隐蔽点，或者带上便携式掩体，可以在需要的任何地点搭建使用。

关于掩体设计，人们已经开展了大量研究工作。因此无须在这方面闭门造车、浪费时间。只需使用经过实践证明的现成模型即可。要求当地企业为掩体提供赞助，并在掩体内放置简易广告牌，作为对赞助商提供支持的回馈。

修建掩体的主要原则是让观察者从掩体内部拥有良好的外部视野，但是掩体外面的鸟类却难以观察掩体内部。最佳设计方案就是设置水平弦月窗，因此掩体内更暗，观察者的水平视野很开阔，但是野生动物只能看到一条很狭窄的暗缝。

图 7.7- 图 7.9 是典型掩体设计和布局的示意图。

通过互联网、观鸟会或狩猎露营供应商，购买摄影用小型掩体或伪装网也很方便。

人行通道也可以屏蔽，因此游客进出掩体时不会被发现。掩体不必太复杂，织物、树篱或茅草等都可以使用。

靠近野生动物的道路应设计为安静模式。人们行走在木板或石路面上产生的噪声可能会干扰附近的鸟类。游客一旦接近野生动物，应告诫他们在任何时候都必须保持安静。

图 7.7　基本的全天候观鸟屋（1）

图 7.8　基本的全天候观鸟屋（2）

图 7.9　在一个典型的公共观鸟屋内的观鸟者

可以修建瞭望塔（图 7.10）监测水体，作为观测点很有用处。鸟类离开地面后，对人类不是太担心。但瞭望塔成本很高，如果维护不当，可能还会存在危险因素。瞭望塔无法移动，视野有限。就中国湿地而言，在大多数情况下，修建瞭望塔并不划算。

图 7.10　瞭望塔

7.1.6　船艇、停泊点和船库

如果水位相对稳定，将简易码头固定在支杆上即可充分满足管理要求。如果水位时涨时消，可以根据铰接斜坡浮箱式码头设计（图 7.11、图 7.12）。游客用船舶停泊点必须安装安全栏杆。

7.1.7　建造鱼梯

很多鱼类在河流上游温度更低、通气条件更好的水体里交配和产卵，但是成鱼往往在河流下游甚至海

图 7.11 固定停泊点

图 7.12 浮动停泊点

洋生活。因此，它们需要沿着河流来回迁徙繁殖。修建大坝、水坝和其他设施切断了鱼类的迁徙路线，除非专门为鱼类迁徙修建通道。我们将这种设施称为鱼梯，它是维持河流上下游连通性的一种有效方式。我们可以通过互联网查看很多设计方案，并且将其纳入全部新建和现有人造障碍物的结构规划和设计方案（图 7.13）。如果保护区水源补给河流修建了大坝，应坚持采用鱼梯这种设施。

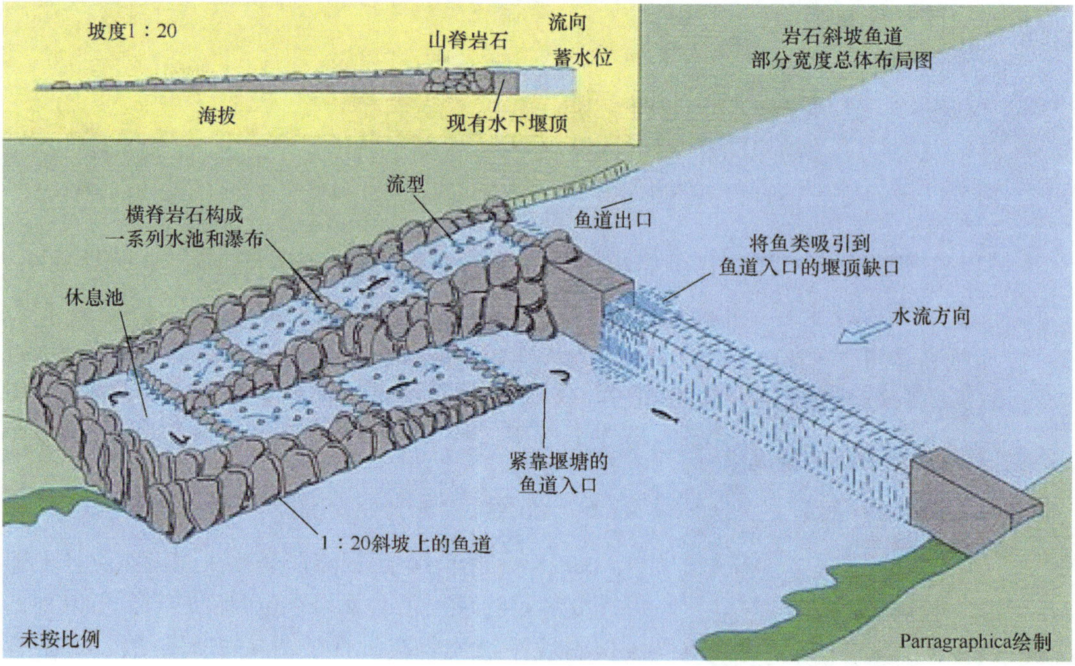

图 7.13 鱼梯的设计

7.1.8 是否修建栅栏？

修建栅栏成本很高，需要经常维护，并且会对野生动物带来严重伤害（图 7.14、图 7.15）。所以即使使用栅栏，也应将其伤害降至绝对最低水平。

图 7.14　死于木栅栏的狐狸

图 7.15　死于铁丝网的羚羊

有些湿地管理方认为需要在每块湿地自然保护区四周修建栅栏，一方面防止人类和家畜进入保护区，另一方面防止野生动物离开保护区。在有些情况下，可能需要修建栅栏，但是一般而言，这种观念已经过时，反而又回到过去的自然保护区隔离状态。目前关于普通保护区（尤其涉及湿地）的理念是必须对保护区（湿地）进行有效规划，将其纳入更为广阔的自然景观中。野生动物需要进入自然景观，能够与外界互联互通。

野生动物的保护需求不会止于保护区边界，保护区周围的社区也不应将保护区视为它们的私家花园。

湿地保护地是野生动物的安全港，很多鸟类、其他生物和鱼类等可以在这里休息、进食或繁殖，但大多数情况下，无法将其限制在保护区内，它们也会四处探险，到邻近的农田、草地、其他湿地，甚至到城区内觅食。

修建栅栏可以防止人类或家畜进入保护区敏感区域，不过使用栅栏防止人类涉足的想法往往徒劳无功。虽然栅栏造价不菲，但是只需使用木棍锯开，就能轻易让栅栏失效（图 7.16）。

图 7.16　人们很容易穿过围栏的两种情况

有时使用电围网，可以防止大型哺乳动物侵入保护区外的农业区。

7.2　如何在树上悬挂标牌

无论您开展研究工作监测植被，抑或只是让游客了解树木名称之类的信息，您可能希望在树上悬挂标牌。不幸的是，这项工作虽然十分简单，但是效果往往不理想，要么树木受到损害，要么标牌很快受损掉落。这是因为**树木一直在生长**。不能在树上钉钉子挂标牌。何况有些钉子对树木有毒。

沿木板人行道，可以悬挂很多公告和树木识别标牌。植物园为树木悬挂标牌已有100多年的历史，它们知道如何开展这项工作，因此没有任何理由做不好树木标识。标牌应做到：①在5m远距离应清晰可辨；②最好使用与自然环境协调的颜色；③固定之后能够维持数年，而且也不会损害树木。管理方似乎总是忘记树木在生长移动这个事实。如果用钉子将标牌固定在树上，它们肯定会在两年内断裂，因为树木在膨胀和弯曲，从而将标牌折断。所以标牌选材要合理（图7.17-图7.21）。

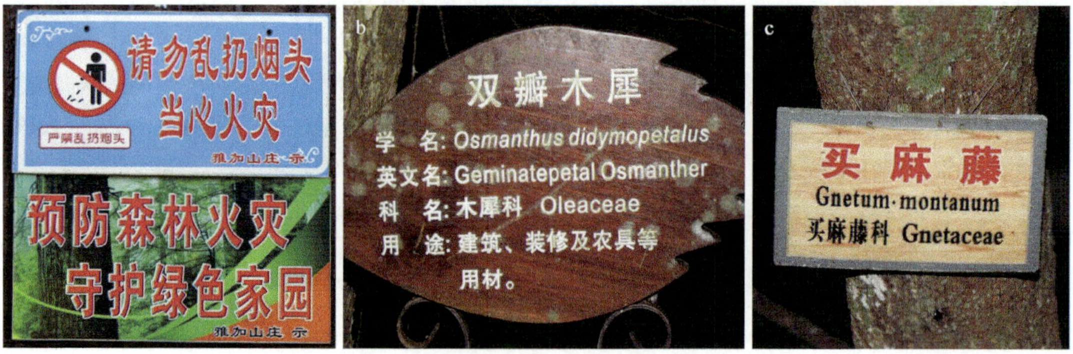

图7.17　3种标牌（1）

a. 标牌内容不错但过于花哨；b. 很棒的独立式标牌；c. 标牌套在软线上

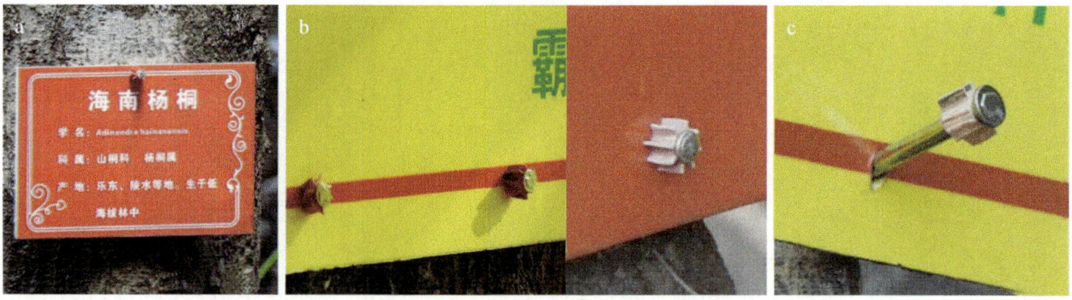

图7.18　3种标牌（2）

a. 字体太小；b. 钉子钉在树上太紧；c. 为树木生长充分预留了空间

悬挂标牌规则

- 避免将公告牌挂在活树上，如果可能，应悬挂在木板路枯木结构或独立支杆上。
- 标牌采用大字号，5m开外清晰易读。公告牌色彩不要太炫目。

图 7.19　树木标牌国际标准实践

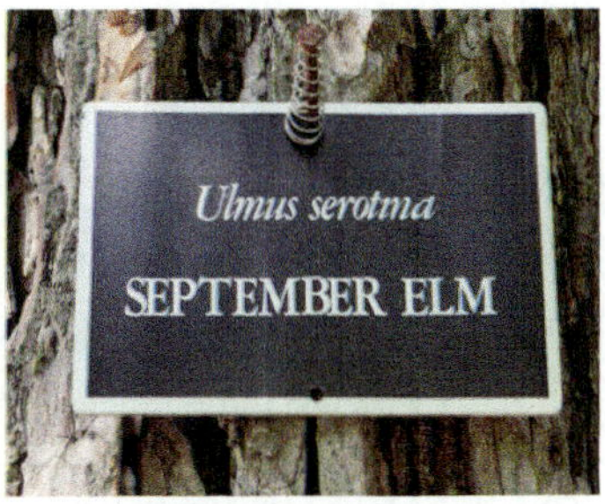

图 7.20　弹簧固定的标签，为树木生长预留了空间

图 7.21　适用于小型植物的灵活标签

- 如果需要将标牌固定在树上，可以使用膨胀式项圈；如果使用钉子，必须用铝材或不锈钢制成。铁对树有害，铜有毒。钉子露出标牌部分的长度能够适应树木几年生长的需要，并且标牌应很宽松，能够沿着钉子滑动，有风或其他情况可以移动。每个标牌只能使用一个钉子。

7.3　使用地图及野外定向

了解保护区内动植物的分布情况，就需要具备良好的空间意识。野外工作人员需要了解保护区的地形和布局。工作人员应携带信息翔实的地图，并知道如何使用地图。如果当地存在任何明显海拔高差，携带的地图还应显示地形特征。巡护人员应能按照地图参考坐标格网或 GPS 坐标系统记录观察结果。

7.3.1　使用地图

地图有各种比例。必须熟悉在用地图的比例和投影。标准地图网格使用十进制或使用米作为单位。

等高线即高程相等各点的连线，它可以显示地形（图 7.22）。通过轮廓线，可以确

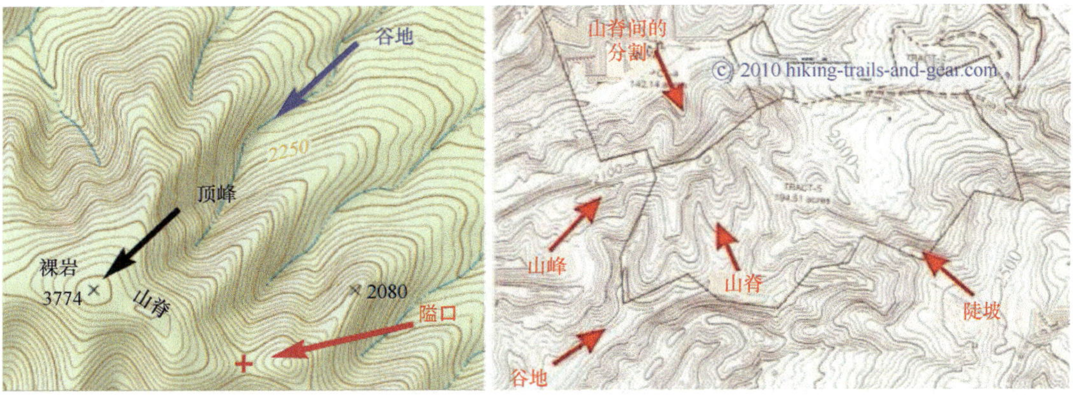

图 7.22　地形图的使用

定两点之间的最浅层路径或最陡峭路径。等高线间距越小，说明地形越陡峭。等高线间距越大，坡度越缓，地形越平坦。一系列 V 形线指向高地（海拔更高），表示山谷、沟壑或峡谷。表示山脊或尖角的等高线像字母 V 或 U，但它们指向低地（海拔更低）。另外一种确定地形的简易方法就是寻找沿 V 形等高线中间或侧面的河流。查看两侧等高线更高的沙漏地形，可以确定它是隘口地形。类似同心等高线的最内侧环线代表山峰。它通常会标注 X 或 Δ 及其海拔。地图添加的颜色或底纹可以区分林区、草地或高山苔原。地图上的其他线条可以表示道路、山径或土地利用边界。这在寻找营地、水源、风景或观察野生动物理想地点时可能非常有用。

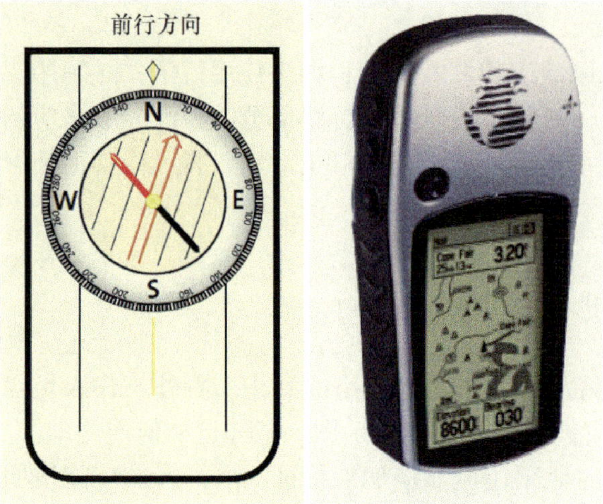

图 7.23　野外指南针和 GPS 装置

使用指南针（图 7.23），可以在任何突出山顶测角，或者在地图上识别标志地形，以便为旅游路线确定方向。在两个不同点测角，可以对地图进行三角测量，从而推断当前所处位置。应将指南针对准磁北。这可能与大多数地图上的正北方略有不同。

掌握指南针使用技巧非常重要，其无法代替。

高精度指南针通常使用镜子或棱镜，即可精确测角。测高仪也很有用，通过它可以确定自己处在哪条等高线。目前很多手表都内置了指南针和测高仪。

巡护人员一旦掌握使用地图和指南针的技能，就可以在景观区内轻松巡逻。但是如果发生突发事件，一场浓雾或暴风雪就会让通常作为参照物的景观模糊不清。使用全球定位系统（GPS）接收器（图 7.23），可以迅速确定地面上的具体位置，通过地图即可导航，无论参照物是否清晰。也可使用 GPS 提前标绘路线，并沿着线路确定路标间距。

7.3.2　使用指南针确定时间

如果有太阳，使用简易指南针可以确定具体时间（白天）。将指南针对准东西方向。让指南针左侧边缘保持水平，提升右侧边缘直到指南针平面指向太阳。在这个角度不会有任何阴影。取一支铅笔，将笔尖绕着指南针边缘四周移动，确定笔尖阴影落在指南针中点时的具体位置。这个点可以视为相对于指南针中点的钟面。

相反，如果有手表，但没有指南针，可以对准手表，确定指南针方位。如果手表有时针（即不是数码显示屏），可以很精确地确定指南针方位。时区可能与太阳时有所出入，但是对于大多数时区而言，完全可以采用以下方式：

将 12 点指向太阳，那么北方也就在时针所指位置与两个合理圆弧中最长圆弧上的

12 点位置的中间。在南半球，将北方改为南方即可。

7.3.3　使用指南针估测距离

如果没有地图或 GPS，但是可以看见远处突出点，这时可以使用指南针估测距离。在该突出点精确测角，然后从该直线侧向移动一个直角（17.4m），再另外测角。相差 1° 计为 1km 距离。如果差值不足 1°，继续侧向移动，直到差值刚好为 1°。那么该点距离为 *n*km，其中 *n*＝测角过程中位移 1° 所移动的横向总距离 /17.4（17.4 即 1/360 × 半径为 1km 圆周周长）。计量 1° 位移使用的指南针必须非常精确。

7.3.4　使用 GPS 应用程序

如果整个保护区都能接受手机信号，巡护人员可以充分使用汽车驾驶员或徒步旅行者在市面上可以购买的各种 GPS 应用程序。也可使用专用手提式 GPS 装置。这样可以在任何地点为巡护人员精确定位。如果应用程序更理想，还可提供海拔。有些测绘软件，如世界自然基金会推出的智能程序，可以在记录野生动物现场观察结果和人类活动迹象过程中使用 GPS。即使有了这些高端技术辅助设备，但是巡护人员仍应携带备份的物理地图和指南针，并接受地图阅读和野外定向相关培训。毕竟会出现手机没电、GPS 装置淋湿或者坠落等诸多情况。

7.4　使用适当设备

多年来，双筒望远镜已经成为野外鸟类学主要观测设备，它有很多品牌和型号。用户可以权衡一下重量和携带便利性。如果是长途旅行，携带大号笨重型双筒望远镜会成为一种负担，但是其在稳定性和放大率方面却有明显优势。重量与惯性相关，因此图像更稳定。由于放大率的影响，导致视角很小，因此很难为观察目标定位。

总之，如果双筒望远镜主要用于固定掩体或视点位置，大号笨重型比较理想；如果只用于移动观测，小型轻便型比较理想。

还有非常轻便的小直径管双筒望远镜，随身携带非常方便；但是太便宜的却不耐用，很容易失准，镜头有限，光线不足，很难保证视野清晰，通常视野很狭窄。很多观鸟者选择的双筒望远镜放大率为 8 倍，视角为 30°。

最好使用高品质、功能强大、带防水套筒的双筒望远镜。应让双筒望远镜保持干燥，妥善保护防止振动，这也很重要。双筒望远镜一旦失准或者镜头充满湿气和真菌，也就报废了。如果继续使用，还会影响视力。

观测镜。观鸟爱好者通常会携带一副观测镜和一个轻便三脚架，它比双筒望远镜的放大倍数更高，视野更稳定。如果需要远距离统计鸟类在水中和岸边的数量或者识别这些鸟类，观测镜几乎就是一种标配。现在已经有了小型摄像机，放在目镜上，它可以将观测镜对折成摄影用大号长焦镜头。使用观测镜意味着需要携带长而笨重的设备，这时只能减少其他装备携带数量。

7.4.1　设备管理

购买光学设备和电气设备需要投入大量资金，而且很难维修或更换，因此必须悉心保管和维护，设备指定使用人应接受适当培训。应遵循以下原则：

- 为所有设备做好记录，包括保管地点、负责人和当前状态相关细节。
- 自然保护区设备只能用于工作，不能用于个人目的。
- 保持设备干燥，在干燥的地方保存设备，做好保护，防止受到雨水和喷雾的影响；如果设备弄湿，应轻轻擦去。
- 将设备放在温暖的阳光下，自然干燥，即可去湿。
- 确保设备处于干燥状态的最佳方法就是在密闭容器放一包硅胶晶体，然后将设备密封在容器内。硅胶在干燥状态呈蓝色，一旦吸水就会变成粉红色。将粉红色硅胶放入无油锅，加热至蓝色后，可以重新使用。
- 避免接触尘、沙和盐。
- 避免极端温度。
- 避免敲击，小心保管，防止坠落。
- 运输过程中，小心包扎起来。

7.5　人 员 培 训

保护区，尤其湿地保护地的管理是一项十分复杂的工程，要求具备广泛的技能和知识：生态、保护、地理、水文、行政、地理信息系统、数据管理、规划、执法、通信、工程、船舶、社会关系等。

受过良好培训的工作人员确实很难招募。我们经常会从各种林业单位借调保护区工作人员，但是却并不具备保护区管理技能。人员招聘也会受到地方政府配额的严格限制，并且聘请专家所需资金也十分有限。最佳解决方案就是提供在岗培训。

宣传册（图 7.24）详细介绍了不同层次工作人员所需要的技能或知识（能力），并为识别工作人员培训需求确定了一个框架。培训需求评估可以为组织相关管理人员培训提供依据。这样有助于为管理机构不同岗位确定适当人选，还可以确定哪里需要另行提供培训。

图 7.24　员工能力标准手册

目前正在开发各种自我导航在线培训模块，也可从网上下载大量有用的培训材料、工具和指南。

但是这无法代替亲力亲为，湿地管理方应努力让附近大学、研究所、非政府组织、客座专家、工程项目或国际项目向工作人员提供外部培训。与其他湿地保护地交换人员也是一种学习经验、吸取教训的有效方式。

管理方还应对其他机构或项目组织的培训课程进行跟踪，并设法让工作人员学习类似培训课程。

自 1990 年来，世界自然基金会一直在香港米埔中心定期提供湿地相关培训（图 7.25）。中国大陆湿地管理人员来香港的培训费用通常由他们承担。

图 7.25　世界自然基金会培训广告

采用能力标准与采用国家认证体系同步进行，确保聘用人员具备工作所需技能。这同样适用现场管理人员，工作人员通常从其他部门借调，他们的生态专业知识有限，难以管理复杂的生态系统。

7.6　如何申请其他资金支持

资金不足经常是一种限制因素，它决定了经济上所能承受的管理和保护水平。虽然政府继续向湿地保护加大投入，但是对于很多省级或县级湿地保护项目而言，实乃杯水车薪。只要处于良好状态的湿地价值被低估，湿地管理拨款就会很有限。那么该从哪些渠道去弥补资金缺口呢?

资金渠道有很多，包括：

- 向国家和省级渠道申请资金
- 向国际项目申请资金
- 针对特定需求，提出研究申请
- 申请生态赔偿基金
- 寻求企业赞助
- 开展合法风投项目（在精心挑选的区域开展可控旅游项目，将影响控制在一定范围内——承载能力）

其他单位也在设法筹资，所以申请资金本身也是一种竞争。如果本项目需求清晰，有说服力，并确保负责、透明地使用资金，就能有效提高竞争力。

如何确定需求：

- 对人员和预算需求进行差异分析，说明本项目确实需要更多资金，配备更多人员，提高管理能力。

- 从经济层面初步估算本项目产生的生态服务价值和其他效益或收入，说明有必要进一步投资。
- 将本项目应用进行调整，使其符合投资方的特定目标和要求。研究项目需要解决具有科学意义的问题，或者需要亟待采取保护措施的问题，同时申请动物保护资金必须适应相关物种需求，而申请慈善基金则需要解决贫困、人权、性别等问题。
- 在申请前，设法获得高层同意。
- 确保申请方案在逻辑、目标、时间和成功指数方面清晰透明。
- 尽力提高地位，提高知名度，从而提升本项目的形象。国内或国际地位越高，获得更多资金的理由越充分，成功率也就越高。

如何证明本项目价值：

- 提供本项目可以盈利
- 本项目能够提供有价值的生态服务
- 本项目受到公众的喜爱、热衷和支持
- 本项目具有很高的休闲养生价值
- 保护环境是中国在国内及在国际层面需要履行的一项义务

任何申请方案应包括以下内容（表 7.1），并确保申请方案与目标投资方的模板或任务书保持一致。申请方案务必篇幅短小，主旨清晰。不要添加太多无关细节或信息，这样反而会喧宾夺主。如果申请方案冗长乏味、说服力不强、翻译错误连篇，审核人肯定不希望在这种申请方案上浪费时间。

表 7.1　典型的申请书格式

章节	说明
标题	标题应简短有力。标题应反映具体目的，并让人联想起投资方的宗旨。标题加入投资方流行术语
目标	明确阐述提出申请的目的
背景信息	有关保护区的背景基本细节。位置、规模、现状、主要功能
保护区的意义	加入的细节内容强调保护区的意义——稀有濒危物种、独具特色，或者详细介绍向社会提供的经济服务
保护区面临的威胁	对当前面临的威胁进行排序评估。不要说得太严重，否则目标投资方误认为保护区已经没有希望，反而认定保护区没有投资价值
需要解决的问题	列举申请人希望消除的主要威胁，说明潜在障碍或原因。一个资金来源无法解决所有问题，但是可以选择与投资方宗旨最相关的问题
融资现状	列举本项目年度开支和需求估值，说明存在的资金缺口
申请理由	说明申请方案在哪些方面符合融资条件及投资方的宗旨
融资事项细节	说明本项目使用额外资金计划开展的各项活动及这些活动产生的结果
逻辑框架	提供一份表格，阐明相关活动与预期结果及 / 或输出之间的逻辑关系，活动本身与确定需要解决的问题之间的逻辑关系

续表

章节	说明
指标	通常作为可验证成绩插入逻辑框架，表明是否正在实现目标
活动日程表	根据项目日程表制定的活动一览表，表中列出已经确定的具体里程碑
汇报日程表	向投资方递交报告的频率和类别时间表，或者应投资方要求设计时间表
详细预算	列出的各类成本或采购成本。如果是合作项目，还应包括任何其他合作方输入内容
背书	担保函、推荐函及与任何合作机构或合作方达成的协议

第8部分　监测和报告

为了确定管理是否有效或充分，哪里还存在问题，我们需要对生存环境健康状况进行监测。这项工作十分重要，也十分艰巨，但是管理方的监测预算和人力都很有限，因此我们需要在选择关键监测要素和低成本方法过程中提高效率。生物学家已经制定了几百种实地调查方法，专门解决不同物种和不同自然条件的具体问题。关于这个主题还有很多专著，但是大多数方法都是某些经典标准措施的变更形式。首先需要确定我们希望解决哪些问题。

什么？湿地管理方需要了解具体处理对象——什么生态系统、什么物种、什么问题。管理方可能还需要查看基本定性调查或者存量调查。

多少？数量统计既困难又耗时。管理方需要选择量化评估涉及的物种或因素。

哪里？现场生物群不会均匀分布。因此在调查过程中，必须确定哪些地点或栖息地类型对不同物种而言十分重要。

哪些变化模式？通常，监控物种的种群走向比了解具体数量更重要。这是评估随着时间流逝对状态产生影响的唯一方式。

为什么？湿地管理方经常需要了解为什么发生变化，但是识别自然过程的起因相当复杂，而且经常无法直接进行。这可能需要进行专业分析和专门研究。

不同目的需要开展不同类型的监测活动（表8.1）。

表8.1　监测需求表

监测类型	改善领域
气候	恢复能力得到改善
污染	识别和关闭污染源
植被	提高规划、分区和栖息地管理水平
植被状态、生物气候学	提高对动物群响应的认识水平
目标物种（数量、繁殖成功率、迁徙物种）	提高生存率和迁入率，减少致死因数
指标物种	生态改变的早期预警
外来入侵物种	对有害物种加强控制力度
迁徙物种	在全球范围提高保护力度

8.1　基线调查

湿地生态系统很复杂，具有动态性质。通常无法准确预测全部物种如何响应各种变化。在很大程度上，管理取决于定期监测。需要观察和记录发生的现象，并且只有在关键物种或栖息地受到变化的威胁，才能确定介入管理。各管理方必须为关键物种、指标物种和问题物种状况及人类活动的物理气候状况和水平的巡逻与汇报，提供充分人力和时间资源。

仅记录变化情况并不充分。重要的是了解哪些条件正在发生变化，为何发生变化。每片湿地最好具备一些生态经验，这样有助于向管理层解释监测结果，并将其转化为建议方案。

监测湿地生物多样性并不是一项孤立活动。因此，《拉姆萨尔公约》强调需要将监测活动纳入一个更大的框架，包括盘点、评估、监测和管理：

- 确定湿地位置和生态特征（基线列表）。
- 评估湿地状态、趋势及面临的威胁（评估）。
- 监测状态和趋势，包括识别现有威胁因素的减少及新的威胁因素的出现（监测）。
- 采取措施（包括现场内外），对导致或可能导致生态特征破坏性变化的任何变化进行调整（管理）。

8.1.1　国家林业局第二次全国湿地资源调查

国家林业局已经开展了全国湿地调查。为了确保该调查也能满足国际和《湿地公约》报告的要求，应保持下列核心字段的统一。

核心湿地调查字段修订版

（根据《湿地公约》生态特征列表进行了调整）

湿地名称：

湿地和流域正式名称/其他识别符（如参考编号）

面积、边界和尺寸：

湿地形状（横截面和平面图）、边界、面积、水域/湿润区域面积（季节最大值/最小值，若相关）、长度、宽度、深度（季节最大值/最小值，若相关）

位置：

投影系统、地图坐标、地图几何中心、立面

地貌：

景观/流域/集水区环境——包括海拔、流域上/下游、与海岸之间的距离（若相关）等

生物地理区域。

气候：

盛行气候类型概述、地带和主要特征（降水、温度、风）

土壤：

地质、土壤和深土层；土壤生物学

水文状况：

水源（地表水和地下水）、流入量/流出量、蒸发量、洪水频率、季节性和持续时

间；流量及 / 或潮汐状况、与地下水的通路

水化学：

温度、浊度、pH、色彩、盐度、溶解气体、溶解或悬浮营养物、溶解有机碳、电导率

生物群：

植物群落、植被带和结构（包括特殊稀有性等相关评论）；动物群落（包括特殊稀有性等相关评论）；现有主要物种（包括特别稀有 / 濒危物种等相关评论）；种群规模和比例（若已知）、发生的季节性及在分布区内的大致地位（如靠近分布区中心地带或边缘地带）

土地利用：

当地及在流域及 / 或海岸带内

压力和趋势：

涉及上述任何特征及 / 或生态系统完整性

土地所有制和管理部门：

湿地及流域及 / 或海岸带关键部门

湿地保护和管理状态：

包括影响湿地管理的法律文本和社会或文化传统；包括根据世界自然保护联盟体系及 / 或本国任何体系确定的保护区分类

生态系统服务：

（相关生态系统服务目录，参见《拉姆萨尔公约》生态特征列表）

管理计划和监测方案：

在湿地、流域及 / 或海岸带内已经到位并经过规划

8.2 选择适当调查方法

为需要调查的物种选择合适的，并且在有限时间和可用人力范围内，在十分紧迫的情况下可以管理的方法。

8.2.1 线样带

鸟类样带采样

每次调查采用确定的同一样带。应尽力坚持在白天鸟类主要活动期间同一时间观察。每个小组应由两名观察人员组成，其中一名观察员查看观测线的左侧，另一名观察员查看观测线的右侧。应缓慢平稳推进，不要偏离样带线。保持安静，记录观测的目标物种全部鸟类。记录与样带线形成的夹角及首次看到的鸟类距离和数量。可以将其计入直角距离，方便以后分析结果。精确记录观测的总距离。在天气状况和能见度报表上添加附注。一年内，随着树叶角度、草本植被高度、雾、水位等发生变化，能见度可能也会有所变化。

注意鸟类并非随机分布，它们成群结队或者在领地内均匀分布。它们使得大多数统计假设都不成立。同样在线样带上，它们与来自样带线的预计分布情况不匹配。接近样线的带条内总是缺乏记录。因为鸟类已经飞走，或者在您之前已经有人在附近散步或者它们知道人们经常在附近（如大堤）散步，必须与您保持安全距离。这意味着需要对数据进行适当校正，并且事实上，需要分别计算每个目标物种，因为它们的行为和可见性都不相同。一般而言，可以将所有鸭类或所有鹅类混在一起。

- 已经测量的长度，可以沿线样带（或现有踪迹）继续前进。从汽车或飞机下来慢速步行。可以在一侧、两侧或预定带宽进行。
- 记录发现时，在与样带线垂直距离预计范围内，观测到的目标物种所有动物（动物可能会首先发现观察者，并在观察者看到它们之前逃离样带线）。

计算密度和种群数量：

- 密度＝观察到的个体数量（观察次数 × 团体成员平均数量）/（样带长度 × 有效带宽 × 调查的面数）
- 指定栖息地种群估计数量＝密度 × 有效栖息地类型面积（假定抽样面积为整个栖息地典型面积）

有效带宽即每次观测时与调查线索之间的平均距离（图 8.1）。它是带宽范围内丧失的观测机会与带宽范围外增加的观察机会相等的特定点。

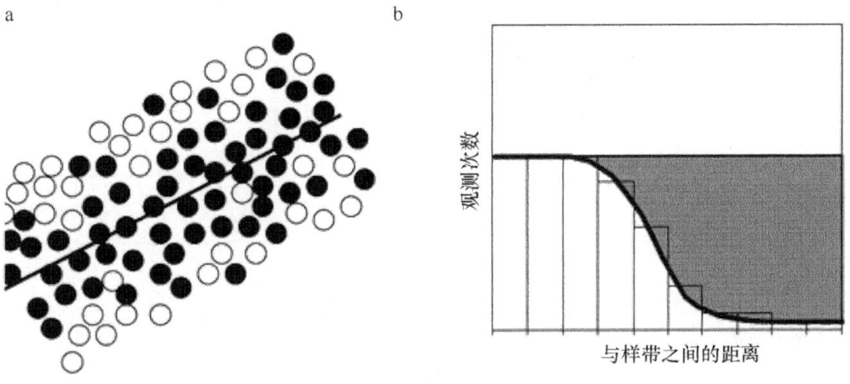

图 8.1　假设线调查

a. 观察员沿调查线推进，以记录所观察到的鸟类情况。黑点代表所观察到的鸟类个体，白点代表未能发现的鸟类。
b. 通过确定在与调查线不同距离上的观察频率，可以计算出平均样带宽度和密度

8.2.2 观测统计

鸟类定点采样

每次调查时采用同一固定点。只能在白天 08:00-10:30 或 15:00-17:00 时段内计数，并且每个站点应在 1h 内彼此反复观测，即一个站点逻辑上可以总是在 08:00-09:00 时段内观测，而远离该站点的另一个站点可以总是在 09:30-10:30 时段内观测。使用观测镜，但观测镜必须安装三脚架。180° 或 360° 扫描目测计数，并记录半径 1km 范围内目

标物种的所有鸟类。如果采样距离低于该值，应注意能见度。

鸟群计数

在黄昏到达前（17:00）做好定位，选择有利地形，从接近鸟窝的天空各个角度进行观测。记录返回鸟窝的所有目标鸟群（白鹭），记录每次返回的时间和数量。添加天气和能见度相关批注。一旦光线太暗，无法识别，应停止记录。

8.2.3　样方

由于没有时间对整个区域进行统计，因此可以使用针对整个区域统一选择的或者随机选择的样方（图 8.2）。现在可以统计选定样方范围内的鸟类或者诸如此类的目标对象，并按比例增加，确定整个区域的估计值。在下例中，20% 样本密度范围内发现6 只鸟，估计总数为 30 只鸟（图 8.3）。实际总数为 33 只。估计值稍微偏低，因为样区忽略了东南角集中区域。如果知道在整个集合中的整体密度或角度范围内，一个物种有若干栖息地，那么可以在首选栖息地类型中，分别抽取样本密度更高（加权后）的每个栖息地。加入每个栖息地小计，估测种群总数量（图 8.4）。

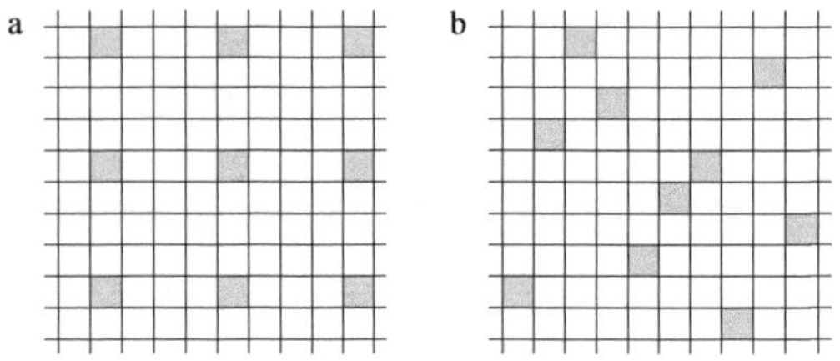

图 8.2　选择系统或随机样地

a. 系统抽样示例；b. 随机抽样——拉丁方格设计。阴影方格即抽调的样方

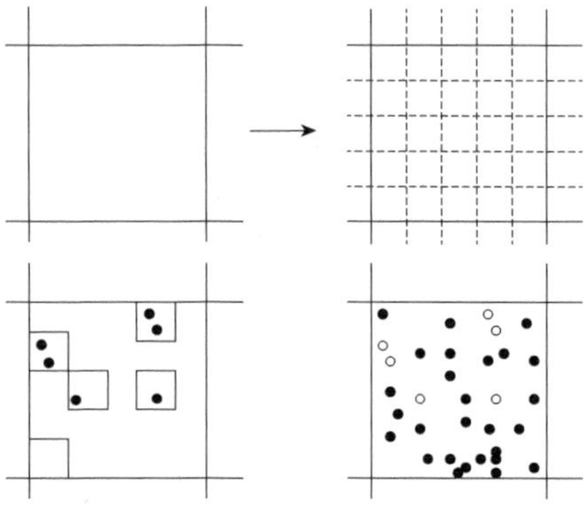

图 8.3　样地与总体状况的比较

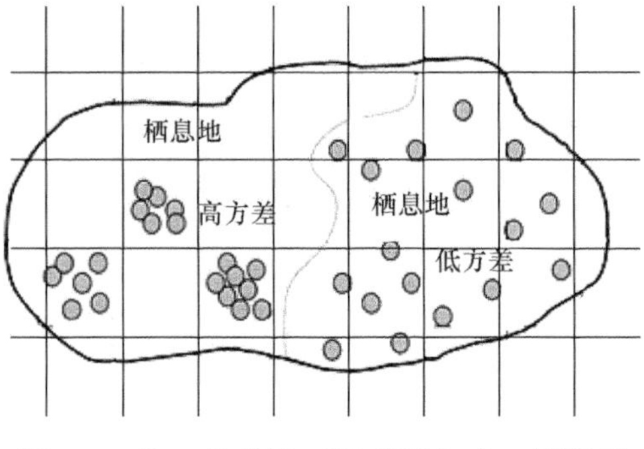

图 8.4　基于物种的不同利用方式，根据不同的生境类型确定不同的样方强度

图示两种栖息地的非匀性方差。图中●代表鸟类

8.2.4　其他物种 / 因素监测法

- 陆生哺乳动物——野外证据（脚印、粪便、喂食迹象）或当地市场影响。
- 普通鸟类——定期对代表性采样点或样带的鸟类进行统计。
- 珍稀鸟类——在已知的最受鸟类欢迎的地方高峰季节统计数量。
- 昆虫——扫样或专用诱捕器，如飞蛾诱捕器（吸引苍蝇、蜉蝣和其他昆虫）。
- 鱼类——渔获量、市场供应和价格。

- 水生哺乳动物——乘船在已知水面上巡查的观测频率。累计偶然观察次数。
- 外来入侵植物，如槐叶萍和凤眼蓝——按月绘制选定样区内地被物示意图，并且对水面富营养化进行测量（氧、氮）。
- 外来入侵小龙虾——克氏原螯虾。每周对市场供应和物价进行监测。
- 栖息地范围——在最低水位期间，根据遥感图像按年绘制地图。
- 栖息地质量——小块永久样区内的树木年度生物量估值、月度估值和草本植物高度测量值。可以使用摄像机技术，测量树木全叶郁蔽度。
- 水质——在选定地点按月提取样本，测试氧、氮、磷和大肠杆菌含量。一旦报告鱼类相继死亡，任何情况均应提取样本。在水塘、浅水、河流和湖泊中央选择采样点。确立各省环境保护部门监测链接方式。

8.2.5　马敬能列表及发现曲线

这是在指定区域，有效评估物种丰度的一种快速稳健的方法。记录人员记录看到（听到）的前 n 个物种，然后新建一份列表。大多数适度至丰富种群区域，建议每份列表采样 n=20 个物种。如果现场种群量很少，可能需要将列表长度减少到 12 个或 10 个物种（图 8.5）。

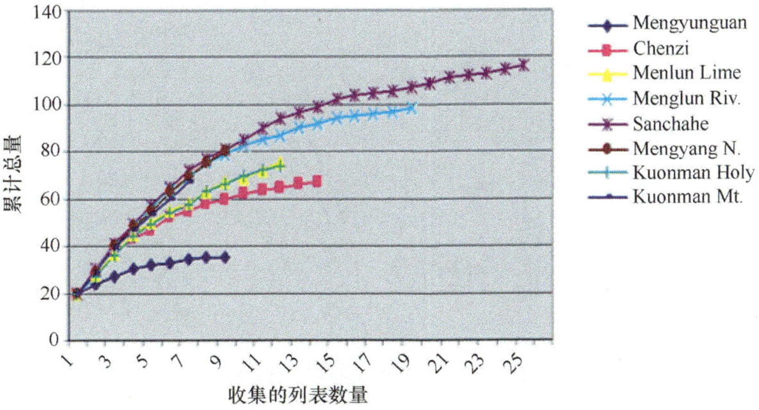

图 8.5　不同地点的物种发现曲线

随着记录人员在调查现场的持续观测，需要填写的这种列表会越来越多。然后可以根据收集的列表总数，标绘发现的物种总体数量。绘制的曲线越陡，表明物种丰度越高。

标绘数据的另外一种方式就是分别在 1、2、3 等列入数量记录的物种数量直方图。这样我们可以通过图表或数学（泊松分布）估计仍待发现的零级物种（图 8.6）。零级

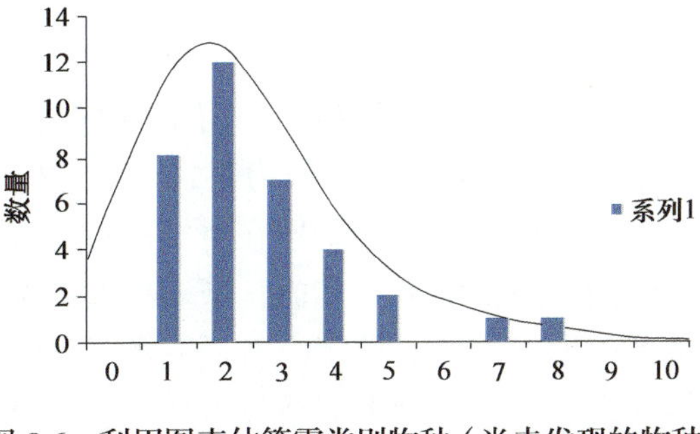

图 8.6　利用图表估算零类别物种（尚未发现的物种）

物种总数与已经发现的总数相加也就是此时此地物种数量的估值。

采用这种方法，可以对各个现场、各栖息地类型和各个季节进行相互比较。如果对所得结果进行更为详细的比较，除了纯丰度外，还可对鸟类群落的质量和构成进行比较。

经过简单修改，即可采样这种方法，对其他分类群（昆虫、珊瑚甚至树木）本地丰度进行评估。

8.2.6　现场绘制草图

工作人员一旦熟悉定期活动鸟类和其他动物，就会注意到何时发现新事物，何时发现不常见的事物。如果携带了摄影机或手机，应拍照留存，方便以后识别。尽量纳入已知尺寸物体（在图中展示硬币、双筒镜等，以便计算陌生物种的具体尺寸）。

如果生物运动速度太快或者距离太远，无法拍照，最好现场画草图（图 8.7）。工作人员应针对熟悉鸟类，练习现场绘图，掌握正确的外形比例等数据，培养相关技能，然后在草图上批注颜色、图案甚至行为。

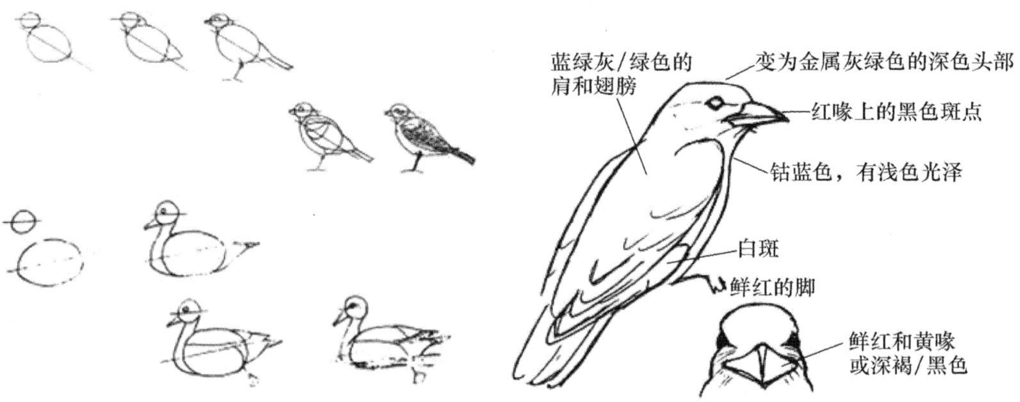

图 8.7　现场绘制草图示例

8.3　实地调查标准工具

8.3.1　使用鸟网

鸟网是一种捕捉鸟类，用于研究、上环或监测的有效方法。不幸的是，猎人也会采用这种方法捕捉鸟类作为食物，对外出售或者保护庄稼。

将各种尺寸细网悬挂在竹竿之间（图 8.8）。飞鸟如果没有看到细网，就会坠入宽松细网形成褶皱的口袋内，或者被缠在网眼内。使用这种捕获的鸟类很容易脱水、饥饿或冻僵，因此必须定时查看这种细网，尽快取下被捕获的鸟类（图 8.9）。如果鸟儿缠在网上，在从网上取下过程中，尤其因为操作不小心，很容易导致鸟儿受伤。鸟儿取下后，应放入软布袋，防止鸟儿挣扎，减轻压力。这些鸟儿应尽快处理（称重、测

图 8.8　被雾网捕获的鸟

图 8.9　研究人员正小心地将捕获的鸟从网中取下

量、上环、拍照等）和释放。

在大多数国家，鸟网受到严格控制，只有获得认可的专家才能受权设置鸟网，移除和处理野生鸟。不幸的是，在中国尚未限制使用鸟网，通过互联网和其他供应商随时可以购买这种鸟网。在中国，由于鸟网使用不当，导致每年鸟类损失数量超过几百万只。尤其是中国一些机场，为了在跑道上减少机鸟互撞事故，使用了这种鸟网（参见 2.4.10 节）。

在中国，对使用鸟网、火箭网及其他动物捕获装置加强管理和控制是一项十分紧迫的工作，应纳入野生动物和栖息地管理条例修订版。

8.3.2　使用火箭网

火箭网（炮网）使用大量小火箭，将一张大网撒向正在进食或休息的鸟群（图 8.10）。它们在国际上比较通行，可以捕获数量较大的水禽群，用于监测或环志。

图 8.10　用来捕获大型水禽的火箭网

但是这些装置对操作人员和鸟类都有一定的危险性，因此应严格控制，只能用于重要研究项目。

8.3.3　鸟类环志

鸟类环志已经成为研究鸟类学的一个重要步骤，拥有几十年的历史。鸟类环志为很多鸟类的运动、分散类型、活动范围、迁徙模式、寿命和健康状况提供了宝贵的数据。鸟类环志是一种单独编号的小型金属或塑料标签附着物，可以固定在鸟类腿部或翅膀上，以便对鸟类进行个体识别（图 8.11）。

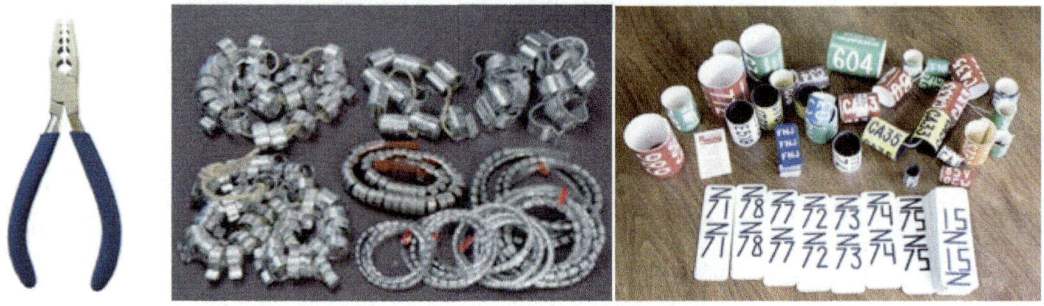

图 8.11　用于鸟类环志的专业设备

大多数国家都制定了全国鸟类环志计划，并且大多数此类计划与为数不多的国际鸟类环志计划保持一致。

在鸟巢内，给羽翼渐丰的雏鸟上环，或者在日常生活或迁徙过程中捕获鸟类。

原环由铝材制成，并刻上唯一数字代码。鸟类一旦上环，需要再次捕获，检查代码和识别数据库中的环志记录。根据不同鸟类的体型，已编号环志采用不同尺寸，并使用配有不同尺寸环孔的专用钳，固定在鸟类下跗骨。

小块区域鸟类研究已经采用了有色塑料环，使用双筒望远镜或射弹观测镜，可以通过配色对已上环鸟类进行个体识别，无须重新捕获，但数量有限。

现在，很多研究项目在鸟类脚环上悬挂彩色编号标签，或者附着在翅膀上，方便远距离识别。

现在有些研究项目还使用小型无线电广播发射机，将其附在鸟类翅膀上，以便对其活动进行监测。

捕获鸟类，为其上环或检查环志的同时，可以对鸟类进行称重和测量，还可检查它们的健康状况。也可提取血标本，进行 DNA 分析等。只有经过充分培训和授权，才能开展环志各项研究工作。鸟类捕获、上环、处理和释放各个环节都会存在伤害鸟类或向鸟类传播疾病的风险，必须将这些风险降至最低水平，并且这些风险必须低于研究成果产生的价值。在中国，鸟类环志由全国鸟类环志中心负责管理，该中心成立于 1981 年，负责全国鸟类环志事务。全国鸟类环志中心接受国家林业局林业研究所、中国林业科学研究院和全国鸟类环志办公室的监督。

8.3.4　无线电跟踪及使用 GPS 跟踪器

使用无线电跟踪对大型哺乳动物进行监测已有几十年的历史。生物学家使用镇静剂捕获哺乳动物，或者让哺乳动物暂时失去运动能力，然后安装无线电传输项圈。未来数月或数年，可以使用两台或两台以上接收器进行三角测量，以便对哺乳动物进行跟踪。通过这些数据，生物学家可以研究很多物种的运动、周期、活动范围、社会互动等。

直到最近，GPS 无线电广播发射机的体积才变得更小，有些型号重量甚至不足 5g（图 8.12）。一直以来，鸟类学家在捕获的鸟类身上安装小型吊带，然后在释放鸟类之前，将无线电广播发射机固定在吊带上，以便观察它们的飞行路径。这在识别迁徙物种迁徙路线和危险区域时特别有用。

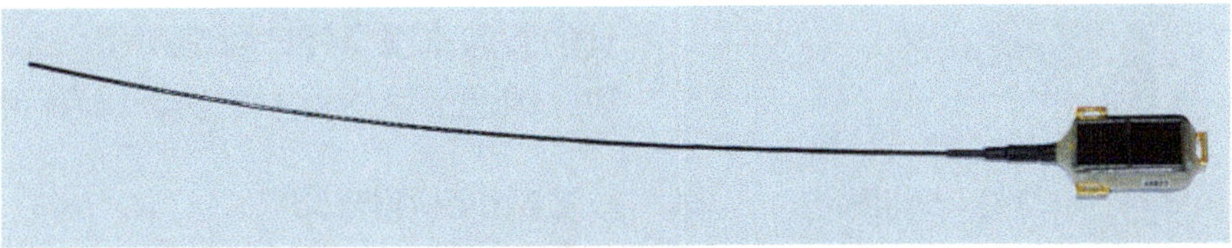

图 8.12　微型 GPS 跟踪器

GPS 跟踪已经用于塍鹬、其他涉禽类和水鸟、鹤类和布谷鸟。北京观鸟会正在采用这种方式开展布谷鸟跟踪项目。这些鸟类在欧洲越来越稀少，通过非洲越冬场可以明显发现这些问题。中国部分地区的布谷鸟数量也在减少，但是我们仍不清楚主要问题在哪里。

8.3.5　使用自动照相机

现在，自动照相机（图 8.13）价格便宜，也十分有效。在一个网格图形内使用若干台照相机，针对可以个体识别的物种，估算它们的密度及活动范围。另外，还可使用这些照相机简单记录胆怯和稀有物种的情况，或者拍摄适当照片，用于出版、网站等。

图 8.13　用于野生动物调查的自动照相机

一般而言，白天这些照相机可以拍摄彩色照片，晚上可以拍摄黑白照片，可以设定为静止状态、视频剪辑、广角或窄聚焦。电池可以工作几周，具体取决于拍摄照片或视频镜头的数量、电池本身质量和温度湿度条件。

通常，对这些照相机进行设定，可以拍摄通过目标开放区域的陆生动物，但是也可使用照相机拍摄栖息在树上的禽鸟和鸟巢等。

设置定点照相机，需要一定技能，必须充分了解目标动物的行为习惯。如果观测人员拥有狩猎经验，他们很擅长为定点照相机确定最佳位置，照相机可以俯视动物经常出没的路线、喂食或饮水地点、气味标记点或休息场所。

使用诱饵、气味吸引剂或制作引导栅栏，引导动物靠近照相机，可以提高成功机会。

偷猎者、好奇的村民、被激怒的大象、熊等可能会盗窃或破坏照相机。所以需要将其放置在人烟稀少的地方，并且尽可能隐蔽好。

8.3.6　使用无人机摄像头

无人机摄像头（图 8.14）越来越受到欢迎，并且价格也比较便宜。如果湿地通过陆路或水路很难开展调查工作，那么在获取湿地影像过程中，上述工具就非常实用。但是它们看起来很奇怪，可能对很多水鸟产生威胁，因此必须注意不要伤害或者吓跑鸟类。

图 8.14　小型无人照相机

8.3.7　识别脚印

留在基地，让自动照相机监测野生动物固然不错。但是无论使用多少照相机，我们只能局限于选定区域。我们需要亲自到现场巡查和监测，这是照相机永远也不会代替的。通过这种方式观测野生动物，可以收集大量相关信息，其中很多信息都是间接的。例如，哺乳动物在泥土或沙地留下很多脚印（图 8.15），通过这些脚印，工作人员可以了解动物相关信息及其出没地点。在识别当地哺乳动物出没路线，甚至它们的粪便或特殊进食迹象、巢穴等方面，向工作人员提供培训十分有用。

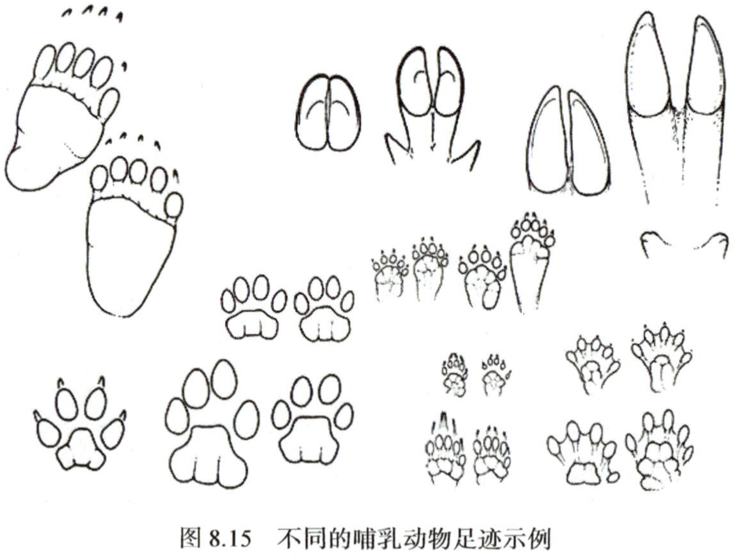

图 8.15　不同的哺乳动物足迹示例

如果无法识别脚印或者需要永久记录，通常需要制作石膏模型。图 8.16 介绍了制作石膏模型的具体步骤。

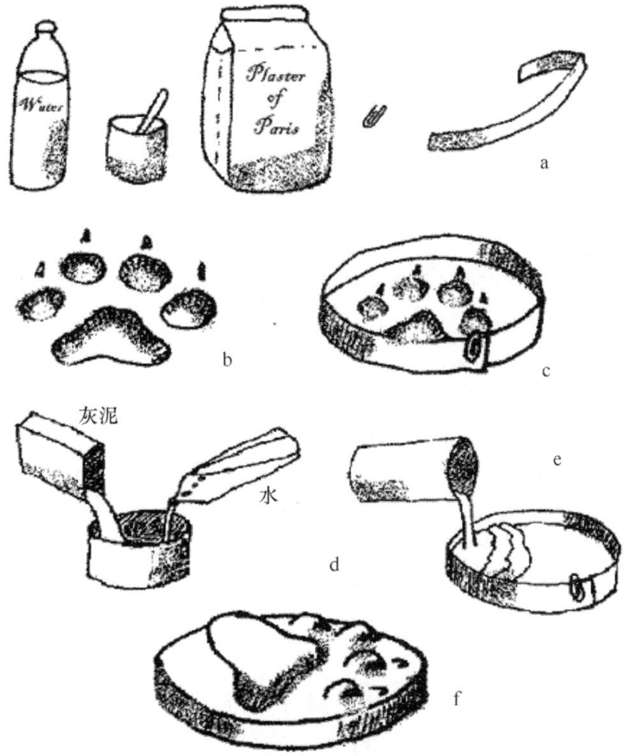

图 8.16　为动物足迹制作石膏模型

a. 所需材料：熟石膏、搅拌盒、水、回形针和纸板条。b. 在软质泥土或湿沙中，找到一个需要制作模型的清晰足迹。小心清除落在上面的所有散叶或棍棒。c. 将纸板条围绕足迹搭建成一道墙，用回形针固定到位，注意不要破坏足迹。轻轻地将纸板条压入周围泥土中，确保石膏不会从纸板条下方流出。d. 此时可以制作石膏。石膏与水的配比大约为 2∶1。迅速搅拌，清除所有结块。在地面上轻轻敲击搅拌盒，除去所有气泡。e. 将石膏倒入事先准备好的模具内。首先从精细部分开始，如爪印。倒入相对较为稠密的石膏，确保模具不会松散。让模具凝固，持续至少 1/2 小时，直到表面无光泽，手感很硬为止。f. 用手探入模具底部，将整个模具取出来。取出模具时，不要使用木棍撬模具下方。否则可能导致模具出现裂纹。在仔细清洗或着色之前，让模具阴干几天

8.4　分析结果和监测数据

调查和巡查报告生成列表、数量和测量值，但是这些数据未经加工毫无意义。需要对这些数据进行分析和展示，但应确保经理、领导、公众和媒体能够理解、领会和遵照执行。

如果没有评估尺度，单纯数量也没有什么意义。例如，一个调查组看见了 200 只绿头鸭，他们提供的报告或许准确地反映了一个事实，但是如果没有比例或语境，这种报告就没有什么实际意义。是太多还是太少？使用这种信息，我们可以获得什么成果？

试比较这段文字：“调查组在 2h 内，从 3 个观测点记录了 200 只绿头鸭。它相当于去年统计数量的两倍，包括之前一个从未记录的区域看见的 50 只绿头鸭。”似乎这种统计忽然具备了某种比较意义。

看看这份报告："我们说A区有200只绿头鸭，但是B区仅有105只。"我们是否应该推断A区的绿头鸭会更多呢？如果不利用搜索方法，我们就不能得出这个结论。如果我们知道A区10个观测点发现200只绿头鸭用了50h，而B区3个观测点发现105只绿头鸭仅用了5h，那么我们必须得出结论：事实上，B区的绿头鸭数量可能更多。原始数据不能进行公正对比。由于A区搜索力度更大，所以A区样本的偏差也更大。如果我们希望将一个样本和另一个样本的数量进行比较，那么我们必须使用可比较单位进行表述，并尽可能多地消除存在的偏差。

8.4.1 偏差最小化

尽力通过以下方式使偏差最小化：

- 复制所用方法的条件、样本区域、季节、天气状况、白天具体时段、观测人员的能力
- 使用稳健方法
- 选择更常见的物种，扩大样本量
- 实现搜索标准化
- 采用适当统计数据

我们需要对可比搜索行为进行量化，如平均遇见率、每步行1km看到的数量、净10天内捕获量等。甚至如果一个区域的观测条件与另一个区域存在较大差异，这些单位都有可能出现偏差。如果是后面一种情况，可能需要使用比例，如A区发现的全部鸭类，其中60%为绿头鸭，而B区仅为40%。

8.4.2 使用清晰图表

对于大多数读者来说，在数量表中列出结果也没有什么意义。设法使用清晰图表展示数据。大多数经理和领导都比较熟悉直方图、饼形图、时距图和地图插图。使用这些工具可以有效揭示大量调查数据的结果。

如果分值可以比较，那么我们只需使用简单的直方图，展示不同时段、地点或物种之间存在的差异。

图8.17即采用这种方式的一个示例（虚拟数据）。

图8.18提供了海南东寨港一年内很多物种的相关数据（真实数据）。但是应注意在泥泞的夏季，数量出现陡升现象。说明出现了正在筑巢的苍鹭和白鹭，这与冬季迁徙类涉禽的峰值完全不同。

这表明在综合不同物种数据时务必小心。图8.19是根据英国多年来鸟类总数量年估计值（图中顶端曲线）获取的真实数据。

数据显示存在一点变化，通常可以认定总体数量的变化幅度不大。但是看看下方曲线，我们发现整个示意图比较符合要求，不过隐藏在里面的是生活在湿地和农田的鸟类数量急剧下降。需要仔细分离数据，以便发现这些细节。

地图上也可进行空间比较。以下示例说明如何展示大面积区域或单一湖泊不同

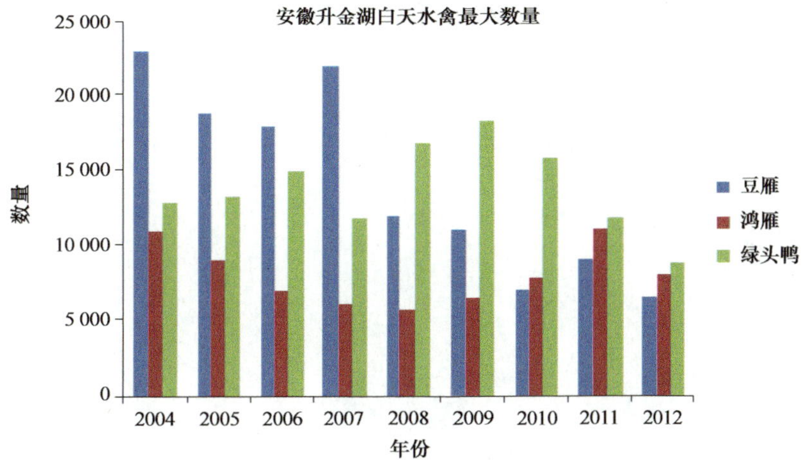

图 8.17　按年份计算的三个物种可比丰度示例

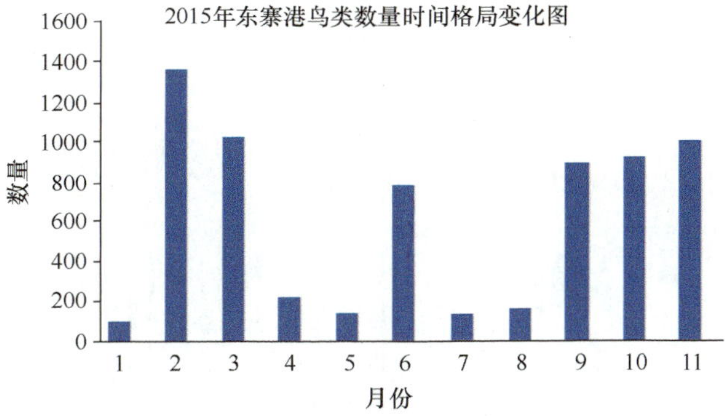

图 8.18　按月份计算的差异性观测丰度图

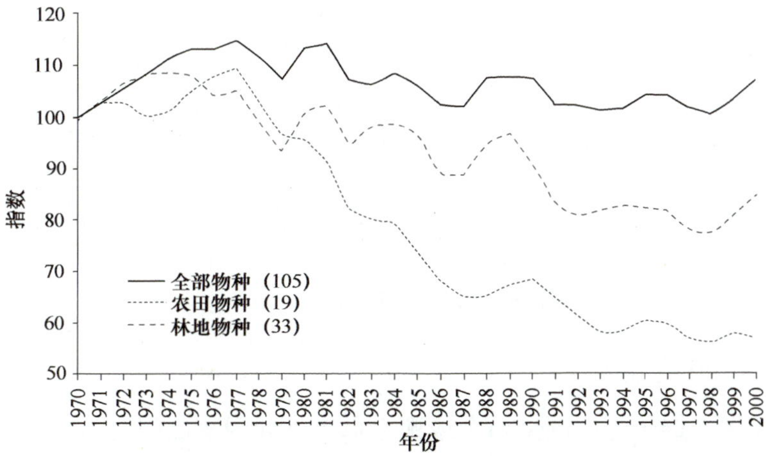

图 8.19　英国政府颁布的生活质量指标显示普通本地繁殖鸟类的种群发展趋势

过去 30 年对英国的所有鸟类差异性监测结果表明，这些鸟类种群数量总体保持平稳，但林地鸟类种群数量有所减少，农田鸟类种群数量尤其大幅减少物种或时间分布的比较数据（图 8.20）。最后一个示例说明人类干扰水平的真实情况（图 8.21）。

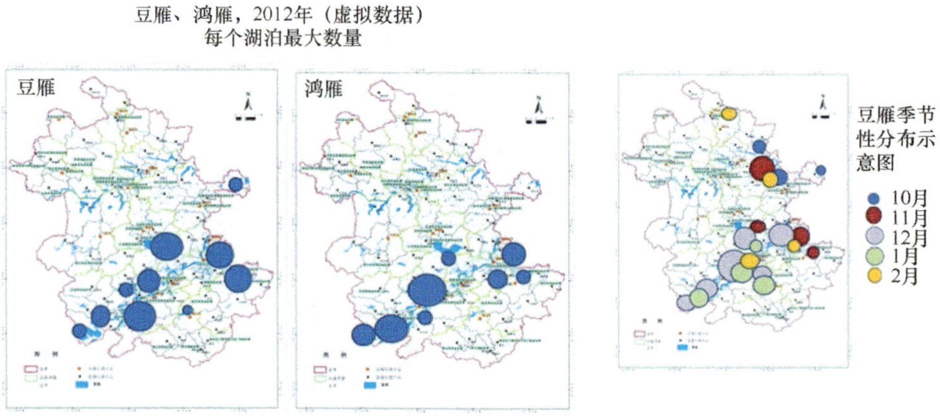

图 8.20　如何展示不同物种或同一物种在不同月份之间可比分布丰度的示例（虚拟数据）

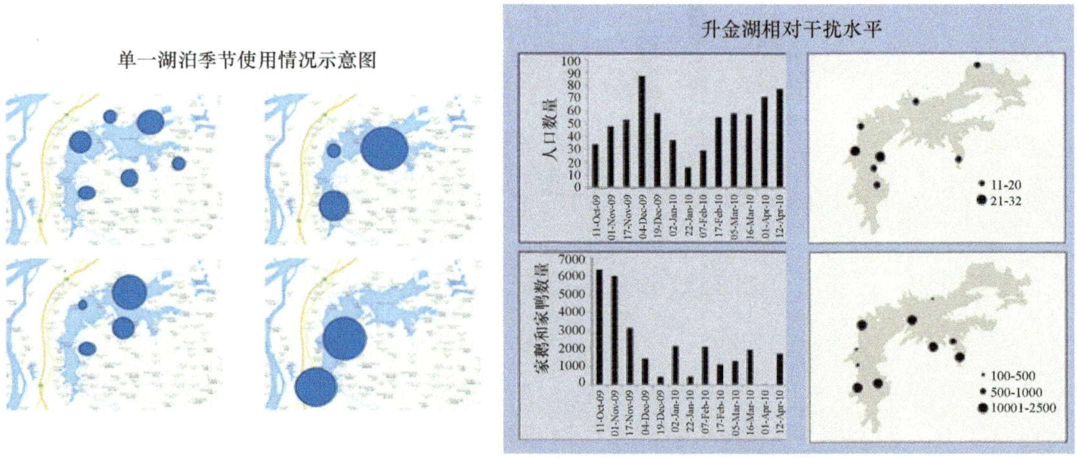

图 8.21　人类活动也需要得到监测——人类对安徽升金湖的利用情况

后一幅图提出了相关性问题。我们可以说明人类干扰水平如何影响雁类在升金湖的季节性分布吗？

这里我们需要进入统计检验最具挑战的领域。

8.4.3　什么是相关性？

相关性（图 8.22）可以测试两个或两个以上不同因素之间是否存在关系。相关性可以是正值、负值或中性值。相关性可以很强，也可以很弱。使用适当统计数据可以提供答案，并且可以得出相关性是否重要的结论。

根据上述结论，统计人员希望说明巧合相关度和不真实关系的概率低于 55。

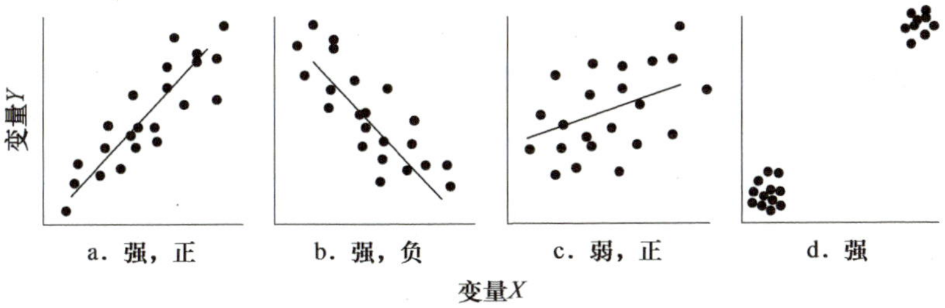

图 8.22 不同类型的相关性

对两个样本直接进行比较（如不同季节、空间对于两个不同物种的用途），如果差异显著，说明两份样本从相同统计数据库提取的概率低于 5%。

警告：相关性可以支持如下论据：一个因素影响另一个因素，但是无法证明存在因果关系。就第三个未识别因素或因素组合而言，上述两个因素可能存在一定差异。

8.4.4 更长时间的数据历史记录

湿地管理方主要负责处理数量、周期和空间分布记录，同时通过其他重要测量值，可以获得在更长时间范围内令人感兴趣的数据，如树木年轮或年龄级结构（图 8.23），从而说明常住种群数量处于稳定、增加或减少状态。

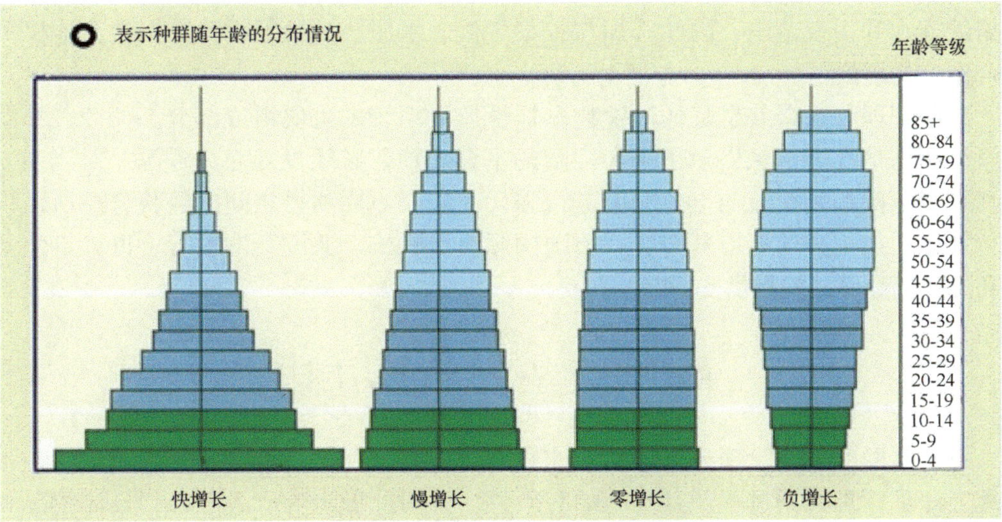

图 8.23 年龄结构示意图（人类数据）

8.5 数据管理、共享和汇报

在监测方面付出太多努力毫无意义，除非产生的数据需要使用、分析和上报，并且可以用于后续策划或管理。

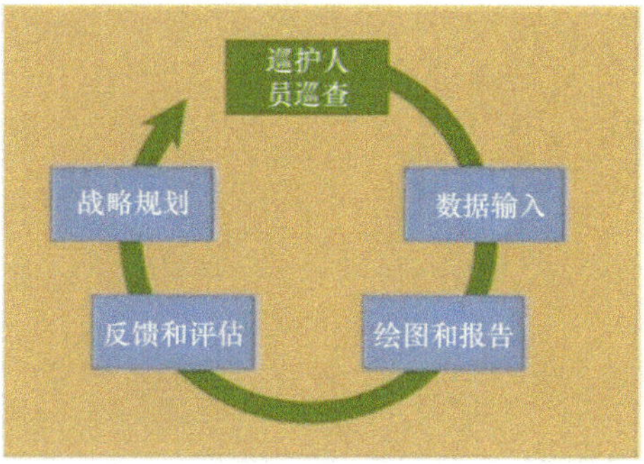

图 8.24　报告周期

管理方、媒体、决策者、政治家无法直接使用原始数据（列表、数量、地图、表格、照片）。必须使用有效格式，对数据进行分析、汇总和上报（图 8.24）。

通过标准统计数据可对受影响因素和诱发因素之间的相关性进行测定，通常这些统计数据即可确定直接关系。但是生态状况很复杂。各种变量并不独立，因此不符合标准统计数据的必要条件。很多变量都在发挥作用，但是几乎没有任何变量处于可控状态，因此需要进行多元分析。若变量明显偏离常态分布，则需要提供非参数统计数据。在任何情况下，必须按照系统且无偏差的方式收集数据，这就需要数据收集或熟练矫正采用严格协议，从而处理现有数据收集模式产生的偏差。

邀请优秀的生态学家设计数据收集协议，核对获得的数据，并对结果进行分析，整个过程必须揭示真相、令人信服。大多数湿地管理局缺少这样的生态学家，因此通常情况下，与当地大学或技术研究所建立合作关系，在研究和监测过程中由他们提供支持，这一点十分重要。

在了解变化过程时，需要考虑很多相关数据集，但却超出了湿地管理局收集此类数据的能力和授权范围，因此必须与收集气候、水质、污染等相关数据的其他机构深入合作、共享数据。

数据管理和数据共享是任何生物多样性监测系统的重要组成部分。如今，互联网为组织和评估此类信息搭建了一个广阔的平台。项目必须投入充分资源，对当地、本省和全国范围沟通交流的上述方面进行强化。应对不同项目之间的经验分享所需设施进行规划，如由全球环境基金会、德国国际合作机构、亚洲开发银行、世界自然基金会和湿地国际等机构提供支持。

8.6　管理有效性跟踪工具

为了协助保护区管理方，包括湿地负责管理方，我们已经开发了大量工具，帮助管理方评估管理策划过程及其实施的有效性，并对此做出响应。考虑到具体情况和需求涉及范围十分广泛，需要采用不同评估方法，因此世界自然保护联盟世界保护区委员会已经制定了一个评估框架［M Hockings、S Stolton、F Leverington、N Dudley 和 J Courrau（2006），*Assessing Effectiveness—A Framework for Assessing Management Effectiveness of Protected Areas*，第二版，世界自然保护联盟，瑞士］。该框架向评估体系开发提供总体指南，鼓励评估和汇报实现标准化。该框架及更为具体的跟踪工具为管理方及各国提供了一些评估机制，帮助它们对履行《拉姆萨尔公约》和生物多样性保护区工作方案公约及其目标的进展情况进行评估。

管理有效性跟踪工具旨在对实现世界保护区管理有效性的进展情况进行跟踪和监测。管理有效性跟踪工具有望成为公园工作人员使用的一种廉价、简单的现场管理工具，随着时间的推移，它还可以提供保护区和管理进展相关的一致性数据。该工具旨在：

- 识别保护区管理有效性的相关进展。
- 提供保护区组合相关基线数据，在汇报和问责过程中提供支持。
- 针对工具开发和政策制定，确定组合趋势和优先顺序。
- 识别具体保护区中的关键管理问题，确定解决这些问题的具体方法。
- 识别跟踪步骤，尤其是现场跟踪步骤。

开发的管理有效性跟踪工具，在实现世界银行 / 世界自然基金会联盟组合过程中，有助于跟踪和监测进度。目前，全球环境基金会所有保护区项目都有义务开发这种工具。几个全国性保护区体系已经用来开发基本管理有效性评估工具。管理有效性跟踪工具已经用于全世界很多陆地保护地，包括一些湿地和国际湿地保护地。

8.7　生态系统健康指数

定义：生态系统健康即适合继续提供安全条件，确保组成物种生存并提供关键生态服务的能力，包括对于气候和其他变化因素的恢复能力。

目标：生态系统健康指数不是一项评估。它是一种反映生物多样性健康状况的动态且不断变化的指标，正如金融指数反映经济表现一样。

- 生态系统健康指数提供了一个基线，根据这种基线可以为维持或实现指定的健康水平设定目标。
- 生态系统健康指数可以用作基于结果的项目成绩和影响指标。
- 生态系统健康指数能够表明项目成功和失败的具体位置，并且允许在整个项目过程中对具体措施进行修改。
- 项目监测与评估管理有效性评分卡会显示生态系统健康指数。

简介：生态系统健康反映了当地维护生物多样性价值和生态功能的能力。区域不同，生物多样性价值和生态功能会存在显著差异。任何指标应包括三个组成部分：①重要生物多样性的栖息地适合性得分；②重要生物多样性的状态；③更为广泛的环境背景。这种得分未必能够反映稳定性。很多湿地变化速度较快，但我们感兴趣的是生物群适应变化甚至在变化过程中繁荣的能力。随着气候和水流形态的变化，这种能力显得越来越重要。采用这种指数的每块湿地应开展基线调查，基线调查也会为后续调查项目选择指标和目标物种。这些指标应包括生活在湿地的关键鸟类、重要的水生动物群——鱼类、软体动物；选定的指标性昆虫；濒危哺乳动物；植被的主要构成；外来入侵物种发生率。

该指数确定了观测时的快照值；根据早期确立用于识别不同指标中存在趋势的基线，该指数可以将现有得分联系起来；可以为了改善每个不同指标，制定合理目标，并根据确定的目标对现状进行比较。

虽然人体还不存在体力严重衰退的迹象，但是我们可以确定一生中威胁人体健康的几种指标（过度饮酒和吸烟的习惯、睡眠不足、未充分接种预防、生活区域存在已知疾病、糟糕的卫生习惯、医疗设施不足等）。同样，我们可以识别外部环境中威胁生态系统健康的几种风险，虽然在物种栖息地或状态条件下可能无法立刻反映这些风险。这些指标包括外部开发威胁等级、有效法律保护的等级、未来应用或预期的人类利用压力等级。

8.7.1 洞庭湖指标物种示例及基本原理

洞庭湖指示物种范例见表 8.2。

表 8.2 洞庭湖指示物种范例

物种	指标	下降原因	采样频率
普通翠鸟	水很清澈，鱼类密度小，人类在浅水区域的干扰较小	水很浑浊，缺少鱼类，干扰严重，预期存在一些季节变化	沿样本水路每月统计一次数量
斑鱼狗	深水区鱼类密度较小	水质较差，鱼类密度较小	乘船沿样带每月统计一次数量
鸬鹚	鱼类密度较大，人类干扰较小	鱼类资源较少，人类干扰较小，缺少栖树	每月在已知栖息地统计一次数量
苍鹭	水道边缘鱼类密度较大	食物减少，人类干扰较大	在采样点和样带每月统计一次数量
夜鹭	农田和湿地生态健康状况	由于过度使用杀虫剂，导致农田中的食物不足	在已知栖息地每月估测一次数量
白鹭	农田和湿地生态健康状况	由于过度使用杀虫剂和污染现象，导致农田中的食物不足	一年内，在样本栖息地每月统计一次数量
黑卷尾	农业用地的生态健康状况，季节性	由于过度使用杀虫剂，导致农田中的食物不足	冬季沿采样路线每周统计一次数量
棕背伯劳	一年内农业用地生态健康状况	由于过度使用杀虫剂，导致农田中的食物不足	一年内，沿样本线路每月统计一次数量
骨顶鸡	湖泊中的植被质量	湖水污染	一年内每月统计一次数量
针尾鸭	普通水禽适合性	人类干扰，湖泊状态下降	整个冬季每月统计一次数量
蜻蜓	一年内的水质和生态健康	过度使用杀虫剂或受到污染	样本水塘和岸边每周统计一次数量
蛾	陆生植物的多样性和健康状况	植物多样性较差，杀虫剂使用率较低	飞蛾诱捕器，每周统计一次数量

8.8 水质指标

经常使用**大型底栖无脊椎动物**作为生物学水质指标，因为大型底栖无脊椎动物数量庞大，比鱼类更容易捕捉，并且比藻类或原生动物更容易识别。如果河流水流量较大，可以使用海斯采样器收集大型无脊椎动物样本；如果水流量较小，可以使用索伯采样工具收集大型无脊椎动物样本。识别和列举大型无脊椎动物，并根据三个相同尺寸样地的平均值，估测每个区域的生物体数量。大型底栖无脊椎动物密度按每平方米河流底部生物体总数量上报。除上述生物体总量外，还应标注多样性指标，分类阶次尤其如此，如蜉蝣、石蝇、甲虫和其他生物体。采用香农指数和蜉蝣目、横翅目毛翅目指数［＝发现的蜉蝣目（蜉蝣）、横翅目（石蝇）和毛翅目（石蛾）物种总数］分别测量无脊椎动物群落的多样性和质量。

有些细菌、病毒和原生动物可能对人类和野生生物有害，从胃肠疾病到轻微呼吸道疾病和皮肤病不等。这些生物体可以通过下水道、排水沟、化粪池、农田径流、动物加工厂，并从生活在水体内或附近的野生生物进入水中。微生物经过土壤过滤，可以进入地下水。

因为无法针对每种潜在致病生物体测试水体，所以通常对数量庞大的物种指标进行测量，如排泄物中的大肠杆菌。海水目前通常使用被称为肠球菌的一组细菌。如果有这种细菌，说明可能存在类似排泄物，据此认定可能存在致病生物体。

测量指标细菌浓度，通常采用每 100ml 水中的生物体数量估值。水质管理方感兴趣的两种指标包括在一段时间内采集的单一样本浓度和一系列样本“平均”浓度。

溶解氧

即氧气溶解于水的浓度（单位：g 氧 /m^3 水）或实际存在的氧气与水的理论溶氧能力之比（单位：饱和百分率）。有时优先使用后者，因为水的溶氧能力会随着温度的变化而变化。

溶解氧是水生生态系统保持健康状态的一项基本要求。如果溶解氧浓度低于 3-4g/m^3，人们需要的很多鱼类（如鲑鱼和三文鱼）就难以生存。幼虫和幼鱼对此更加敏感，它们对溶解氧浓度的要求更高。如果溶解氧长时间保持低浓度，会导致成鱼窒息死亡；由于鱼卵和幼鱼对溶解氧浓度更敏感，一旦溶解氧长时间保持低浓度，会导致鱼卵和幼鱼窒息死亡，进而降低鱼类繁殖生存能力。如果水体状况发生变化，导致水生幼虫和其他食物死亡，鱼类可能会因为饥饿而死亡。溶解氧浓度下降有利于厌氧（无氧）细菌活动，它们会产生有害气体或恶臭，而这通常与水体污染有关。

如果大量可生物降解有机物质，如污水或食品加工废弃物进入地表水，可能导致氧气快速消耗。

细菌使用氧气分解有机物质。如果污染物含有机废物，就会向细菌持续提供食物，从而加快细菌活动，导致细菌数量增加。一旦水体受到污染，细菌消耗氧气的速度大大超过氧气补充速度，从而导致水体含氧量下降。

其他因素，如温度和盐度也会影响水中溶解氧的含量。如果炎热天气持续时间太长，会减少氧气含量，即使水体清洁，也可能导致鱼类死亡，因为温水持氧能力低于冷水。高温加快细菌活动，细菌消耗更多氧气，进而导致氧气枯竭。

温度

温度是反映水质的一个基本因素，温度能够对水生生物产生巨大影响。如果水系总体温度发生变化，群落构成也可能会发生变化。冷水鱼，如三文鱼对温度变化非常敏感，一旦温度超过约20℃，就会受到生理应激的影响。鱼类如果进入局部温水区，也可能会受到影响。

水温受到很多因素的影响，包括气温剧烈波动、河道和湖边形状变化、悬垂植被减少、阴天，而最重要的则是水流改变。废弃物排入水体，如果污水处理温度明显不同于背景水温，也可能会影响温度。

pH

酸碱度，氢离子和氢氧离子在水中的含量，是生物体内很多化学反应的驱动因素。酸碱度的标准指标为pH。pH等于7表示中性，低于7表示酸性，高于7表示碱性。

在热带红树林地区，一旦硫酸盐暴露并经过氧化，酸性偏高可能会累积到危险水平。最安全的做法就是让这些土壤一直保持湿润状态。

浅层地下水（通常为井水，深度不足30m）为弱酸性，pH低至6.0。这是因为雨水（雨水本身为弱酸性）携带（植物根系和微生物产生的）二氧化碳进入产生碳酸的深层地下水。弱酸性水具有一定的腐蚀性，能够将来自管道和水泵的金属，尤其是铜溶入水体，进而影响物种成分。

导电性

即水的导电性。因为导电性会随着水中离子（尤其是带电粒子）数量的增加而上升，所以水体导电说明存在溶解物质。这些物质可能是自然存在的矿物质（阳离子和阴离子），或者因为人类活动导致水体存在污染物。通常使用手提仪器在野外测量导电性。常用计量单位是：毫西门子/平方米，即mS/m^3。

营养物

导致水质下降最频繁的营养物质是氮和磷。由于这些物质在环境中以很多形式存在，因此从事水质研究的科学家常常采用不同方式对它们进行测量。例如，水中氮元素可能会固定在动植物组织中，这时将其称为“有机”氮。这种氮元素最终分解成无机物：硝酸盐（NO_3）、亚硝酸盐（NO_2）或氨（NH_3）。营养物质来源包括化肥、排放污水、动物和食品加工废弃物及城市雨水。

营养物是水生群落维持健康状态的必要构建模块，但是过量营养物（尤其是氮化合物和磷化合物）会过度刺激水草和藻类的生长。水草和藻类生长速度远远超过本地水下水生植被，因此可能覆盖水生动物栖息地，最终导致水生动物窒息死亡。过量水草和藻类的分解可能导致氧气耗尽。地表水养分浓度高被称为“富营养”。湖泊高浓度营养物质产生的不利影响尤其显著，其中营养物质通过相同水体实现循环，逐渐聚集在一起。

氨

氮元素在水体中存在的一种形式就是氨（NH_3）。氨是维持生命的一种必需营养物，不过一旦超过特定浓度，对水生生物有剧毒作用。如果氨的浓度太高，鱼类会死亡。氨在水中的化学形式比较复杂；其毒性取决于未电离分子的比例（即以 NH_3 而非 NH_4^+ 存在），它反过来与水的 pH 和温度相关。通常根据总氨（NH_4^+）分析水样，了解氮浓度。根据总氨相对于 pH 和温度的浓度，计算未电离氨（NH_3）。一般而言，地表水中的氨氮浓度大约不应超过 1.5g/m^3。水体中氨的来源包括动物粪便径流、农业用地肥料、污水排放和食品加工废料，如冻库废水废料。

硝酸盐

硝酸盐来源和氨来源相同。这是因为硝酸盐在土壤和水体中形成的一个重要原理就是氨的分解（或“矿化”）。作物残茬通过土壤过滤分解，有时会在地下水中产生硝酸盐。一旦硝酸盐超过植物生长所需营养物，就会发生这种现象。农业系统中硝酸盐的一个重要来源是固氮植物，如豆类，它们可以从空气中获取氮元素。牲畜排尿也是地下水中硝酸盐的一个重要来源。

硝酸盐测量结果为硝酸盐（NO_3）或以氮元素（NO_3-N）存在的硝酸盐。典型饮用水中硝酸盐的标准最大容许值为 50mg/L，但是有些地区几乎不能经过河流稀释，并且雨水将氮元素从土壤带入地下水，允许超过上述值。一旦地下水位上升，从土壤剖面底层冲刷氮元素，地下水中的氮含量也会增加。

悬浮沉积物

沉积物由各种尺寸颗粒物构成，包括细黏土颗粒、淤泥、沙和砂砾。在水质环境中，人们最关注的颗粒物是细黏土和泥沙。水柱内沉积物通常指的是悬浮沉积物，一般测量指标为浓度：g/m^3。

一旦沉积物沉淀，可能会严重改变水生群落。沉积物可以堵塞和损害鱼鳃，使水底的鱼卵和幼虫窒息，并填满鱼类产卵所在砾石之间的空间。悬浮泥沙和沉积物降低水体透明度，妨碍水生植物的生长。

沉积物也可携带其他营养物和有毒重金属，它们附着在沉积物上，然后进入地表水。这里的污染物可以和沉积物一起沉淀，或分离，并可溶于水。

雨水冲刷所有表面，带走泥沙和土壤颗粒，但尤其是植被已经受到破坏的地面泥沙和土壤颗粒。因此，水土流失及诸如开挖土方、清除植被和耕作之类的活动可能会导致泥沙流入地表水，强降雨之后尤其如此。牲畜践踏河床、水边或岸边，导致大量泥沙进入水体。

浊度

浊度是与悬浮沉淀物相关的一种水质指标。它对光进入水体后被水中悬浮颗粒物散射的程度进行量化。悬浮物质数量越多，光散射越强，浊度越高。光散射颗粒物可能是有机物（如藻类和其他植物或动物残屑），也可能是无机物（如细泥或黏土）。透光性下降导致植物生长速度减缓，反过来减少无脊椎动物食物来源，最终减少鱼类食物来源。

测量浊度时采用一种特殊测光表，通常使用浊度单位（NTU）。如果浊度单位小于 25NTU，视为水生生物可以接受，但是只有当浊度单位远低于这个值时，才会影响水体外观。

阳离子

阳离子是一组在水中溶解时呈正电荷的物质。在地表水和地下水中存在的主要阳离子有钙离子（Ca^{2+}）、镁离子（Mg^{2+}）、铁离子（Fe^{3+}）和锰离子（Mn^{2+}）。其他阳离子包括钾离子（K^{+}）和钠离子（Na^{+}）。存在这种阳离子主要因为水流经过区域的地质特征。例如，水流经石灰岩，它的钙、镁和碳酸氢盐浓度就会很高。这些阳离子会提高水的硬度。地下水流经深色（铁镁质）火山岩，一般会含大量铁和锰。

泥炭矿床出水的铁和锰含量可能也很高。任何铁或锰元素在土壤中都会处于溶解状态，但是当地下水升到地面，一旦氧气进入，就会从溶液中析出。如果锰元素含量高，野生生物可能会面临健康风险。

阴离子

阴离子是一组在水中溶解时呈负电荷的物质。从水质角度看，主要阴离子包括硫酸盐（SO_4^{2-}）、氯化物（Cl^-）、硝酸盐（NO_3^-）和碳酸氢盐（HCO_3^-）。通过阴离子，可以显示水源或水质影响因素。例如，地下水氯化物浓度很高，说明盐水已经进入水系。

毒性有机化学品

毒性有机化学品即含碳复合化合物，如多氯联苯（PCB）、二噁英和杀虫剂滴滴涕。这些化合物经常在环境中持续存在、日积月累，因为它们在自然生态系统中难以分解。因此，这些化合物经常称为持久性生物累积性污染物。很多这些化合物对人类具有致癌作用，容易造成食物链顶端附近的其他捕食者，如鸟类和鱼类出生缺陷。

有毒烃类用于石油产品、冰箱、杀虫剂、溶剂、推进剂和清洁剂。因为渗漏，或者在处理过程中，有毒烃类能够污染水源。由于烃类经常储存在地下储罐，一旦储罐破裂，很容易污染地下水。

农药可以对水造成污染，包括杀菌剂、杀虫剂、除草剂和生长调节剂。农药一旦喷洒在路面或农田等区域，通常会经过雨水径流进入地表水。由于喷洒过量或喷雾偏差，或者直接撒入河道控制水生杂草或杀死蜗牛（血吸虫病宿主）时，偶尔会在河流中发现农药。有些农药活动能力很强，能够通过土壤渗入地下水。

重金属和准金属

高密度金属（如铅、锌、铜和铬）及准金属（如砷）在环境中自然存在。但是人类活动（如工业生产过程和采矿）已经改变了这些金属物的分布，从而对地表水和地下水产生污染。在地表水中，金属通常存在于容易附着的沉积物中。在此食物链之上的动物（如鱼类）有可能让金属累积，从而增加金属物质的浓度。

8.8.1　监测水质

为了确保人类安全，由环境保护部门负责对水质进行测量和监控，此项工作技术要求很高，需要使用各种仪器和实验设备。但是，生态学家和湿地管理方可能希望独立开展某些简单的基础类测试项目。表 8.3 介绍了开展此项工作的具体方式。

表 8.3 典型湿地水质监测项目

要素	方法	频率	设备 / 物品	注释	安全水平
氧	嗅一嗅甲烷或硫磺气味。在样瓶中收集水	每月一次	清洗集瓶；标准化学实验室	对鱼类和其他动物健康状况	＞3-4g/m^3
温度	在不同深度测量	每月一次或出现极端天气时	温度计	低温时，水含氧量更高	温带＜20℃ 热带＜32℃
pH	从大片湿地若干采样点采样	每月一次	pH 计，石蕊试纸	由于二氧化碳过量或石灰溢出，导致 pH 上升	5-8
浊度	目测标准试样沉淀物的深度	每月一次	采样瓶		
大肠杆菌（细菌）	从几个采样点采集水样，并保存在无菌瓶中	每月一次	送往实验室	生物性污染一般应对措施低水平为正常	
氮（N）				淡水富营养化最佳指标	＞10mg/L
磷（P）				淡水富营养化最佳指标	
汞（Hg）、硫（S）、铅（Pb）、铜（Cu）		如果靠近排放源，每年或每月一次	送往实验室	重金属毒性很强	
有机化合物	目测表明是否存在油液收集样本，存放在洁净瓶内	每年一次，除非怀疑再次溢出	送往实验室	大多数有机化合物毒性极强	
氯和其他卤化物				消毒和有毒	
钠、钾	通过品尝可以大致估测海水盐度	确定几个点，每月检验一次	单纯伏特计即可测量导电性	在一些淡水系统，盐水可能是一个问题	

附　　录

附录 1　相关国际组织

名称	范围	网站
湿地国际	一个全球组织，旨在维持和修复人类和生物多样性所需要的湿地及其资源	www.wetlands.org/
英国皇家保护鸟类协会	设在英国的一个慈善机构，旨在为鸟类和所有野生动物创造一个健康的环境，为人类创造一个美好的世界	www.rspb.org.uk/
英国鸟类学信托基金会	一个独立慈善事业研究机构，综合专业和公众科学，旨在利用野生生物，尤其是鸟类种群改变证据，向公众、舆论影响者、环境政策制定者和环境决策者提供相关信息。该机构主要关注英国，但也参与很多全球项目	www.bto.org/
美国奥杜邦鸟类协会	其使命是维持和恢复自然生态系统，重点关注鸟类和野生动物，保护人类和地球生态系统。它有很多全球合作伙伴，旨在向迁徙美国境外的鸟类提供帮助	www.audubon.org/
英国斯林布里奇野生鸟类与湿地基金会	世界最大、最受推崇的湿地保护机构之一，它在全球开展工作，为野生生物和水禽保护并改善湿地	www.wwt.org.uk/
国际鹤类基金会	向重点影响鹤类的人群和地域大力提供支持，专门保护全球所有 15 种鹤类	www.savingcranes.org/
东亚—澳大利西亚迁飞区伙伴关系	保护迁徙水禽及其栖息地及迁徙水禽依赖人群的谋生方式	www.eaaflyway.net/
中国香港观鸟会	观赏并保护香港鸟类和自然环境。1999 年设立了香港观鸟会中国保护基金会，支持观鸟者和鸟类学者在中国大陆进行观鸟推广，开展研究工作	www.hkbws.org.hk/

附录 2　湿地生物多样性公约 / 计划

名称	年份	范围	网站
《生物多样性公约》	1992 年	一个全球协定，它关注生物多样性的方方面面：基因资源、物种和生态系统	www.cbd.int/
《联合国气候变化框架公约》	1994 年	联合国环境与发展会议通过的国际环境公约	newsroom.unfccc.int/

续表

名称	年份	范围	网站
《联合国防治荒漠化公约》	1996年	建立全球合作关系，逆转和防止沙漠化/土地退化，缓解干旱对受影响区域产生的影响，支持扶贫和环境可持续性	www.unccd.int/en/Pages/default.aspx
《湿地公约》	1971年	通过国家行动和国际合作，保护并合理使用全部湿地，为实现可持续发展做出应有贡献	截至2010年，已有160个国家加入了该公约，超过1900处湿地纳入了《国际重要湿地名录》，总面积超过1.86亿 hm^2。www.ramsar.org/
《保护野生动物迁徙物种公约》	1979年	一个保护和可持续利用迁徙动物及其栖息地的全球平台	www.cms.int/
东亚—澳大利西亚迁飞区伙伴关系	2006年	保护迁徙水禽及其栖息地及迁徙水禽依赖人群的谋生方式	www.eaaflyway.net/
《濒危野生动植物种国际贸易公约》	1975年	各国政府间国际协议。旨在确保野生动植物种国际贸易公约不会威胁这些物种的生存	www.cites.org/
全球入侵物种计划	1997年	一个生物多样性计划，旨在防范和管理外来入侵物种	www.diversitas-international.org/activities/past-projects/global-invasive-species-programme-gisp/

附录3　最佳实践指南

附录3.1　IUCN世界保护区委员会最佳实践指南系列

最佳实践1：国家保护区体系规划
最佳实践2：保护区的经济价值
最佳实践3：海洋保护区指导方针
最佳实践4：土著民族及保护区
最佳实践5：保护区融资
最佳实践6/最佳实践14：评估效果：保护区管理评估框架
最佳实践7：实现和平与合作的跨界保护区
最佳实践8：保护区可持续旅游
最佳实践9：世界自然保护联盟第五类保护区保护陆地/海洋景观管理指导方针
最佳实践10：保护区管理策划指导方针
最佳实践11：土著和本地社区及保护区：实现公平、加强保护
最佳实践12：森林与保护区：世界自然保护联盟保护区管理分类使用指南
最佳实践13：保护区可持续融资
最佳实践14：参见最佳实践6

最佳实践 15：关键生物多样性区域的识别与缺口分析：综合保护区系统目标

最佳实践 16：自然圣境：保护区管理人员指南

最佳实践 17：保护区工作人员培训：规划与管理指南

附录 3.2　国际湿地合理利用手册系列

1．湿地合理利用概念和方法
2．国家湿地政策
3．法律与制度
4．禽流感与湿地
5．合作关系
6．湿地更紧密经贸关系安排
7．参与技巧
8．水资源相关指南
9．流域管理
10．水资源分配与管理
11．管理地下水
12．海岸管理
13．编目、评估及监控
14．数据与信息需求
15．湿地名录 http://www.ramsar.org/sites/default/files/documents/pdf/lib/hbk4-15.pdf
16．影响评估
17．标注国际重要湿地
18．管理湿地
19．应对湿地生态特征变化
20．国际合作
21．2009—2015 年《湿地公约》战略计划

《湿地公约》第三号技术报告《生物多样性公约》技术系列 27：评估湿地价值 https://www.cbd.int/doc/publications/cbd-ts-27.pdf

附录 4　湿地野外作业建议

附录 4.1　赤脚医生

至少应向全体野外工作人员提供健康、安全及如何处理普通伤口和问题的相关建议。应严格遵循以下规则。

保持干燥

天气可能发生变化，可能淋湿衣服，气温可能下降。衣服淋湿导致热损耗严重因

此应保持干燥。应随时戴上防水披肩或穿上夹克衫。带一套干衣服，装入塑料袋，如果野外作业路程很远，导致衣服淋湿，可以及时更换。

保暖

应务必确保衣服不能太薄、不能太少。如果刮风、衣服淋湿、晚上作业或者夜间外出，会感觉更加寒冷。随身携带干衣服，放入塑料袋，可能会派上用场。如果身体开始发抖，应待在掩体内，避免身体暴露在外，并且大家可以挤在一起。

穿上救生衣

无须冒险。即使游泳健将在划船前进时，也可能会掉入水中。穿上救生衣，并确保同一条船上的任何人都要穿上救生衣。

小心缝隙

很多船只事故发生的原因是有人陷在船只和码头或另外一艘大船之间，或者受到两者挤压。骨骼一旦受到挤压，就很难康复。因此上下船时必须特别小心。

带一瓶安全饮用水

带足水，确保能够维持一天使用。如果山地河流水很清澈，可以放心饮用，或者将水瓶盛满，但是必须意识到中国大多数河流水污染非常严重，尤其是矿区附近或者工业设施或人口聚居地下游。长江下游流域的水体也滋生了东方血吸虫病寄生虫，宿主是水蜗牛。如果直接饮用这种水，人体可能会感染这种疾病。

使用地图

应务必确保路线清晰。如果远离主要道路或显著地标，应随身携带地图。地图对野外步行、勘探、监测等作业而言却至关重要。也可使用手机数字地图（如下），但是在野外无法保证都能接收手机信号。

手机

如果手机有信号，携带的手机最好充足电。如果陷入困境、迷路或发生事故，可以使用手机呼叫其他人。如今，手机使用全球定位系统和地图应用程序越来越普遍，可以帮助自己确定具体位置。

割伤和刮伤

用水或酒精清洗伤口，保持干燥。如果有碘酒或消毒粉，应涂上碘酒或消毒粉。用清洁干净的纱布包扎伤口。纱布不要太紧，以使伤口能够呼吸。尽快让伤口干透，如晒太阳。

昆虫咬伤和螫伤

不要刮擦伤口，否则会感觉更痒，更容易受到感染。涂酒精、桉树油或抗组胺乳膏。如果引来很多螯蝇，应戴上网状帽。

蛇咬伤

不要惊慌，缓慢移动，将心跳频率降至最低水平。如果认为蛇有毒，可以用刀片张开伤口，加快血流速度；也可吮吸伤口，使用橡胶半球形成的真空或者使用吸收材料清除部分毒素。无须使用止血带。

为蛇拍照或者将蛇带走，方便医生识别。通过电话，告诉其他人自己所处位置。

缓慢移动到最近道路、河流交叉点等，呼叫车辆或船舶帮助就医。

溺水和人工呼吸

如果巡逻员因为溺水、事故后昏迷、摄入药物或毒物（如蛇咬导致毒素进入人体）而停止呼吸，那么在处理伤情前最紧迫的任务就是要设法让他重新呼吸。

下面介绍几种人工呼吸法。最常用的方法是“嘴对嘴”或“抢救呼吸”法，这也是急救培训中广泛传授和认可的方法。通过这种方法，将来自施救者肺部的空气直接送入伤者口腔。来自施救者肺部的空气就像一个紧密气封，压入伤者肺部，然后观察每呼入一次空气，伤者胸部是否能够自行隆起。

人工呼吸过程中应按下列步骤操作：

1．查看伤者口腔内有无堵塞物（舌头、呕吐物等），托起颈部，让其头部后仰。

2．捏住伤者鼻子，封住伤者口腔，然后直接向伤者口腔内呼气。

3．松开鼻子和口腔周围密封部位。

4．观察伤者胸部是否自行隆起。

5．查看伤者颈部是否跳动。

6．如果伤者胸部尚未自行隆起，从第一个步骤开始重复操作，直到专业人员到达现场。

人工呼吸应按正常呼吸频率持续进行，大约每分钟 12 次，直到伤者完全能够自行呼吸。即使感觉伤者能够自行呼吸，也要密切注意，因为伤者可能会反复，这种现象非常普遍。

四肢断裂

如果可能，应设法就医。单纯骨折，只需将肢体拉入正常伸位，然后将断裂部位两侧捆绑在硬质夹板上。

如果是严重粉碎或扭曲性骨折，最好留在原地等待就医。应务必保暖。

制作夹板

夹板是一种刚性支架，防止受伤部分移动，保护受伤部位，避免再次受伤。伤者在送往医院就诊过程中，通常使用夹板稳定断骨。如果肢体严重紧张或扭伤，也可使用夹板。如果本人或同伴在野外受伤，可以就地取材，如用棍子临时制作一副夹板，并使用线绳、绷带、衣物碎片或纤维树皮固定。

- 在直接挤压伤口固定夹板前，应做好伤口止血处理，然后缠上绷带、纱布或布条。
- 如果身体部位需要上夹板，不要尝试移动该部位，否则容易产生二次伤害。
- 上好夹板，确保夹板固定在受伤部位上方连接部位和下方连接部位。然后将夹板套在受伤部位一侧的肢体上，避免直接将固定夹板的绳子置于伤口上方。充分固定夹板，确保身体部位保持静止状态，但不能太紧，否则会导致伤者血液循环中断。
- 每隔几分钟查看一下夹板周围，确定血液循环是否存在下降迹象。如果末端开始出现苍白、肿胀或呈现淡蓝色，应松开固定夹板的绳子。

- 伤者如果感觉很虚弱，呼吸短促，可能会出现休克。在这种情况，将伤者放下，但受伤部位不要受到影响。如果可能，应抬起伤者腿部，让其头部略低于心脏。

使用止血带

止血带是一种临时捆绑用品，旨在限制血液流入和流出身体特定部位。使用止血带可以防止身体深度切口过量失血，或者防止蛇咬伤后毒素扩散到身体其他部位。必须学习如何使用止血带，这点十分重要，因为使用不当（或者缠绕的止血带太长）可能会引起极端并发症，如组织坏死和截肢。具体办法是切断动脉内来自心脏的强血流，而不是血液流入心脏的浅静脉。

但是如果伤口太严重，外加压力无法止血，这时（并且只有在这时）才能考虑使用止血带。

- 如果未得到控制，伤口流血将最终导致休克和死亡。
- 必须使用止血带时，应让临时使用的绷带留在伤口上，因为血流减缓有助于凝血。
- 如果可能，应抬升伤口。通常施加压力及减少血管内血液流动承受的重力，两者共同作用足以止血和形成血块。

运送受伤人员

受伤人员应被尽快送往诊所或医院，但是如果受伤现场和公路有一定距离，可能需要运送伤员。肩负法是一种最简单的方法，但是在此之前，必须首先确定伤员任何肢体是否断裂，脊柱是否受伤。运送伤员前，断肢可能需要上夹板。如果在移动伤员颈部或背部过程中，伤员感觉非常痛苦，应使用担架，避免移动这些部位，但同时应添加垫子，避免在运送过程中出现横向移动。

可以临时制作担架，使用两个结实的木棒，长度约为2.5m，将这两根木棒分别穿过两件夹克衫袖口。夹克衫底角共同构成一个悬带，非常结实。将伤员放在悬带上，但要特别注意，不要让伤员身体弯曲或摇晃，用力支撑患者颈部。如果有毯子，效果比两件夹克衫更好。将两根木棒放在毯子上面，间距大约为70cm。将毯子较长两端折叠起来，缠绕圈数越多越好。然后将伤员放在毯子上。在伤员体重作用下，毯子“固定”到位，在运送过程中不会展开。

附录5　湿地管理适用的生态原则

根据《开发和规划人员生物多样性原则》复制（John Mackinnon，2002，BWG/CCICED，中国北京）。

翻印的湿地指南原则如下。

建立具代表性的保护区系统

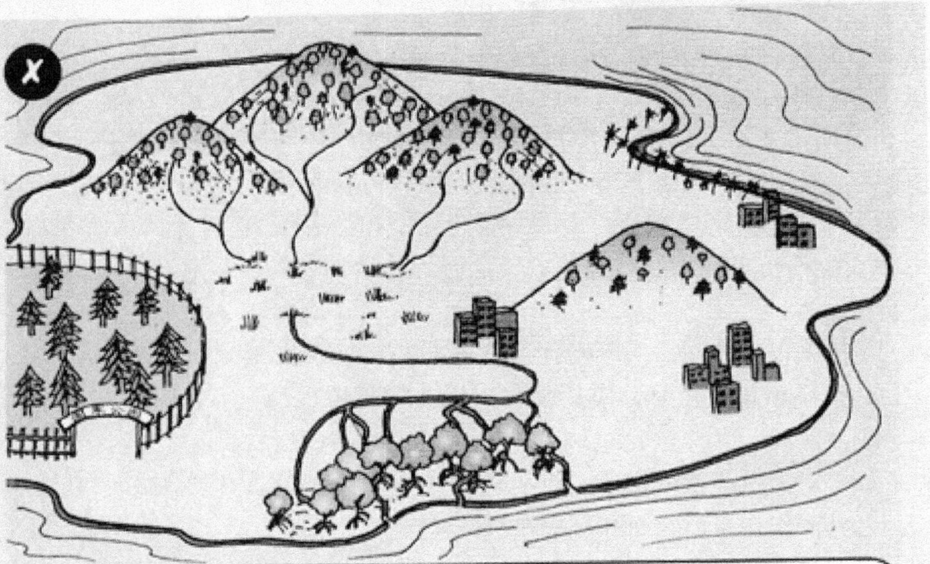

应该建立一个保护区体系，它应该具有代表性，保护着中国所有自然生态系统、栖息地和各种海拔范围内的典范地区。地点应选择在物种丰富或当地特有种集中分布的地区。面积要足够大，以实现可持续性发展。关键物种的种群要达到能够自我维持的数目。大的、连通的、变化多的地点，比小的、孤立的或分散的地点要好。

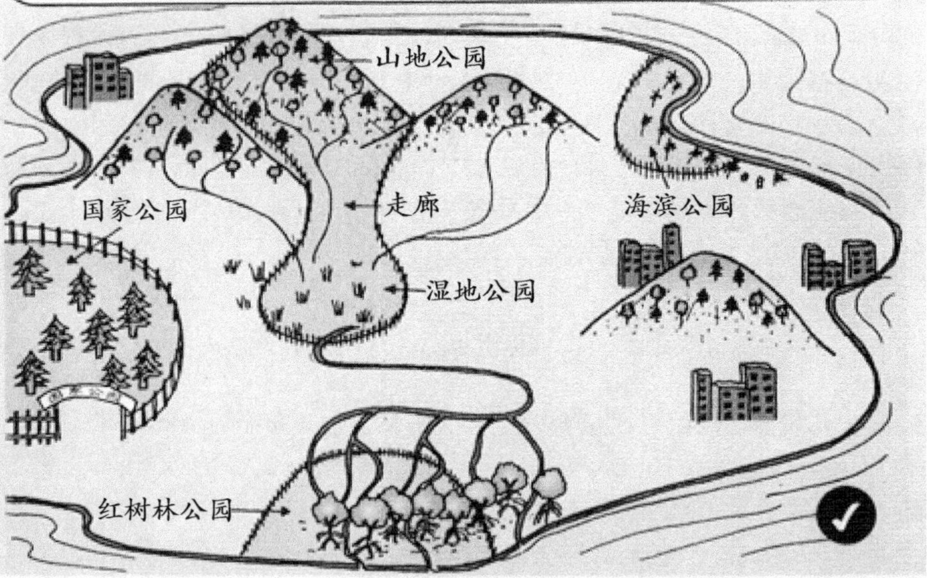

禁止在重点湿地内开垦农田

围湖造田和开发破坏了湖泊的生态系统，降低其蓄水（海绵）功能，使重要的生物多样性丧失，破坏微气候和当地湖泊河流连通体系的生态平衡。为了改善人类的健康水平、提高生产力和环境质量，就必须维护湖泊和天然水系的连通渠道，不要在重要地区开垦农田。保护中国健康的水文功能比生产更多的粮食作物要重要得多。

保持淡水系统的连通性

在水系通道设立人为屏障将影响当地鱼类和其他水生动物的迁徙。必须保持堰坝和其他屏障水体间的连通性，才能保护水生生态系统，使成年鱼类能够到达繁殖区，以及鱼苗能够到达采食栖息地。

维护水系统的排蓄水功能

湖泊在干旱季节是水的来源地，而河流却是雨季的排水渠道。保护湖泊河流的这种功能和防止淤塞是相当重要的。必须在集水区和所有必要的地方保持良好的森林覆盖以便减少河流上游土壤侵蚀，挖走多余的泥土和沙砾，避免河道淤塞。

保护水域和湿地的生物多样性

提高湿地生态系统的连通性，降低污染，避免过度捕鱼和水流量减少的威胁。管理好湿地生态系统以提高生产力、生物多样性、娱乐和旅游潜力，也有助于调节小气候。通过建立保护区或控制捕鱼、狩猎和植被采集来加强保护这些生态系统的重要组成成分。

把捕鱼量限制在可持续性范围内

过度捕鱼减少了单位捕鱼量，破坏了鱼类的多样性，降低了鱼的平均大小和使不受喜欢的鱼类扩散。如果传统的捕鱼方法已经使捕鱼量过高，那么希望通过更大面积和更现代的捕捞方法来达到增加捕鱼量的目的是不可能的。小规模的传统捕鱼方法能在社会上更加公平地分享利益。

保持水系统的平衡

✗

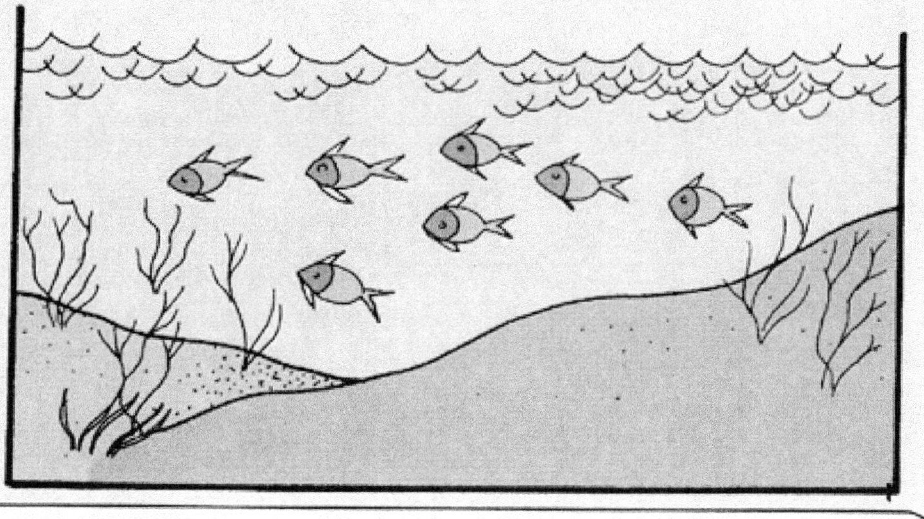

不要提倡通过引入或过度养殖，使某些物种形成优势。这会导致某些食物类型的过度消费，其他物种的丧失，生态平衡的破坏，最终导致产量的降低。可以通过保护多样性，来达到平衡的和可持续性的渔业管理。生态系统中物种越多，系统越稳定。

✓

合理利用水资源

在季节性干旱地区，过度使用水资源将降低水位、破坏植被和加剧荒漠化。在任何地区必须确定可持续性发展的利用限度，发展只能在允许范围内进行。特别注意控制使用深井。

保护风景区的自然景观

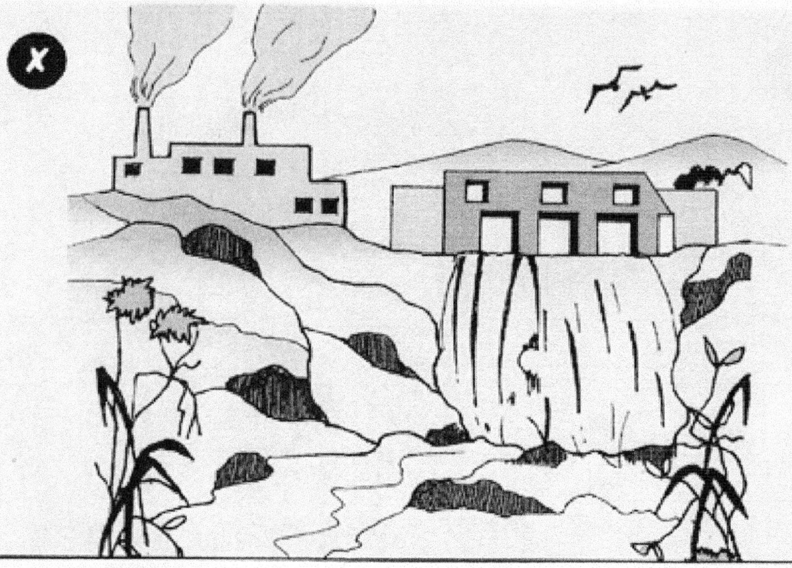

有植被环绕的湖泊和瀑布不仅能吸引游客，而且对生物多样性也有着重要的意义。它们既是室外娱乐、踏青、旅游的绝好去处，也是人们陶冶情操的地方。一定不要以牺牲这些不断增值的宝贵财富为代价，制定那些会造成污染的发展计划。中国有大量并不十分脆弱的地区能为这类发展提供广阔的空间。

需要正确处理废水

城市生活和工业废水对当地的生物多样性有相当大的负面影响，不仅破坏了河流和湖泊，还使我们的生活环境质量降低。要设计和使用足够的废水处理设施，使城镇居民享受美丽、富于生物多样性和健康的环境。必须严格设置污染标准，对造成污染的单位和个人处以高额罚款。

保护城镇内湖泊和水道周围的天然植被

湖泊河流周围的植被有助于滋养和净化生态系统，并可创造出优美、清洁、健康的环境。但这些水系统若被不能生长植被的混凝土填塞，则会造成水体超营养作用，形成无生命的、不健康的和没有吸引力的厌氧性水体。

尽量减少挖泥所带来的负面影响

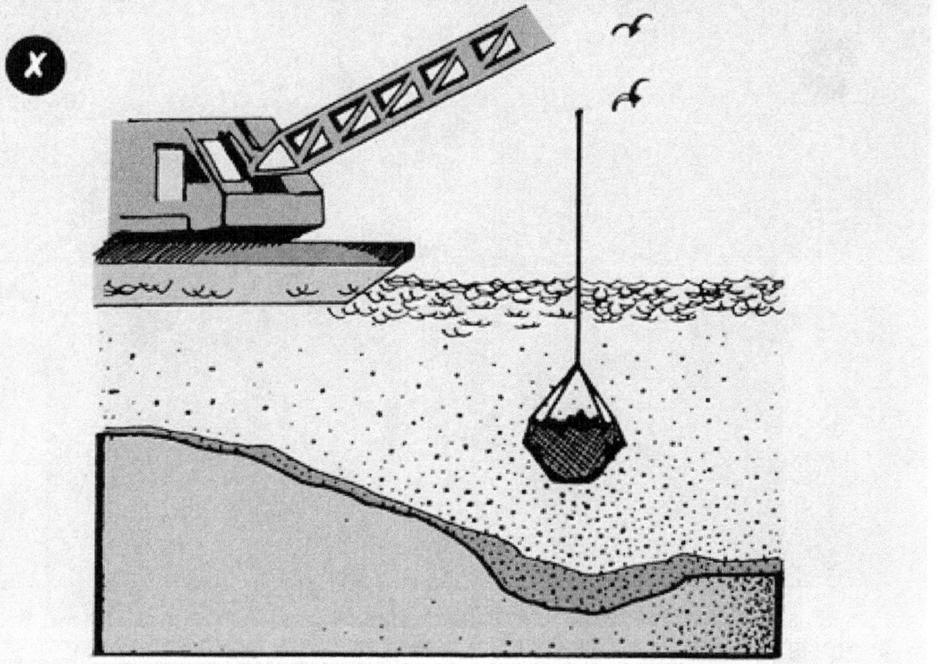

河床湖泊淤塞会降低其蓄水、排水的功能，因此有必要进行挖泥。但挖泥也会带来负面影响，包括降低水质、改变水的流速和流向以及导致洪水和侵蚀等。挖泥量应最小化，弃泥应安置在适宜的地方，使其不会加大侵蚀程度。

将泛滥平原保留为开放地区

随着气候不稳定性的增强，在河流易泛滥的平原开发建设会造成生命和财产的损失，甚至会影响到多年未遭受洪水的地区。易遭洪水的地区应该保留作为开放地区、娱乐地区、野生动植物避难所或农田。

创建人造池塘或湖泊

很多废弃的采石场和矿井都可以转化为宝贵的湖泊或池塘，其他池塘也可以很容易地纳入当地周密的发展规划中。如果这些池塘适合于鱼类生长，其娱乐和生物多样性价值更会提高。深水与浅水区应该结合在一起。浅水区有利于鱼类和两栖动物的生活，芦苇生长区可为秧鸡和鸣禽所用。如果在安全的岛屿上保留矮树丛，则可作为白鹭和其他水鸟的栖息所和繁殖地。

适度开发湖滨地带或海岸线

不正确的开发容易毁坏湖滨地带或海岸线的自然环境质量和生物多样性价值。大量建设码头区会破坏植被和生物多样性，增加腐烂物的污染危险。房屋应集中建设并远离这些地区并使负面影响降至最低。应该建设少量的公用码头和入口处，而不是建设大量的码头和上岸平台。

保护河边和沼泽森林

沼泽和河边森林在蓄留洪水、避免河口淤积、阻拦污染源、维持干旱季节水流量、保持船只畅通以及保护生物多样性等方面起着决定性的作用。伐除这一地带的森林会导致洪灾、河堤倒塌、河流通道淤积以及生物多样性损失。

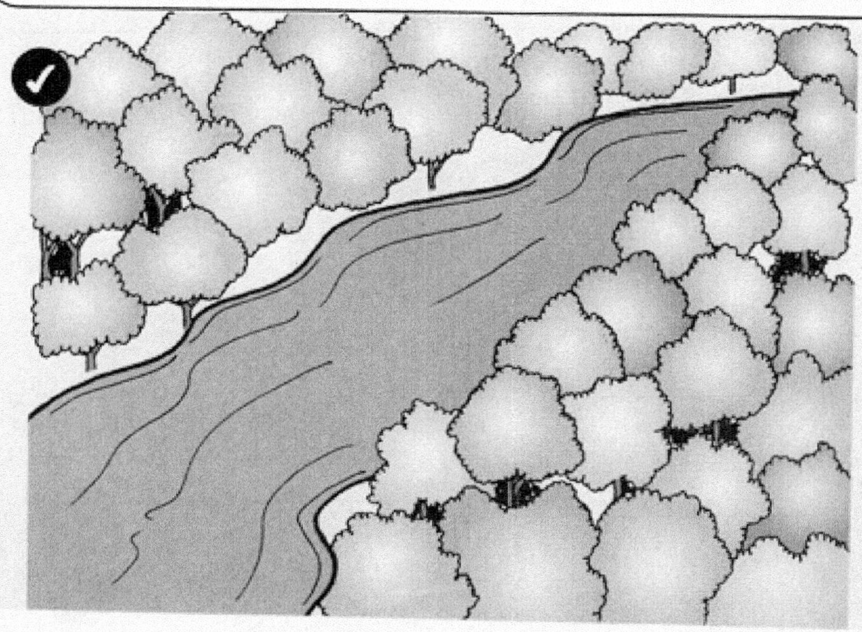

正确使用桥梁和堤道

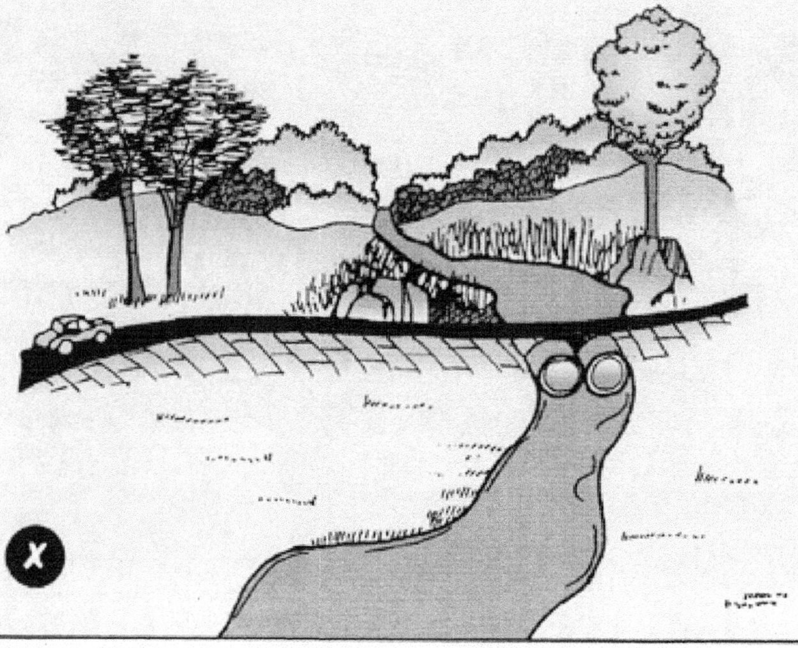

应在湿地和潮水地带建立桥梁和堤道，使水循环不受阻碍。桥梁比管道的作用大得多。途经沼泽和潮汐地的堤道可以保护天然栖息地和生物多样性，这比填充阻塞的建筑方式破坏性要小得多。

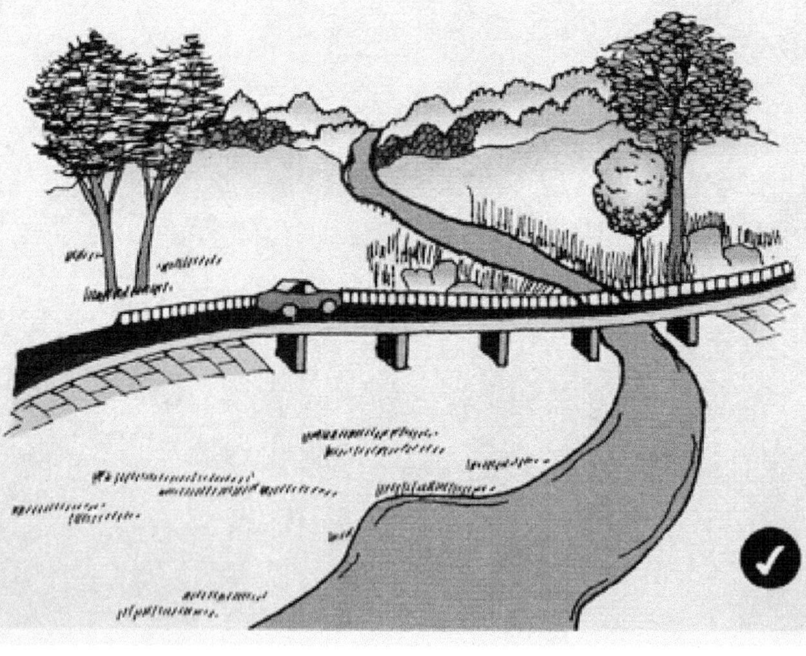

限制沿码头的建设

水道的堤岸非常富于变化。天然植被对于保护堤岸的结构有十分重要的作用。房屋建设要远离堤岸，以免造成堤岸的侵蚀、下沉、洪水破坏等危险，以及对环境和生物多样性的破坏。

保护红树林和沿海植被

沿海植被对稳定泥泞多沙的地层起着重要作用。红树林、番薯属（*Ipomoea*）和鬣刺属（*Spinifex*）沙丘植被，以及浓密的露兜树属（*Pandanus*）植被和海滩森林不仅促进了新农田的开垦，还保护着沿海地区不受海水侵蚀、台风和移动沙丘的影响。它们还是野生动植物的庇护所。红树林是鱼虾的繁殖地，而且能够固着沉积物和防止附近珊瑚礁被遮盖。必须防止不必要地清除沿海植被。如果已经被破坏，就应鼓励人们恢复之。

限制潮间带的开发

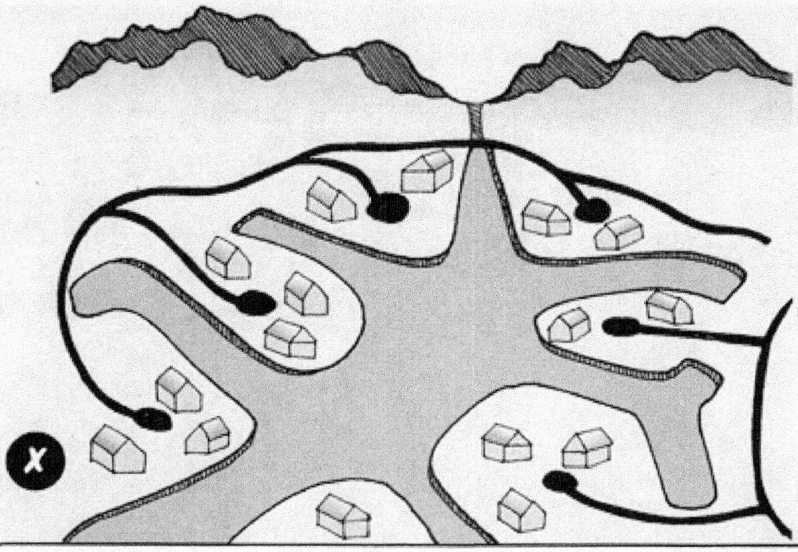

潮间带非常富饶，是重要的净化区、鱼虾繁殖地和大量食用甲壳类的栖息地，也是娱乐休闲场所和重要的生物多样性来源地。同时它也很脆弱，污染、土壤流失或潮汐运动都会毁坏该地带。该地带一旦被污染，采自这里的食物会威胁人类的健康。要严格限制在潮间带内或可能影响该地带的开发，这些活动应该在内陆进行。

建立综合的海洋地区发展计划

由于不同部门在海洋和沿海地区的利益有很大的交叉，因此有必要建立综合的发展计划，确保由某个部门发起的开发不会危及其他部门的需要和利益。

保护珊瑚礁

珊瑚礁是高生产力的生态系统，有很多有价值的功能，如鱼苗的育苗地、旅游胜地、水资源净化、海岸保护等。很多珊瑚礁有机体富含大量蛋白质，正在被越来越多地制成药物。这些利益不应被没有必要的破坏和开发所损害。必须保护珊瑚礁，禁止进行珊瑚礁开采、炸鱼、毒鱼、违规的船只停泊、淤泥或其他污染物排放等活动。

为海洋哺乳动物提供特殊保护

海豹、鲸和儒艮等珍稀海洋哺乳动物的觅食和繁殖栖息地极易遭到污染、人为干扰和非法猎捕等的影响。开发计划必须保证不给海洋动物种群集中地增加更多的威胁。

避免在沿岸和潮汐海域进行不必要的开垦

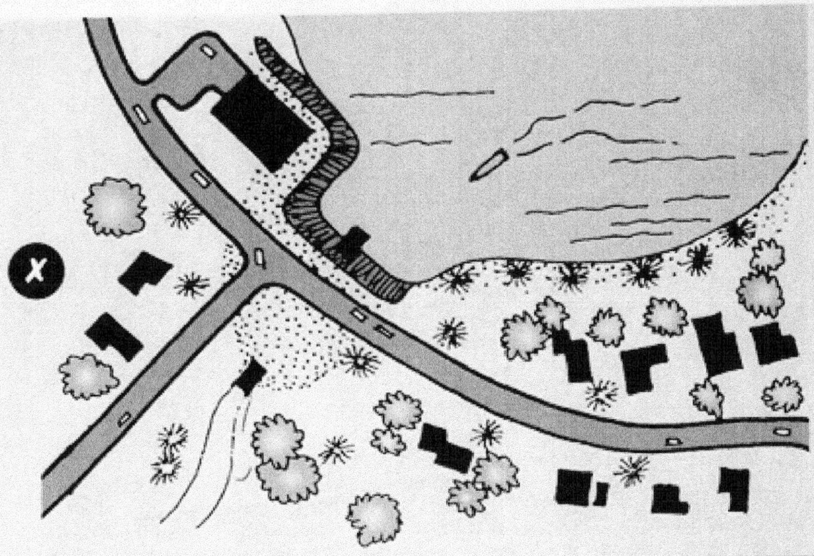

一般来说，沿岸和潮汐海域不应该被填起来或者用其他方式进行开垦。被填起来的潮汐海域往往会受到洪水的侵袭，也可能导致侵蚀问题。而且，填塞会改变水和沉积物的流动，破坏野生动植物的栖息地和高产的浅水区。一旦这些情况确实发生了，我们就必须建造人工暗礁或栖息地来弥补对生物多样性和其他功能造成的损失。这些工作既困难又花费昂贵。

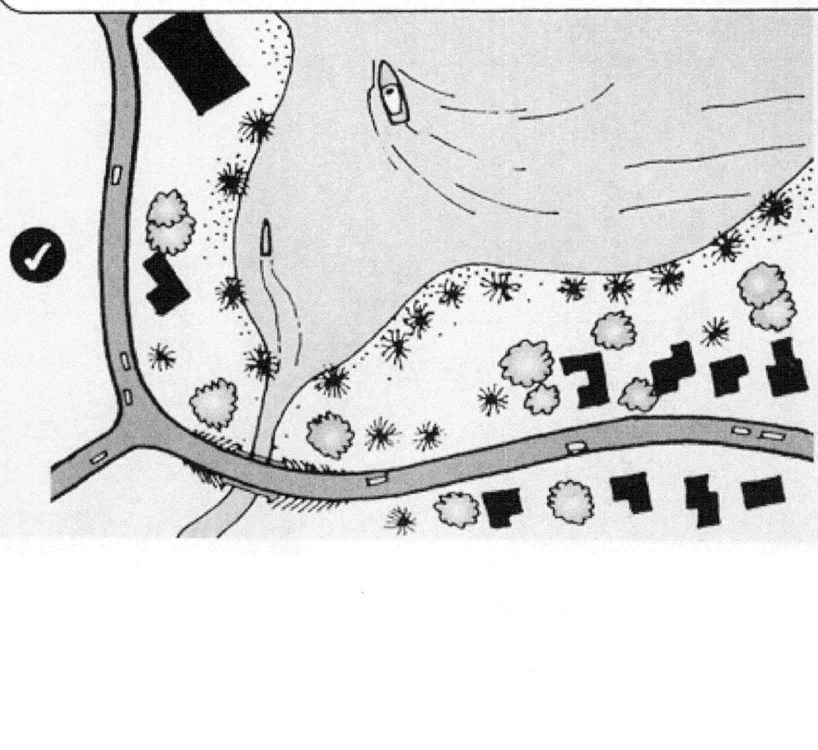

防止赤潮和其他污染的影响

海洋经常被当作巨大的垃圾存储地，但是这些污染物会对海洋生态系统造成严重的破坏。珊瑚在淤泥中会窒息。由于重金属的污染，食用这些鱼类、贝类和虾类对人类的健康构成了威胁。其他废弃的化学物质则导致有毒藻类的大爆发，这些藻类对食物链中的鱼类和软体动物释放出大量的神经毒素，甚至威胁到人类的健康，这种现象被称作“赤潮”。必须对河流和沿岸工厂的污染源头进行净化处理，减少污染的扩散。

消除防水壁

建设开发应远离海岸线，以免沿着被冲蚀的海岸线形成防水壁。在防水壁不可避免的地方，可以在防水壁和海水之间保留由天然植被组成的缓冲带，这将有助于保护防水壁和野生动植物的栖息地，并提高生物生产力。

注重风景质量

沿海地区发展规划应注意保护景观的价值。尽可能地使这些景观对大众开放。可以通过建立缓冲地带或仔细规划建筑物的规模和地点，来减少建筑物对景观的视觉破坏，保护沿河口和沿海地区的景色。

使用桩子

在潮汐带建立码头时，打桩比填方好。沉积物和野生动植物可在桩子底下自由活动，而填方只会成为阻碍，并改变流速和侵蚀过程。另外，打桩的建筑和保养成本低，而且可以被生物分解。

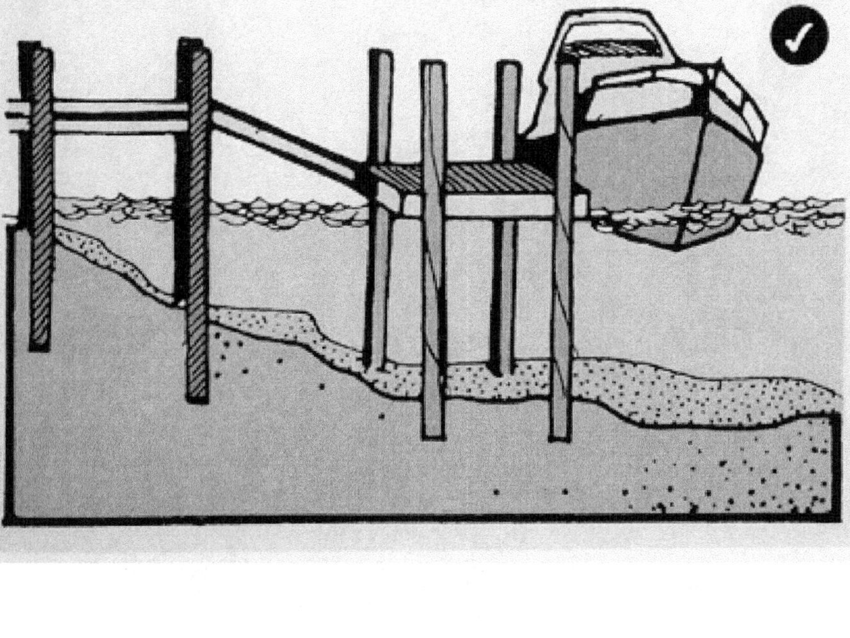

保护天然排水模式

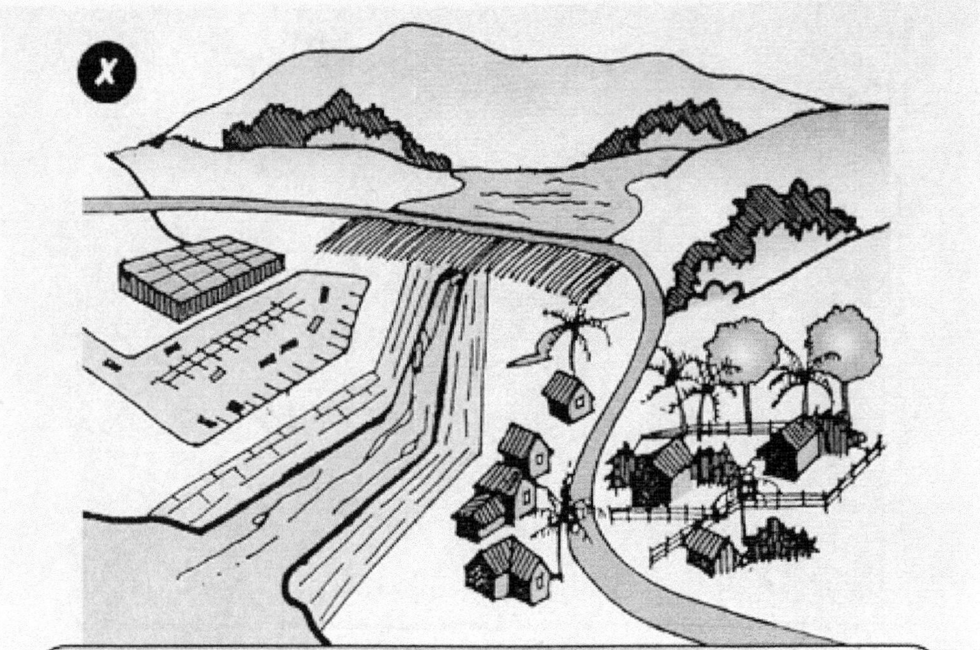

保护沿岸地区的天然排水模式。开凿运河和改变沿海水流的流向会改变或转移沼泽、潮汐带和其他浅水区的水流流向，从而增加污染、改变盐度、减少河口的生物活动。同样的道理，沿岸地区也不应建设堤坝和水电站。

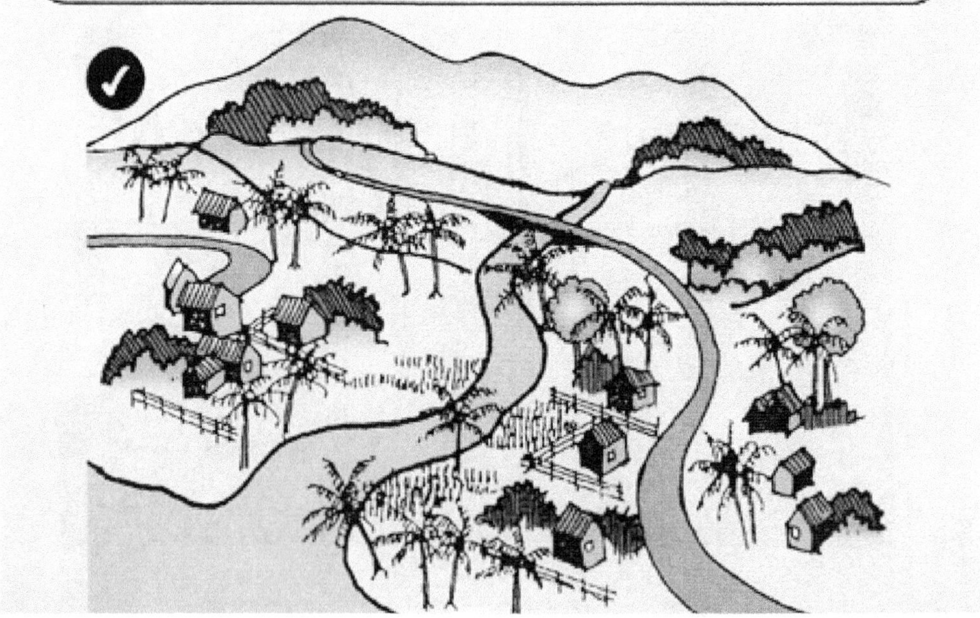

避免沿沙质海岸的开发

沙质海岸的建设要十分小心。这些地区十分脆弱，如果失去天然植被的覆盖，很容易被风暴侵蚀而破坏。人类和机械的过度使用会增加侵蚀的严重程度。洪水、坍塌、水供应和废物排放等问题随之而来。开发应该在远离海岸的更坚实的地表进行。

保护被保护地区的“荒野”价值

建设、分区和制定规章制度要以保护被保护地区的“荒野”价值为前提。这就是说游客们在这里应感觉到其身处真正的大自然中，而不是在人造环境之中。应特别注意要尽量减少噪音和影响视觉效果的建筑，严格禁止采集植物，损伤树木，以及乱扔垃圾和乱涂乱画等。

限制建设进入保护区的公路

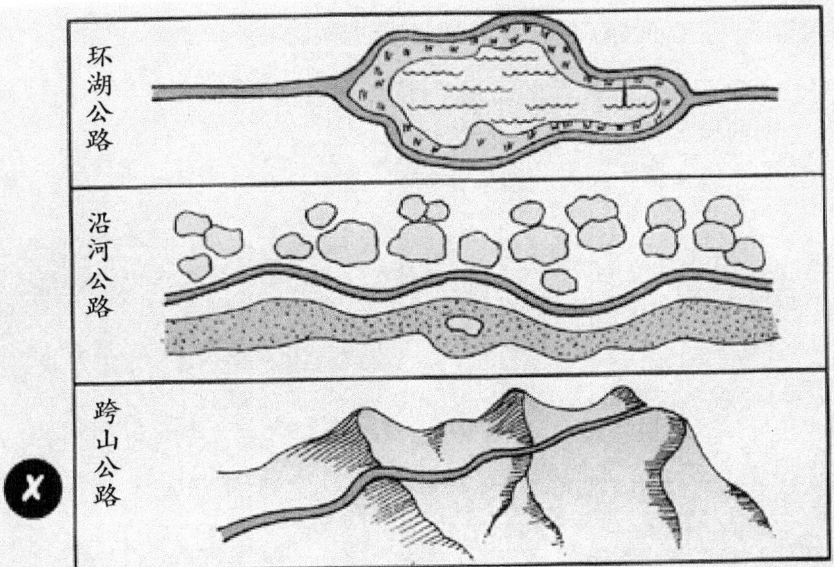

进入保护区的公路虽然便于管理人员工作，但更为滥用和掠夺保护区内的树木及野生动植物资源提供方便。此外，还具有经济效益的诱惑力，使人们想在保护区内开垦农田，所以必须抵制这种诱惑，不要开设进入保护区的公路。难以进入保护区可以避免不适宜的发展，是保护自然最好的方式。公路要建在负面影响最小的地方。

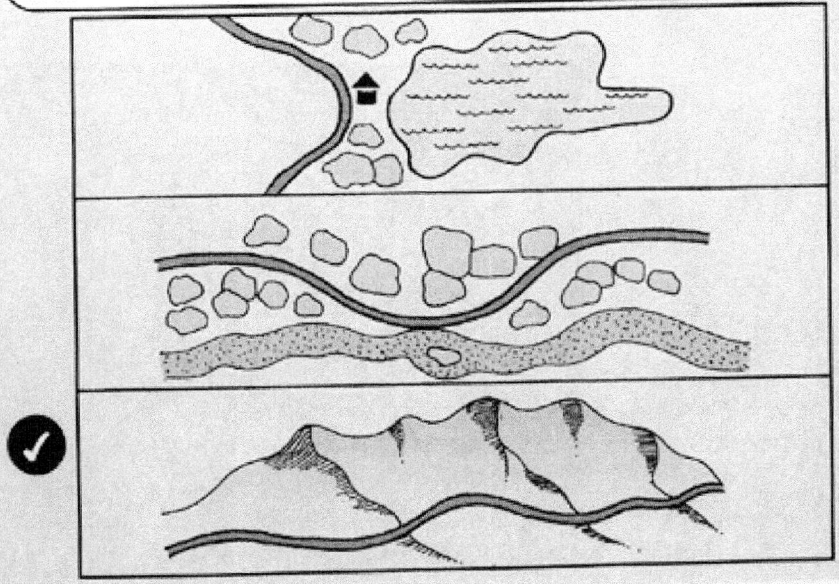

附录6　特定指标监控的具体内容

附录6.1　监测天蛾（天蛾科）和蚕蛾（天蚕蛾科）

飞蛾大多数体型较小，种类繁多，很难区分。因此只需监测这两科体型较大、容易识别和特征分明的蛾类。

天蛾外形像战斗机，翅膀突出并向后延伸。

蚕蛾体型较大，翅膀很宽，呈圆形，上面有斑点或新月状图案。

夜间放置带水银蒸气灯的简易飞蛾诱捕器，每周一次。南方地区一年四季均可，北方只能在夏季进行。飞蛾诱捕器应固定在醒目位置，远离其他明亮灯光。可以在白天查看诱捕的飞蛾，但只能在夜间释放，否则会被鸟类捕食。可以使用旧蛋盒，让飞蛾隐藏在诱捕器底部下方。

典型自制飞蛾诱捕器。商业供应商可以提供专业模型。

附录 6.2　监测蜻蜓

通过水里或周围的蜻蜓幼虫和成虫，能够很好地了解湿地状态。选择几个理想采样点、小水塘，或者几段沟渠或河流，重复采样。统计每次采样发现的个体和物种总数（为采样点设定时间限制）。首先不同物种使用平行分类名称即可（物种甲、物种乙等）。随着时间的流逝，我们就可以识别主要物种，并且能够使用确切名称。注意雄性蜻蜓的颜色可能不同于雌性，因此要按配对组合观察，了解哪些是相同物种。首先了解一些主要科别。

差翅亚目——蜻蜓（强壮有力，翅膀呈扁平状）

蜓总科：体型较大，身体长而直。

大蜓总科：体型很大，身体为黑色，具黄色斑纹，眼睛为绿色，沿边缘接触头顶。

蜻总科：体型更短，腹部很宽，呈扁平状。通常停息时翅膀前伸（最大科）。有些物种的翅膀为彩色。

裂唇蜓科：体型很大，身体为黑色，具黄色斑纹，但眼睛完全不接触。后翅弯曲而非后边有角。雌性后翅可能为彩色。

春蜓科：体型更小，身体为黑色，具黄色斑纹，但眼睛不接触。后翅弯曲而非后边弯曲。腹端展开。

均翅亚目——豆娘（体型细长，眼睛有一定间距，通常翅膀保持垂直状态）

色蟌总科：翅膀很宽，身体和翅膀呈金属色。

鼓蟌科：体型较小，“鼻子”突出。

细蟌科：体型细长，翅膀窄而短（最大科）。

扇蟌科：类似细蟌科，但雄性腿部略微展开。

原蟌科：深色，体型细长，翅膀较短。

昔蟌科：体型较大，粗壮，身体很短，翅膀交叉成 X 形。

幽蟌科：体型中等，粗短，翅膀略微分开。

山蟌科：体型较大，粗壮，翅膀呈半开状。
洵蟌科：体型较大，呈金属色，末端较宽，一直延伸到腹部，翅膀很小。
丝蟌科：体型较小，属于豆娘畸变种群，腹部上方翅膀呈扁平状。

附录 7　选择适当的统计测试方法（改编自 www.graphpad.com）

各种统计测试方法及其应用

测试原因	数据类型			
	测量值（来自正常种群）	排名、得分或测量值（来自非正常种群）	二项（两种可能结果）	存活期
描述一组	平均值 ±SD	中位数，四分位差	比例	卡普兰 - 迈耶生存曲线
将一组与一假设值进行比较	单样本 t 检验	威尔科克森检验	卡方或二项检验	
比较两个未配对组	未配对 t 检验	曼 - 惠特尼 U 检验、克鲁斯卡尔 - 沃利斯检验	费歇尔检验（大样本采用卡方检验）	对数排名检验或曼特尔 - 亨塞尔检验
比较两个配对组	配对 t 检验	威尔科克森检验	麦克尼马尔检验	有条件比例风险回归法
比较三个或以上不匹配组	单因素方差分析	克鲁斯卡尔 - 沃利斯检验	卡方检验	考克斯比例风险回归法
比较三个或以上匹配组	重复测量方差分析	弗里德曼检验	柯克兰 Q 检验	有条件比例风险回归法
对两个变量之间的联系进行量化	皮尔逊相关	斯皮尔曼相关	列联系数	
从另一个变量测量结果预测数值	简单线性回归或非线性回归	非参数回顾	简单逻辑回顾	考克斯比例风险回归法
从几个测量的或二项变量预测数值	多元线性回归或多元非线性回归		多元逻辑回顾	考克斯比例风险回归法

非参数检验法评估

比较测量结果，需要选择恰当的检验方法，这就需要在两个检验系列进行选择：参数型和非参数型。很多统计检验方法依据的假设条件是从常态分布提取数据。这些检验项目被称为参数检验。上表第二栏列出了常用参数检验项目，包括测试和方差分析。

无须对总体分布进行相关假设的检验项目被称为非参数检验。大多数非参数检验项目将结果变量从低到高排序，然后对排序结果进行分析。表中第三栏列出了这些检验项目，包括威尔科克森检验、曼 - 惠特尼 U 检验和克鲁斯卡尔 - 沃利斯检验。这些检验项目也称为任意分布检验。柯尔莫哥洛夫 - 斯米尔诺夫检验可以对完全非算数变量得分进行比较，如栖息地偏好。

在参数检验和非参数检验之间进行选择：简单案例

在参数检验和非参数检验之间进行选择有时比较容易。如果确定从遵循常态分布（至少大致为常态分布）的种群中提取数据，应明确选择参数检验。下列三种情况，应明确选择非参数检验：

结果是一种排名或得分，并且种群明显不正常。示例包括学生班级排名、阿普伽新生儿健康状况评分（按0-10分值测量，所有得分均为整数）、疼痛视觉模拟评分（在连续量表基础上测量，其中0表示无疼痛，10表示疼痛难以忍受）及影片和餐厅评论家通常采用的星级评分（*好；***** 极好）。

有些数值超出范围，表示测量的指标太高或太低。即使种群很正常，也无法采用参数检验对这些数据进行分析，因为我们并不知道全部数值。利用这些数据进行非参数检验比较简单。指定数值太低，是为了测量任意极低值；指定数值太高，是为了测量任意极高值。然后进行非参数检验。由于非参数检验只需了解数值相对排序，因此即使不能准确掌握全部数值也无大碍。

数据就是测量值，并且确信种群不是常态分布。如果不是从常态分布提取数据，应考虑是否能够将这些数值转换，从而实现常态分布。例如，可以求所有数值的对数或倒数。进行特定转换，通常存在生物或化学（及统计）方面的原因。

在参数检验和非参数检验之间进行选择：复杂案例

通常确定样本是否源自常态种群并不容易。需要考虑以下几点：

如果收集很多数据点（超过100个），可以查看数据分布情况，并且可以明显确定这种分布大致呈钟形。可以采用公式统计测验法（柯尔莫哥罗夫-斯米尔诺夫检验，本书不再深入阐述），测试数据分布是否明显不同于常态分布。如果数据点太少，很难通过检验确定数据是否正常，并且这种公式测验几乎无法区分常态分布和非常态分布。

我们还应查看以前的数据。请务必记住，至关重要的是整个种群的分布，而不是样本的分布。在确定种群是否正常的过程中，应查看所有可用数据，而不是当前实验中的数据。

应考虑分散源。如果分散源太多（没有任何一种源头是这种分散现象的主要决定因素），可以大致确定是一种常态分布现象。如果存在疑问，让一些人选择参数检验（因为他们不确信违反了正常假设条件），让其他人选择非参数检验（因为他们不确信符合正常假设条件）。

在参数检验和非参数检验之间进行选择：它重要吗？

无论选择参数检验或非参数检验，它重要吗？答案取决于样本量。我们需要考虑4种情况：

大样本。使用非常态种群数据开展参数检验时，产生了什么现象？中心极限定理确保如果样本量大，参数检验可以发挥良好作用，即使种群不正常也是如此。换言之，只要样本量大，参数检验就能应对与常态分布之间存在的偏差。问题是无法界定样本究竟多大才算够大，因为它取决于特定非常态分布的性质。如果每组至少有24个数据点，可以放心地选择参数检验法，除非种群分布确实很奇怪。

大样本。使用常态种群数据开展非参数检验时，产生了什么现象？如果来自常态种群的样本量很大，非参数检验可以发挥良好作用。P值往往要大得多，但差异较小。换言之，非参数检验比大样本参数检验弱，但差距很小。

小样本。使用非常态种群数据开展参数检验时，产生了什么现象？我们不能使用中心极限定理，所以 P 值可能不准确。

小样本。如果采用常态种群数据进行非参数检验，P 值往往太高。如果样本量较小，非参数检验就缺乏统计功效。

因此，大数据集不存在任何问题。通常很容易确定数据是否来自常态种群，但是这并不重要，因为非参数检验和参数检验本身都很有说服力。小数据集则进退两难。很难确定数据是否来自常态种群，但是这非常重要。非参数检验和参数检验本身都不具说服力。

单侧 P 值还是双侧 P 值？

如果需要开展多次检验，必须选择希望计算单侧 P 值还是双侧 P 值（等同单尾 P 值或双尾 P 值）。我们可以在检验条件下对单侧 P 值和双侧 P 值之间的差异进行评估。计算零假设 P 值，即两个种群平均值相等，并且两个样本平均值之间存在的任何差异都是一种偶然现象。如果这种零假设为真，那么单侧 P 值也就是两个样本平均值与按照假设条件仅通过偶然因素确定的方向观察（或后续）所得值存在差异的概率，即使整个种群平均值实际相等。双侧 P 值也包括样本平均值与反向所得值存在差异的概率（即另一组的平均值更大）。双侧 P 值是单侧 P 值的两倍。

能够肯定（并且在收集任何数据之前）平均值之间不存在任何差异或者这种差异的方向可以事先确定（即已经确定哪组的平均值最大），那么采用单侧 P 值比较合适。如果不能在收集数据之前确定任何差异的方向，采用双侧 P 值更合适。如有疑问，应选择双侧 P 值。

如果选择单侧检验，应在收集任何数据之前进行，并且需要明确实验假设的方向。如果数据遵循另外一种方式，我们必须愿意将这种差异（或关联）归于偶然因素，无论数据如何引人注目。如果我们对按照“错误”方向运行的数据感兴趣，即使有一点兴趣，我们应该使用双侧 P 值。

配对检验还是未配对检验？

比较两组时，需要决定是否进行配对检验。如果比较三组或四组，使用“配对”不太恰当，因此只能使用“重复测量”。

如果个别数值不配对或者不相互匹配，应通过未配对检验，对各组进行比较。如果数值表示一个主题的重复测量结果（干预前后）或者匹配主题的测量结果，应选择配对检验或重复测量检验。对于不同时间反复开展的实验室实验项目，配对检验或重复测量检验也比较合适，每种实验均有自行控制机构。

如果一组中的数值与另一组中特定数值的关联性超过与后者任意数值的关联性，应选择配对检验。如果在收集数据之前，主题匹配或配对，只有选择配对检验才合适。不能根据正在分析的数据进行配对。

费歇尔检验还是卡方检验？

在分析两横两纵列联表时，可以进行费歇尔检验或卡方检验。费歇尔检验是最佳选择方案，因为它任何时候都能提供精确 P 值。卡方检验计算更简单，但是只能获得近似 P 值。如果计算机正在进行计算，应选择费歇尔检验，除非对卡方检验更熟悉。如果列联表中的数字很小（大约小于 6 的任何数字），应明确避免选择卡方检验。如果数字更大，卡方检验和费歇尔检验提供的 P 值比较类似。

卡方检验计算近似 P 值，而耶茨连续性校正可以让近似值更精确。如果不采用耶茨校正，P 值就会太低。但是，校正过度，得到的 P 值就会太高。统计学家为耶茨校正提供了各种建议。如果样本大，耶茨校正几乎不会产生任何差异。如果选择费歇尔检验，P 值很精确，就不需要耶茨校正。

回归还是相关?

线性回归与相关比较类似，因此容易混淆。有时进行这两种计算才有意义。如果每种情况同时测量了 X 和 Y，并且希望对它们的关联性进行量化，应计算线性相关。如果可以假定采集的 X 值和 Y 值均来自常态种群，应选择皮尔森（参数）相关系数。否则应选择斯皮尔曼非参数相关系数。如果改动了 X 变量，请勿计算相关系数（或者它的置信区间）。

只有当其中一个变量（X）高于或产生另一个变量（Y）时，才能计算线性回归。如果改动了 X 变量，应明确选择线性回归。确定哪个变量为 X、哪个变量为 Y 十分重要，因为线性回归计算在 X 和 Y 并不对称。如果交换这两个变量，可以得到一条不同的回归线。相比之下，回归计算在 X 和 Y 是对称的。如果交换标签 X 和 Y，仍然会得到同一相关系数。

附录 8　巡护和监测报告样表

<table>
<tr><td colspan="4">站点：</td><td colspan="8">日期：　　年：　　月：　　日：</td></tr>
<tr><td colspan="7">巡护人员姓名：</td><td colspan="5">通讯员姓名：</td></tr>
<tr><td colspan="7">调查方式：1. 步行；2. 驾车；3. 驾船；4. 其他</td><td colspan="5">路线：</td></tr>
<tr><td colspan="7">地点：</td><td colspan="5">天气：</td></tr>
<tr><td colspan="3">起始时间：</td><td colspan="4">结束时间：</td><td colspan="5">时长：</td></tr>
<tr><td>部门：</td><td colspan="3"></td><td colspan="3"></td><td colspan="3"></td><td>栖息地描述</td><td>备注</td></tr>
<tr><td>海拔 / 经纬度</td><td colspan="3"></td><td colspan="3"></td><td colspan="3"></td><td></td><td></td></tr>
<tr><td>名称</td><td>编号</td><td colspan="2">包括</td><td>编号</td><td colspan="2">包括</td><td>编号</td><td colspan="2">包括</td><td rowspan="2"></td><td rowspan="2">性别比</td></tr>
<tr><td></td><td></td><td>成虫</td><td>幼虫</td><td></td><td>成虫</td><td>幼虫</td><td></td><td>成虫</td><td>幼虫</td></tr>
<tr><td></td><td></td><td></td><td></td><td></td><td></td><td></td><td></td><td></td><td></td><td></td><td></td></tr>
<tr><td></td><td></td><td></td><td></td><td></td><td></td><td></td><td></td><td></td><td></td><td></td><td></td></tr>
<tr><td></td><td></td><td></td><td></td><td></td><td></td><td></td><td></td><td></td><td></td><td></td><td></td></tr>
<tr><td></td><td></td><td></td><td></td><td></td><td></td><td></td><td></td><td></td><td></td><td></td><td></td></tr>
<tr><td></td><td></td><td></td><td></td><td></td><td></td><td></td><td></td><td></td><td></td><td></td><td></td></tr>
<tr><td></td><td></td><td></td><td></td><td></td><td></td><td></td><td></td><td></td><td></td><td></td><td></td></tr>
<tr><td></td><td></td><td></td><td></td><td></td><td></td><td></td><td></td><td></td><td></td><td></td><td></td></tr>
<tr><td>总计</td><td></td><td></td><td></td><td></td><td></td><td></td><td></td><td></td><td></td><td></td><td></td></tr>
<tr><td>植物</td><td colspan="3"></td><td colspan="3"></td><td colspan="3"></td><td></td><td></td></tr>
<tr><td></td><td colspan="3"></td><td colspan="3"></td><td colspan="3"></td><td></td><td></td></tr>
<tr><td></td><td colspan="3"></td><td colspan="3"></td><td colspan="3"></td><td></td><td></td></tr>
<tr><td>人类活动</td><td colspan="11"></td></tr>
</table>

附录 9　改善中国湿地保护所需要的改革方案与实践变更汇总

- 保护区分类和法规改革
- 湿地保护地分区改进方案
- 湿地周围景观划定红线
- 缓解大坝和调水影响
- 水位控制优化
- 生态系统健康指数监控利用
- 解决污染源问题
- 迁徙中途停留关键区域网络维护
- 加强本地社区共同管理力度
- 增强意识
- 相关培训能力标准的采用
- 渔场、电力公司、水资源等提高主流化水平
- 推广破坏性较小的生态旅游类型
- 降低气候变化影响的活动策划
- 采用更多自然再生法
- 湿地内停止植树
- 红树林恢复方法改进
- 数据共享

参考文献

保护地课题组．2004．利用保护地为中国农村提供经济效益——中国保护地体系评估及建立合理化体系的建议．北京：中国环境与发展国际合作委员会（CCICED）.

丁长青．2010．朱鹮．中国鸟类．1（2）：156-162.

李世豪，屈若搴．1937．中国渔业史．上海：商务印书馆.

李思忠．1981．中国淡水鱼类的分布区划．北京：科学出版社.

汪松，马敬能．1993．拯救中国生物多样性的紧急建议．提交给中国环境与发展国际合作委员会（CCICED）的报告．中国生物多样性，1（1）：2-13.

解焱，汪松，Peter Schei．2004．中国的保护地．北京：清华大学出版社.

Appleton M, Texon G, Uriarte M. 2003. Competence Standards for Protected Areas Jobs in South East Asia. Laguna: ASEAN Regional Centre for Biodiversity Conservation.

Athanas A, Ghersi F, Phillips A, et al. 2001. Guidelines for Financing Protected Areas in East Asia. Gland: IUCN.

Bailey JA. 1984. Principles of Wildlife Management. New York: John Wiley & Sons.

Bibby CJ, Burgess ND, Hill DA, et al. 2000. Bird Census Techniques Second Edition. San Diego: Academic Press.

BirdLife International. 2001. Threatened Birds of Asia: The BirdLife International Red Data Book. Cambridge: BirdLife International.

Bowman ME, Eagles PFJ, Tao TCH. 2001. Guidelines for Tourism in Parks and Protected Areas of East Asia. Gland: IUCN; Waterloo: University of Waterloo.

Brown J, Mitchell N, Beresford M. 2005. Protected Landscape Approach (The): Linking Nature, Culture and Community.

Christ C, Hillel O, Matus S, et al. 2003. Tourism and Biodiversity: Mapping Tourism's Global Footprint. Washington DC: Conservation International.

Davey AG. 1998. National System Planning for Protected Areas. Gland: IUCN.

de Sherbinin A, Lacko A, Jaiteh M. 2012. Evaluating the Risk to Ramsar Sites from Climate Change Induced Sea Level Rise. Ramsar Scientific and Technical Briefing Note No.5. Gland: Ramsar Convention Secretariat.

Diamond JM. 1975. The island dilemma: lessons of modern biogeographic studies for the design of natural reserves. Biological Conservation, 7: 129-145.

Dudley N, Hockings M, Stolton S. 2000. Evaluating Effectiveness: A Framework for Assessing Management of Protected Areas. Cambridge: IUCN.

Eagles PFJ, Haynes CD, McCool SF. 2002. Sustainable Tourism in Protected Areas: Guidelines for Planning and Management. Gland: IUCN.

Guidelines for Protected Areas Management Categories.1994. IUCN World Commission on Protected Areas and the World Conservation Monitoring Centre.

Hamilton LS, Sandwith T, Sheppard D, et al. 2001. Transboundary Protected Areas for Peace and Co-operation. Gland: IUCN.

IUCN. 2000. Financing Protected Areas: Guidelines for Protected Area Managers. Gland: IUCN.

Kelleher G. 1999. Guidelines for Marine Protected Areas. Gland: IUCN.

MacKinnon J. 2002. Biodiversity Principles for Developers and Planners. Beijing: BWG/CCICED.

MacKinnon J, MacKinnon K, Child G, et al. 1987. Managing Protected Areas in the Tropics. Journal of Applied Ecology, 25 (1): 369.

MacKinnon J, Mengsha G, Cheung C, et al. 1996. A Biodiversity Review of China. Hong Kong: WWF International.

MacKinnon K, MacKinnon J. 1991. Habitat protection and re-introduction programmes//Gipps JHW. Beyond Captive Breeding: Re-introducing Endangered Mammals to the Wild. Symposia Zoological Society of London No. 62. Oxford: Clarendon Press: 173-198.

Meffe GK, Carroll CR. 1994. Principles of Conservation Biology. Sunderland: Sinauer Associates.

Middleton J, Thomas L. 2003. Guidelines for Management Planning of Protected Areas. Gland: IUCN.

Riney T. 1982. Study and Management of Large Mammals. New York: John Wiley & Sons.

Schemnitz SD. 1980. Wildlife Management Techniques Manual. Washington DC: The Wildlife Society.

She ZG, Lin JX, Peng YG, et al. 2005. A preliminary study on mangrove and aquaculture system. Chinese Journal of Ecology, 7: 837-840.

Soulé ME. 1996. Viable Populations for Conservation. Cambridge: Cambridge University Press.

Sutherland WJ. 1997. Ecological Census Techniques: A Handbook. Cambridge: Cambridge University Press.

Sutherland WJ. 2001. The Conservation Handbook: Research, Management and Policy. Oxford: Blackwell.

Worboys G, Lockwood M, De Lacy T. 2001. Protected Area Management: Principles and Practice. Melbourne: Oxford University Press.

Part One Introduction to Wetlands

1.1 Definitions

What is biodiversity?

Biodiversity is the name we give to the total richness of living things on our planet—the genes that make up so many millions of different species and their varieties, the species themselves and the functioning ecosystems that these species form through their interactions with other species and with their physical environment.

What is a wetland?

Wetlands include all lands that are permanently or regularly submerged by shallow water. They include all lakes and rivers, underground aquifers, swamps and marshes, wet grasslands, peat-lands, oases, estuaries, deltas and tidal flats, mangroves and other coastal areas, coral reefs, marine waters less than 6 m deep and all human-made sites such as fish ponds, rice paddies, reservoirs and salt pans.

Wetland assessment: The identification of the status of, and threats to, wetlands as a basis for the collection of more specific information through monitoring activities.

Wetland monitoring: Collection of specific information for management purposes in response to hypotheses derived from assessment activities.

1.2 Why are wetlands important?

Wetlands contribute to human welfare in many ways both economic and also non-economic. Valuable wetland functions include nutrient cycling, sediment retention, maintenance of important biotic communities, conservation of important genetic resources, water storage and purification, carbon fixation, flood control and coastal protection. Wetlands can also be grazed, provide a sustainable harvest of plants and animal products, have recreational and scenic values and thence enhance property values.

ECOSYSTEM SERVICES PROVIDED BY OR DERIVED FROM WETLANDS

Services	Comments and Examples
Provisioning	
Food	production of fish, wild game, fruits, and grains
Fresh water*	storage and retention of water for domestic, industrial, and agricultural use
Fiber and fuel	production of logs, fuelwood, peat, fodder
Biochemical	extraction of medicines and other materials from biota

Continued

Services	Comments and Examples
Genetic materials	genes for resistance to plant pathogens, ornamental species, and so on
Regulating	
Climate regulation	source of and sink for greenhouse gases; influence local and regional temperature, precipitation, and other climatic processes
Water regulation (hydrological flows)	groundwater recharge/discharge
Water purification and waste treatment	retention, recovery, and removal of excess nutrients and other pollutants
Erosion regulation	retention of soils and sediments
Natural hazard regulation	flood control, storm protection
Pollination	habitat for pollinators
Cultural	
Spiritual and inspirational	source of inspiration; many religions attach spiritual and religious values to aspects of wetland ecosystems
Recreational	opportunities for recreational activities
Aesthetic	many people find beauty or aesthetic value in aspects of wetland ecosystems
Educational	opportunities for formal and informal education and training
Supporting	
Soil formation	sediment retention and accumulation of organic matter
Nutrient cycling	storage, recycling, processing, and acquisition of nutrients

*While fresh water was treated as a provisioning service within the MA, it is also regarded as a regulating service by various sectors.

Many of these benefits contribute directly into the market economy, others are freely enjoyed by society and do not enter market economy but their economic value can still be estimated on a case by case basis and is usually very large. Global estimates can also be made. According to Costanza et al. 1997, the dollar value of wetlands ecological services worldwide was estimated to be \$19 trillion per year. In a revised evaluation using updated values and areas for 2011 (Costanza et al. 2014), these estimates were raised up to \$50 trillion per year. The following unit values were used:

Wetland types	Ecosystem services / (\$/ha/yr) (2007)	Ecosystem services / (\$/ha/yr) (2011)	Main valuable functions
Estuaries	31,509	28,916	Staging areas for migrant birds, edible molluscs and crustaceans, fish nurseries
Seagrass/algal beds	26,226	28,916	Wave containment, food for humans and biodiversity
Coral reefs	8,384	352,249	Coastal protection, fisheries and tourism
Tidal marsh/mangroves	13,785	193,843	Coastal protection, fisheries, soil capture, water cleansing, biodiversity
Swamps/floodplains	27,021	25,681	Water cleansing, storage, biodiversity, products
Lakes/rivers	11,727	12,512	Water channels, storage, energy, fisheries, biodiversity

The main reason for raised evaluation for coral and mangrove services is better appreciation of their role in coastal protection following disastrous tsunami events of 2004 (Indonesia) and 2011 (Japan).

Japan tsunami sweeping in 2011

The values are generally much higher than either forests valued at $5,382 (tropical) and $3,137 (temperate) or croplands valued at $5,567 per ha per year. (Note: It makes no economic sense to plant trees on wetlands. This immediately reduces their ecosystem value.)

These high numbers may be difficult to explain or grasp. It is sometimes easier to convert these values into scales that decision makers can more easily understand.

Reservoir equivalents

So a given forest catchment of 200km^2 may deliver ecosystem services valued at $60,000,000 per year. But this number is difficult to comprehend by local planners and decision makers because these numbers are not captured in normal accounting of expenditures and production. It may make more sense to express the service in terms of something more tangible such as saying this is equivalent to a water reservoir of 5 billion cubic metres!

Treatment plant equivalents

One of the main functions of marshy wetlands is their valuable service in cleansing water. The marshy vegetation causes metals and other pollutants to become trapped into the soil and sediments to be deposited also. They act as sewage treatment plants. Plants and bacteria, which are powered by the sun, break down pollutants and cleanse wastewater.

It is possible to calculate the effectiveness and efficiency of this cleansing service and translate it into the equivalent number and size of artificial sewage treatment plants for which we could estimate an economic cost for construction and maintenance. Expressed in these terms it may be more meaningful to impress local planners and decision makers of the need to leave such healthy wetlands undeveloped and not drained.

Recreational equivalents

Some values are harder to calculate in purely monetary terms including much of the scenic, recreational, spiritual and cultural values. If one of the services delivered by a wetland site is recreation, one can calculate its potential economic earnings in terms of visitor numbers or tourism revenues. But much of its value is non-economic in the form of happiness, well-being and health. It may be more comprehensible to government planners to express this function in terms of recreational equivalents of 10 municipal parks etc.

1.3 How does biodiversity contribute to wetland function?

We benefit from the great biological richness of wetlands in many ways:

- We derive many renewable resources such as fish, crustaceans, shellfish, timber, fuel-wood, plant and fence poles, animal fodder, edible plants, game meat, fungi and honey
- High biodiversity increases the efficiency and health of wetland function, greatly adding to its ecosystem service delivery in the form of soil formation, nutrient recycling, water storage and purification, carbon fixation, flood control, storm protection
- We derive many benefits in the form of beautiful scenic and recreation areas, living field laboratories, culturally sacred sites and inspirational heritage
- Biodiversity serves as indicators of environmental health and serves as an early warning system of dangerous changes

1.4 Wetlands inventory

Wetland inventory consists of the collection and/or collation of core information for wetland management, including the provision of an information base for specific assessment and monitoring activities. Inventory can take place at a range of scales from site-level to catchment to national. Detailed guidelines on conducting such inventories are provided by Ramsar at http://www.ramsar.org/sites/default/files/documents/pdf/lib/hbk4-15.pdf.

The handbook recommends 13 steps in conducting any inventory.

- **Clarify objectives**—What is the inventory needed for?
- **Review existing information**—Published, unpublished, maps, remote sensed imagery, photographs, earlier inventories, etc.
- **Review possible inventory methods**—Consider which approaches will provide the data needed to achieve the objectives identified.
- **Define scale and resolution for inventory**—Depends on size of area and types of base maps available.
- **Pull together a core baseline dataset**—You will not need all information available, select the key elements most relevant to the objective.
- **Select suitable habitat classification**—Use or adapt from existing mapped vegetation, soil types, land units, or clear types distinguishable from remote sense images, etc.
- **Select preferred methodology**—There are dozens of different methods. select the most

appropriate for local conditions, available manpower, etc.

- **Develop data collection protocols and management system**—Define data sets required and meta-databases of those sets. Plan data flows and data sharing protocols.
- **Establish timelines for the collection of different data**—Scheme what datasets need collecting at what intervals and by who.
- **Assess cost effectiveness of methods and options**—You always want more information but must be selective given time, funds, man-power and equipment available.
- **Establish reporting procedures**—Raw data needs collation, analysis, interpretation and combining with other data before it can be usefully reported.
- **Establish review and evaluation process**—Determine procedures for checking and approving different datasets. Consider who has access to what sets of data.
- **Plan pilot study**—Run a trial to see if the process is workable and adequate to meet the objective. Modify as necessary before undertaking full-scale inventory.

The State Forestry Administration has now completed two generations of national wetlands inventory of China. In the first inventory, conducted up to 2003, all wetlands larger than 100 ha were listed with pertinent details of wetland type, ownership and legal status, biology and physical characteristics. From 2009-2013 the second inventory was conducted at even more detailed scale; all wetlands greater than 8 ha being included and all sites also being mapped. The new data show China supports wetlands with a total area of 53.60 million ha, accounting for 5.58% of national land. Compared with the previous national wetlands inventory, 3.39 million ha of the wetlands has been lost, even with the addition of many smaller wetlands not counted in the first inventory. Natural wetlands cover 46.67 million ha, accounting for 87.08% of the total of national wetlands. But natural wetlands have reduced by 3.37 million ha. From wetlands distribution, Qinghai, Heilongjiang, Tibet, Inner Mongolia contain the largest areas of wetlands.

1.5 Wetlands for multiple use

Even though many of the most biologically important wetlands in China have already been registered and listed as protected areas at either national or provincial level, this does not mean they can be managed solely for conservation. All serve as multiple use areas and are expected to deliver varied services to other sectors of society. State Forestry Administration has issued its own regulations for the protection and management of wetlands but other sectors also apply their own regulations e.g. fishery, sand dredging, water extraction, pollution discharge, shipping and there may be a complicated overlap of different interests, user rights, ownership and regulations.

Cooperative systems

With so much regulatory overlap (see section 2.3 below) and often conflicting interests by different sectors, harmonious use of such multiple use areas, as most wetlands are, requires high levels of cooperation and coordination between different people, agencies and sectors. Creating systems of such cooperation is complex and delicate and needs to be built up on the back of pre-existing cooperative systems such as irrigation systems where rules already determine who can take how much water, when and where from the overall water systems have already been organized for many generations. Similar cooperation systems also have to apply to access to shared fishery resources, forest resources, grazing rights and other use of shared commons.

1.6 The dynamic nature of wetlands

Wetlands are by nature opportunistic habitats. They are very dynamic. They never were and never will be stable. Lakes form, fill up with sediment and become dry land. Coastlines change eternally and sea level is rising fast. Glaciers expand in winter and shrink in summer but all are now shrinking fast with climate change and as they melt one by one, so the streams of their melt water dry up also. Rivers change course carving deeper and deeper river courses but occasional storms and floods block the drainage of floodplains and create new wetlands. Wherever depressions occur or drainage channels become blocked, so new wetlands form, become colonized by wetland plants, wetland animals and evolving in nature. The factors that create suitable conditions for wetlands may be natural geological or geographic events but they can equally be the result of human activity. Agriculture and engineering may accelerate or slow down drainage and thus create or destroy or at least greatly modify the condition and nature of wetlands. Thus many of the world's great wetlands are in fact artificial lakes, reservoirs or lands with deliberately reduced drainage. Natural origin is not as important as natural species composition. It is in human hands to create, destroy, preserve or modify our wetlands to ensure the services we need. And there is the secret of good wetland management-knowing what services you need from a given wetland and managing it to ensure the delivery of those services required or the preservation of what biodiversity you might want to preserve or encourage. And whilst protection of biodiversity *per se* is a noble objective that enriches our planet, we are ultimately a selfish species and put our own development needs ahead of the needs of wildlife but we must also recognize that one of the most valuable services of biodiversity is to ensure the ecological health and indicate the health of our wetlands.

Part Two Basic Wetlands Knowledge and Theory

2.1 Distribution and types of wetland in China

Wetlands includes all lands submerged permanently or periodically under shallow water (generally taken to be less than 6m). This includes many different types of wetlands.

Type	Description	Distribution	Significance
Marshes and swamps	Poorly drained and generally wet areas with herbaceous or tree cover interspersed with streams and ponds	Relatively flat, poorly drained sites all over country including many in highlands	Very important often peaty water storage. Breeding habitat of many water birds and some important mammals
Meadows	Open fields of grasses and herbaceous plants, periodically flooded so generally used for grazing rather than agriculture	Alluvial flood plains of major rivers, often beside rivers or around lakes	Can be useful grazing areas for wildlife and domestic animals. Variable habitat for many birds, mammals etc. and important for purification of water-borne pollutants
Lakes	Large natural water bodies of freshwater or partly saline (on plateau)	Scattered all over China	Important fisheries, water bird breeding and wintering sites, human transport, natural water storage; some important aquatic mammals
Reservoirs	Artificial lakes, held behind dams for water storage and irrigation	Scattered all over China	Often relatively sterile but can become important fisheries and hence support water birds and other aquatic biota
Ponds	Small, slow flowing natural or artificial water bodies	All over China	Support rich pond life of diatoms, dragonflies, amphibians, small fish and other biota
Rivers and streams	Natural network of water channels leading from upper catchments to the sea	All over China but more in humid zones	Vital connections for aquatic creatures dispersal. Rich habitats for fish, crustaceans, birds, otters and important for oxygenation of water
Ditches and canals	Artificial water connections to divert water from or return water back to natural water system. Some made to drain wetlands for agricultural conversion	All over China	Many are polluted and rather sterile but they can support rich biota similar to natural rivers and streams
Coast line	Rocky, sandy, muddy beaches and cliffs with marine lagoons and coral reefs	All coastal areas, many habitat types	Many habitat types support different biota, southern corals support rich biota and fisheries, coastal protection and recreation opportunity
Intertidal flats	Mozaic of mud, sand, islets and freshwater and sea water. Various vegetation types	All coastal areas	Very important for shore birds, mollusks, crabs, fish spawning
Mangroves	Vegetation type of trees and associated plants capable of withstanding regular inundation of sea water	Tropical alluvial lands exposed to tidal action	Very rich biota of crustaceans, fish, worms, mollusks, shorebirds, canopy birds, insects, mammals etc.

2.2 Wetland PA system of China

Two main categories of protected wetlands currently include nature reserves with primary objective of protection of species and ecosystems and also wetland parks primarily established to protect scenic wetlands and facilitate visitor enjoyment. According to the 2nd national wetlands inventory, there are 577 wetland nature reserves and a further 468 wetland parks in China, giving a total of 23.24 million ha of protected wetlands, an increase of 5.25 million ha of wetlands protected since 2003. There are current plans to reform the protected area system in China and we may soon see more flexibility of management options through the adoption of a wider range of categories taken from the IUCN international PA categories.

2.2.1 International versus Chinese categories of PAs

Seven categories (IUCN)		Four categories (suggested in China)	
Ia	Strict Nature Reserve	Category I	Strictly protected area
Ib	Wilderness Area		
II	National Park	Category III	Natural park
III	Natural Monument or Feature		
IV	Protected Landscape/Seascape		
V	Habitat/Species Management Area	Category II	Habitat and species management area
VI	Protected Area with Sustainable Use of Natural Resources	Category IV	Sustainable use area

2.3 Wetland laws and regulations

Several different ministries have issued their own laws and regulations that relate to use of wetlands and waterways. The State Forestry Administration is mandated by the State Council to be responsible for wetlands protection (wildlife, protected areas and wetland reserves, catchment protection), but other agencies regulations also apply (see following table):

Agency	Functions and responsibilities
Ministry of Environmental Protection	pollution issues, standards and overall coordination of biodiversity issues
Ministry of Water Resources	dams, reservoirs, diversions and water extractions, flood defenses
Ministry of Agriculture	fisheries, use of chemical in agriculture, reed harvesting
State Oceanic Administration	marine fisheries, mariculture, coastal wetlands and marine nature reserves
Ministry of Transport	ships, boats, channels and bridges
Ministry of Mining	Gravel and sand dredging for construction materials

This overlap of mandates results in many contradictions with Ministry of Agriculture handling issuance of fishing and reed cutting permits, the Ministry of Water Resources controlling sluice levels and water take off, Ministry of Environmental Protection responsible for monitoring pollution and discharge into waterways and Ministry of Transportation organizing shipping lanes and communications. All may operate in the same wetland. Conflicts of interest even exist within each Nature Reserve management which must balance objectives of visitor access, resource use and species conservation.

Regulations issued by one agency can only operate within their own mandate of responsibilities. Other agencies with different mandates may not feel obliged to follow. Thus Forestry may declare no mining can take place inside their protected areas but the Department of Mines feel quite free to issue exploration licenses and open up mines across the entire landscape.

1994 Regulations on Nature Reserves define and prescribe the establishment of nature reserves with three zones—strictly protected core area, relatively unmodified surrounding buffer zone and man-modified experimental zone. The regulations follow the MAB design but that was established for studying the relationships between Man and his Environment not for nature protection *per se*. The Chinese regulations are confusing in that in most international schemes, the buffer zone is a highly modified external zone and not a well protected internal zone. The regulations also provide little flexibility for design or management options so that although in reality many nature reserves in China face rather high levels of human use and modification, there are no suitable zones and only one category for nature reserves.

The need to overhaul these regulations, establish more permissible zones and a range of protected area categories based on the IUCN classification system has been repeatedly proposed but still never agreed upon. Meanwhile regulations for other designations of protected areas—forest parks, wetland parks, national parks, state key scenic areas etc. are being developed in rather uncoordinated manner by different agencies at different levels (national, provincial, county etc.).

2.4 Threats to Chinese wetlands

Chinese wetlands are facing many threats.

2.4.1 Loss of habitat

Loss of wetlands is more than just the sad loss of our fascinating natural heritage but constitute a threat to the lives, health, safety and welfare of hundreds of millions of our fellow humans, a threat to trillions of dollars worth of land and property and a threat to the health of the great oceans on which we also depend so much.

Lakes: Comparing the data from the first national wetlands inventory published in 2003 with the latest inventory, the shrinking trend of lakes in China has been obvious, decreasing by 58.91 million hm^2, or by about 7.05%. Data from the *National Comprehensive Water Resources Plan* shows than since 1950, 231 lakes have shrunk to varying degrees among the 635 lakes with an area of more than 10km^2; the total area of lake shrinkage is about 13,800km^2

(including 4,300km^2 of dry area).

Marshes and terrestrial wetlands: Compared with the first national wetlands inventory published in 2003, natural terrestrial wetlands have decreased by about 13.5 million hm^2, or by 28%.

Intertidal zone: The destruction of Asia's intertidal zone constitutes one of the fastest losses of biodiversity globally. This is the vital habitat for the survival of millions of birds of several hundred species, but also the nesting beaches for endangered sea turtles, the breeding areas for Asia's seals and homes of thousands of invertebrate crustaceans, worms and mollusks. The clearest evidence of high number of globally threatened species dependent on these habitats is seen among the birds, and particularly waterbirds, with most globally threatened species among the waders (e.g. spoonbills, cranes), but also waterfowl and seabirds.

The intertidal zone with its sand and mud bars, beaches, and mangroves deliver a great list of valuable ecological services that we are discarding too causally in favour of near-sighted development goals driven by the lure of short-sighted economic ventures touted by wealthy investors. Although the total area involved is very small, it is very fragile and it is vanishing fast.

Permafrost: Extreme north regions of NE and NW China have peaty permafrost wetlands where a layer of underground ice prevents water drainage and stunts growth of trees resulting in a special and biologically interesting wetland type. The peat itself is a large carbon store which needs to be kept wet to avoid decomposition and release of greenhouse gases of CO_2 and methane. These areas of permafrost are fast shrinking as a result of human disturbance combined with global warming.

2.4.2 Pollution

The development of China's industry, agriculture and construction has raised millions of people out of poverty but has come at a high cost in the form of pollution.

- Since the 1980s, chemical weed and pest control has been increasingly introduced and endorsed by local government extension services. By using these poisons, the farmers gain more harvest, but at the same time more and more aquatic species have been adversely affected.
- A recent government report indicates that agriculture is responsible for 43.7% of the nation's chemical oxygen demand (the main measure of organic compounds in water), 67% of phosphorus and 57% of nitrogen discharges which result in eutrophication (see section 2.4.5 below).

2.4.3 Water diversions

Much of China faces water shortages and everyone wants more water. Water is diverted from natural waterways for agricultural, industrial and domestic uses. Even small streams and water sources in forested mountains are diverted into pipes or channels and effectively removed from the natural ecosystems.

Much larger water diversions result from schemes to take the water from south to north and from the thousands of dams that now block large flowing rivers converting them into a series of reservoir cascades.

These diversions result in loss of water from many former wetlands and also barriers to the free movement of water animals—fish, amphibians etc. that need to reach cool water upstream breeding streams to complete their life cycles.

For the purposes of generating electricity and for irrigation purposes, a number of big dams will be constructed along the Lancang-Mekong River, especially in the upper Lancang-Mekong sub-region. The big dams will form a big risk for the survival of some migrating fish species, such as the catfish *Pangasius sanitwongsei*. Solving this problem is not just as simple as to build one additional fish channel, as dams have considerable impact on the balance of the river system as a whole. Nowadays, many scientists do not agree with the idea of building more dams along the Lancang-Mekong for generation of electricity to further boost the economic development of the Lancang-Mekong region. The impacts of the dams will not only be felt that much by local communities of Xishuangbanna, but more dramatic in the downstream countries. It is likely that even the Tonle Sap lake of Cambodia will be disturbed in its eternal cycle of low and high water levels which has created one of the most outstanding freshwater ecosystems in the world with exceptional biodiversity and productivity.

2.4.4 Loss of water table

In many areas the general water table has been lowered as a result of agricultural or industrial extraction by bore pumps or diversions from upstream sources.

China's shallow groundwater extraction grew from 55.7 billion m^3 in 1980 to 108.1 billion m^3 in 2008, the increase in the north China accounting for 90% of the total. The extraction in many areas has exceed the maximum limit for extraction, leading to a continuous decline in groundwater levels to bring about many environmental and geological problems. Groundwater over-extraction is concentrated in the plains in the north China.

The planting of some water greedy tree species such as willows and poplars can locally lower water tables and compete with wetlands for needed water.

2.4.5 Salination

Salination occurs when freshwater bodies become contaminated by high levels of salt. This may be from the sea or from salts in inland soils and land or result from irrigation schemes.

Estuaries are places where the freshwater flowing down a river meets the tidal sea water. Salinity levels range from pure freshwater to pure sea water and there is a brackish intermediate zone which ebbs and flows up and down the estuary with the tide. As freshwater flows reduce because of high upstream water use or water diversions and as sea levels rise with climate change so the extent to which salt travels up rivers, onto coastal lands and into coastal wells and drinking supplies increases.

Inland, we see osmotic forces drawing salts up and out of underlying soils and strata as the salt concentrations of the surface water increases with evaporation. Thus in arid regions of the world with high levels of solar energy we tend to find saline marshes and lakes well inland. In China many saline lakes can be found on the Tibet-Qinghai Plateau and other desert regions. The area of such saline areas is also expanding with climate change.

Some species of plant (e.g. mangroves and some algae) with high salt tolerance can thrive in saline conditions as can many plankton, crustaceans, mollusks and worms. This saline habitats often support other specialized biota—waders, ducks etc. that can feed on these species.

2.4.6 Eutrophication

Eutrophication is the process in which lakes receive excessive nutrients (phosphorus and nitrogen) and sediment from the surrounding watershed and become more fertile and shallow. Too much phosphorus stimulates rapid blooms of algae and some toxic phytoplankton. The algae can, increase turbidity, block waterways and may choke the habitat of fish, birds, etc. When the algal bloom decays, excessive bacterial action depletes the oxygen and may lead to death of fish and other creatures whose decay also leads to spread of toxic and dangerous bacteria and viruses.

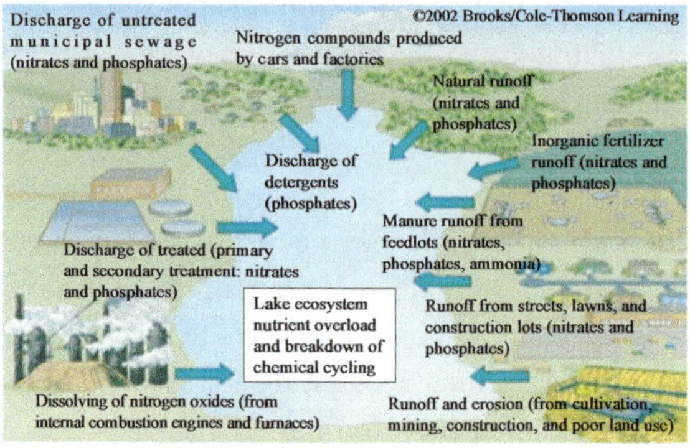

Phosphorus, nitrogen and other nutrients can enter a lake through poorly managed or failing septic systems, phosphorus detergents, phosphorus fertilizers, agriculture, and animal feedlots. In China, excessive use of fertilizer and poor treatment of sewage are the main sources of excess nutrients that lead to increase of phytoplankton in water bodies and often toxic algal blooms. Increasingly these periodic toxic algal blooms have spread into the neighbouring seas.

Too much nitrogen in water causes similar algal blooms in coastal marine waters, destroying fish nursery areas, sea grass beds and prompting plagues of jelly fish or other faunal changes. Huge algal blooms now affect the east coast of China annually, whilst almost all China's freshwater lakes are now classed as eutrophic. A recent government report indicates that agriculture is responsible for 67% of phosphorus and 57% of nitrogen discharges.

It is nearly impossible to turn a eutrophic lake back into an oligotrophic lake but eutrophication can be slowed by reducing nutrient and sediment addition to water bodies.

Whenever erosion occurs along a lakeshore or a stream bank, extra soil gets washed into the water bodies causing cloudier water with less efficient photosynthesis of water weeds and sediments settling on the lake bottom making it mucky and less stable. A lakeshore inventory is recommended to identify the areas where erosion is occurring and take steps to stabilize the area. The most effective stabilization technique is a shoreline buffer of native plants.

2.4.7 Loss of oxygen

Aquatic animals (fish, insects, crustaceans, worms, mollusks) need oxygen to breathe; oxygen levels in water bodies are a critical control of how much fauna each site can support. Oxygen is released into water by the photosynthesis of aquatic plants and also by physical agitation of moving or falling of water (e.g. artificial aerators). Cold water can hold more oxygen than warm water. Activities that increase temperature such as loss of shade, slowing water flow and changing climate will all tend to reduce oxygen levels in a given water body as will loss or damage to the aquatic plants due to pollution, eutrophication, siltation or physical over-grazing. Oxygen is consumed in the processes of decomposition of dead materials.

High faunal richness can only be retained if oxygen levels are kept high and the manager can avoid or minimize factors that would reduce oxygen level.

2.4.8 Alien invasive species

International agencies have labeled the problem of alien invasive species (AIS) as second only to loss and damage to habitat as the greatest threat to biodiversity worldwide.

Alien invasive species can be described as non-native species that get a foothold in a new area and spread causing damage to local environment or local species.

China is particularly vulnerable to AIS because it has become a huge global trader with many people and goods moving around the world and it offers potentially invasive species such a variety of environmental conditions that a high proportion of arriving species can succeed in finding a foothold somewhere in the country. The landscape of China is undergoing such fast and dynamic changes that there are many open areas for new species to invade.

Awareness and species recognition skills are generally low so new species often spread to a high degree before they are noticed as dangerous.

Problems arise when introduced species expand aggressively damaging or outcompeting native species, damaging the physical environment or using up limited nutrients or resources at unsustainable rates.

Some invasive species already causing huge economic losses include the golden apple snails that invade the rice fields of southern China, longhorn beetles from Russia that destroy forests in northern China, water hyacinths that block waterways and reduce oxygen levels in many lakes and waterways over much of the country, fire ants that have become a pest in southern

China and growing list of agricultural weeds such as *Lantana*, *Eupatorium*, etc.

China's wetlands are being increasingly degraded by spread of AIS—both plants and animals. Some of these AIS have colonized by accident but many were deliberate introductions if efforts to add horticultural decoration, commercial crops or new food items.

Some aquatic species, especially the African tilapias, the South American tambaqui, northern carps, and the South American golden apple snail (*Pomacea canaliculata*) were introduced into S China in the last decades. The economic nor ecological impacts of these exotic species were never evaluated carefully and may result in some local aquatic species' competition or even extinction.

Several hundred species are now recognized to be damaging to Chinese wetlands. Removal or control of such species costs large budgets and labour. The follow lists only a few of the major problem species now degrading Chinese wetlands. Section 5.9 describes ways to reduce AIS.

Common name	Scientific name	Region of origin	Area affected	Problems caused
Water hyacinth	*Eichhornia crassipes*	S America	S & C China	Floating weeds choke lakes and waterways impeding boats, reducing oxygen and damaging fisheries
Water lettuce	*Pistia stratiotes*	S America	S & C China	Can spread very fast blocking waterways and reducing oxygen levels at expense of fish and crustaceans
Alligator grass	*Alternanthera philoxeroides*	S America	Most of China	Invades aquatic and terrestrial habitats. Smothers other herbaceous plants. Very expensive to control
Atlantic cordgrass	*Spartina alterniflora*	N America	E & S coastal zone	Dense thickets smother native coastal flora
Lantana	*Lantana camara*	S America	Most of China	Thickets displace native species
Prickly pear	*Opuntia*	S America	S China	Thickets displace native species
Mimosa	*Mimosa pudica*	S America	S China	Spiny thickets spreading
Zebra mussel	*Dreissena polymorpha*	Russia via USA	SE China coastal	Blocks drainage channels; source of avian botulism
African giant snail	*Achatina fulica*	East Africa	S China	Destroys crops and orchards
Golden apple snail	*Pomacea insularum*	S America	S & C China	Destroys aquatic vegetation and rice crops
Louisiana crayfish	*Procambarus clarkii*	South USA	S & C China	Weakens flood dykes, competes with local species
Mouth- brooder, Tilapia	*Oreochromis mossambica*	Southern Africa	SE China and estuaries	Damages benthic substrate, eats other fish and their foods
Cane toad	*Bufo marinus*	S and C America via Australia	S China	Eats native amphibians
Bullfrog	*Rana catesbeiana*	N America	S & E China	Outcompetes and eats smaller endemic amphibians
Mink	*Martes vison*	N America	NE & NW China	Outcompetes local sable and otters, damages fisheries, squirrel populations etc.
Nutria	*Myocastor coypus*	S America	S & E China	Large grazer transforms habitat

2.4.9 Over-fishing and hunting

Hunting for birds, mammals, reptiles and amphibians is a major threat to wetlands biodiversity. Everything that can be sold or eaten is a target for hunters or poachers and some very short-sighted and greedy methods of collection are sometimes employed. Half of China's wetlands have no specific protection and even those wetlands that are officially 'protected' in nature reserves and inadequately patrolled and protected and so vulnerable to farmers or fishermen setting traps and even using guns in more remote areas.

Snares, spring traps and mist nets are regularly set by farmers and poachers to catch birds. Some prized species can be sold to the pet trade, others sold to markets for food or eaten at home.

Where ducks, geese or cranes feed in farmers fields, the farmers sometimes retaliate by spreading poisoned grain or bait to kill the birds. Poisoned birds are even then sold into the market as food and some incautious purchasers have died as a result.

Wetland mammals such as otters, beavers, water deer are also hunted and trapped. China also has an insatiable appetite for snakes, turtles and frogs for the table.

Migratory species face the gauntlet of nets and hunters all along their migration routes and may be killed far from any wetland protected area.

Impact of destructive "modern" fishing methods electricity, explosives, and chemical poison were gradually adopted by farmers over the last decade. Although these methods are all forbidden by the fishery law of P.R. China, many people are practicing it, not only in small waters and tributaries, but in the bigger rivers and lakes such as the Lancang River to catch big catfish species.

The promulgation of the Chinese fishery law and the rules and regulations for local river conservation, a lot of effort by the local Government has gone into aquatic resource protection and control illegal fishing methods. But as rural enforcement is almost absent, these fishing means are controlled efficiently only in very few places.

2.4.10 Wind turbines, cables, buildings and airport

China's larger airports constitute a huge hazard for migrating birds. Each airport has mist nets set to protect planes from bird strike. These nets kill many birds per year. These are not the birds that pose a hazard to planes (large raptors, swans and geese or large flocks of gulls). Other countries do not use nets but have other bird scaring techniques such as dogs, trained falcons, loud noises etc. Airports close to important Ramsar wetland sites are a particular concern e.g. Pudong and Haikou.

Despite local government assurance that they pose no risk, large wind turbines do kill quite a lot of larger birds. Most countries try to avoid placing wind farms close to nesting colonies of important birds and RSPB have issues special guidelines for limiting the damage to birds of such installations.

Light houses and power lines also pose a hazard to large birds. Many countries place easily visible objects on power lines to warn birds not to fly into them. In 2005, APLIC and the U.S. Fish and Wildlife Service released national Avian Protection Plan (APP) Guidelines. The APP

Flocks fly past turbine and dead raptors and cranes killed by turbines

Guidelines offer resources to help utilities manage avian/power line issues.

Many powerlines in China have too little spacing between cables, so large birds can contact both cables and become electrocuted. The crossbars need to be wider to hold the cables further apart.

2.4.11 Airports and mist nets

Tall light-houses and large airport buildings with large glass windows take their toll of naive migrating birds that fly into the windows thinking they can fly through to where they see light at the other side. Kunming airport kills several birds a day in this manner during the migrating season.

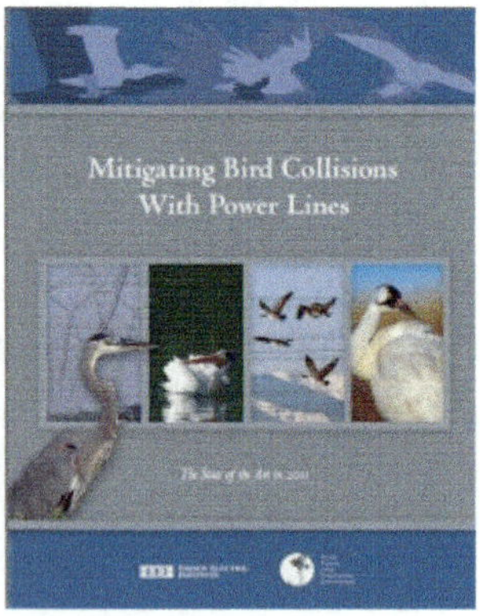

International guidelines on power lines and warning lights for attachment to cables

But far more destructive is the use of mist nets by Chinese airports as a standard way of limiting bird strike to jet planes on landing and take off.

Airport uses mist nets to reduce air-strike

Most airports in China set at least one row of mist nets down the entire length of the main runway. The nets are not very effective in clearing birds and no other countries use this method. There are many more advanced and less destructive methods in general use internationally including dogs, hawks, high-pitched vocalisations and flashing objects. Many airports internationally use remote controlled model planes made to resemble hawks, or use small drones to clear birds off runways.

Bird deaths due to mist nets on runways is probably small in comparison to the numbers being killed by farmers protecting crops, fish farmers protecting their tanks and illegal hunters trying to catch birds. But it is difficult to control public use of nets when they see the government using these dangerous devices so freely.

2.4.12 Climate change

Climate change is irrefutable. The data stand for themselves. Recent years have seen the warmest climates ever recorded. The oceans are warming, the surface is warming, glaciers and polar ice sheets are melting, and sea levels are rising. Extreme weather events are becoming more frequent and more severe worldwide. Weather patterns are becoming more erratic.

Science has proved that much of the change in our climate is related to raised CO_2 levels in the atmosphere. Carbon levels in the atmosphere are higher than at any time in the past 30,000 years and are largely due to human activities. The UN convention on climate change is trying desperately to secure international agreements for countries to limit their carbon emissions to levels that will contain mean global warming to only 2℃ before the end of this century above pre-industrial average. But most climate scientist are pessimistic that this will be achieved. Several regions of the world including the plateaus and high peaks of China are warming faster than the global average and have already risen more than 2℃.

Effects on wetlands and biodiversity will be dramatic and are already being felt. Indeed many of the consequences of global warming are now irreversible and will continue even if we could totally stop carbon emissions.

The wetland manager must expect and plan for more storms, typhoons, floods and droughts: more heat-waves and yet more cold spells; more changes to the distribution of vegetation zones and species, changes to the seasonality of flowering and fruiting patterns, insect emergence and the timing of migrations. It will be a great challenge to retain wetland ecosystem resilience in the face of such change.

2.4.13 Rising sea level

As seas warm, they expand. As polar ice melts it adds to sea level rise. Most estimates now range from 1.0m to even 2m of sea level rise. This will enormously impact coastlines, coastal wetlands, mangroves and small islands. Many of the world's largest cities are in the Asian coastal zone and liable to inundation, including Shanghai, Tianjin and Xiamen.

The largest effects will be noticed in Hainan, Jiangsu and in the Yangtze estuary. In Hainan almost all existing mangroves will be submerged, but new mangrove areas could be established along the current high tide line. In the Yangtze estuary, large areas of land will be lost to the sea and discharge through some of the lower Yangtze valley lakes may be greatly transformed.

The large coastal wetlands of Yancheng in Jiangsu can be expected to be flooded and washed away relatively early.

Villages damaged and boats washed onshore following tsunami

2.5 Need to understand the total ecosystem

An ecosystem is a community of species that are adapted to living together as well as with their physical environment. Each species needs the ecosystem and the ecosystem needs all its component species. Within any ecosystem we can recognize energy flows or food chains and a variety of trophic levels.

Most ecosystems derive their primary energy from sunlight through the process of photosynthesis by green plants. Chlorophyll in the plant leaves combines sunlight with water and CO_2 to make sugars and release oxygen. The oxygen is used by animals in their own respiration and most animals are herbivores that eat some of the green plants and derive their food energy from their digestion. Other animals called carnivores may in turn derive energy by eating the flesh of other animals—their prey species.

We call these different levels in the food pyramid trophic levels. A simple ecosystem may have only three levels but a complex ecosystem may have many.

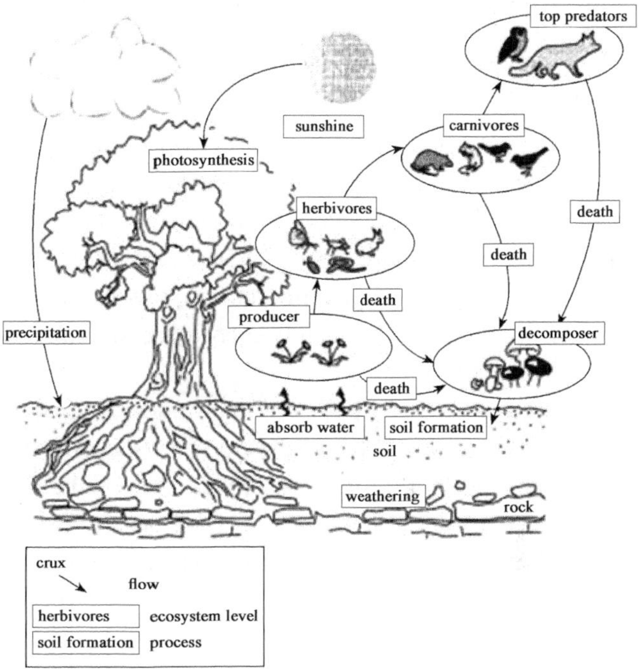

In addition to feeding on plants, some animals may serve other vital roles within the ecosystem. Some species act as pollinators, others may serve as seed dispersal agents. Some species perform important control or balancing roles. Thus the fox eats the rats, which prevents them from becoming so numerous that they eat all grain seeds which would cause their own food supply to collapse.

Out of sight in the soil are other important components of the ecosystem. Worms collect and digest the dead leaves and help create and improve the fertility of soil. Other decomposers such as fungi, isopods, nematodes and bacteria break down dead materials—thus recycling nutrients and essential minerals, maintaining soil fertility for the green plants so they can trap more solar energy and power the whole system.

Even diseases play an important role in controlling the balance of the ecosystem. The more common a species gets, the more liable it is to get infected by disease from another of its kind. Diseases and also parasites thus act as density dependent factors, hitting a species harder when it gets too numerous but having little effect when the density of that species is low.

A natural ecosystem depends on having all these functions—primary producers, basic herbivores, high-end carnivores, pollinators, seed dispersers, decomposers, nitrogen fixing plants, regulating diseases. The more complex the ecosystem the more stable it is and the more efficient and productive it can be at trapping and converting solar energy.

The wetlands manager must see his wetland as part of a larger ecosystem. And must ensure that the basic functions of surrounding areas are maintained to preserve water quality, soil fertility, health of vegetation and safety of wild animals to maintain high productivity of the system and high delivery of essential ecosystem services.

The manager may need to extend protection to other ecosystems way beyond the limits of his own wetland boundary since the wetland may depend on pollinators and insect feeders to come from the wider surrounding landscape; the wetland certainly will depend on water sources from way beyond its boundary and may need the shelter of trees outside of the wetland boundary for protection from storms or as roosts and nesting sites for many wetland species. We call this broader approach to protected areas management the 'landscape' approach.

2.5.1 Diversity is security

Simple ecosystems consisting of few species are very vulnerable. Unusual drought, flood or emergence of a disease or parasite can easily cause the ecosystem to collapse with loss of its ecosystem services. Complex ecosystems are better balanced, more efficient in using energy sources and delivering ecosystem services and more adaptable and more resilient in times of change. Given the dynamic nature of wetland ecosystems and the current fast changing climate, maintaining high biological diversity in wetland sites is the best way to provide security to site health.

The same argument applies to man-made ecosystems such as forest plantations or farms. Monocultures are risky and diversity is security. In spring of 2008 unusual snowstorms hit southern China doing massive damage to bamboo, forests and crops. Some bamboo groves were almost totally destroyed. Farmers that grew bamboo in mixed forest systems found that they suffered far less damage than those that managed bamboo alone as a monoculture.

2.6 Special needs of migrating species

Migratory birds are species where a substantial proportion of the global or a regional population makes regular cyclical movements beyond the breeding range, with predictable timing and destinations.

Migratory Waterbirds are defined broadly as migratory birds ecologically dependent on wetlands. These include populations of shorebirds (e.g. cranes, herons, ibises), Anatidae (ducks, geese and swans), pelicans and seabirds (for example divers, cormorants, gulls, frigatebirds, tropicbirds) and a few other groups.

Large numbers of migratory waterbirds often congregate at staging sites (typically, few such sites) for re-fueling during their journeys, especially before crossing large ecological barriers. Consequently, the loss of staging wetlands may have critical impacts on successful journeys and survival of migratory waterbirds.

Sites that contain or are defined as wetland sites might also be important stopover sites, breeding sites or wintering areas for other migratory birds not specifically regarded as wetland species.

In addition to birds, there are migratory species of mammals—whales, dolphins, reindeer, bats etc.; migratory fish—salmons, sturgeon, eels, sharks, tuna and many pelagic species; amphibians and even insects that may be relevant to a given wetlands management needs.

Migrating fish racing upstream to reach cool water breeding sites

2.7 The need for connectivity

Connectivity is an essential quality for ensuring that meta-populations of component species in individual wetlands remain genetically in touch with their respective species total populations. Without such connectivity, isolated populations will suffer from genetic inbreeding and will have no capacity to recolonise the site in the event of any stochastic or demographic disasters.

Many species need to undertake seasonal migrations between different wetlands or different habitat types. This may be as small a movement as frogs coming out of a surrounding forest to mate and spawn in a pond or larger scale movements of marine fish such as salmon that need to access cold freshwater to spawn in upper rivers.

At the top end of the migrating spectrum are the many migratory water birds that travel each year from their arctic breeding areas in the north to southern winter feeding areas in the sub-tropics or even as far south as Australia and New Zealand. Such species need en-route stopping points to refuel or rest in what is called in passage.

Loss of these migratory stepping stones puts great stress on migrating species forcing them to overcrowd remaining sites and face longer journeys between increasingly rare and distant stopover sites.

Dams, weirs and other barriers constitute severe obstacles to species that need to travel up and down water courses.

Small isolated habitat patches fail to protect viable populations and suffer from dwindling species richness as one after another of the original species community becomes locally extirpated and fails to become re-established by recolonisation.

2.8 Achieving sustainable use

2.8.1 'Sources' and 'sinks'

Any area where a fishery resource grows faster than natural mortality and any human off-take can be classed as a 'source' area. It generates a growing and spreading population that can disperse to seed, repopulate or boost neighbouring populations that may be great distance apart. Conversely any area where mortality plus harvest levels exceed growth capacity, local populations will decline as a drain or 'sink' area dependent on regular fresh influx from neighbouring source areas to maintain its own productive level.

Most wetland localities have now become sink areas, so identifying and offering heightened protection to 'source' areas such as key breeding areas is very important. We can recognize key 'source' areas in streams, lakes, mangroves, coral reefs and other marine areas.

2.8.2 Optimal / maximum sustained yield

A natural body of water will have a maximum carrying capacity for a given fishery resource depending on the water quality, food availability and various mortality factors (predation, disease etc.) present. Humans can exploit each fishery resource by taking off a yield dependent on search effort. It the yield remains low the resource is still able to approach its natural carrying capacity, but if the yield taken is too high (overfishing) then the off-take exceeds the growth rate and the population will crash. The concept of maximum sustainable yield (MSY) is the highest level of off-take at which the resource can still show maximum growth rate. Sustainability is achieved because at this point harvest level = growth rate. This is usually at a level between 30%-50% of the carrying capacity. Many fisheries around the world have used this approach to determine suitable quotas to maximize exploitation benefits. The formula for maximum sustained harvest (H) is one-fourth the maximum population or carrying capacity (K) times the intrinsic rate of growth (r) ($H=\frac{Kr}{4}$).

However, use of this approach is criticized by ecologists and conservationists and has been seen to lead to many population crashes because the formula does not take into consideration such key factors of age structure, reproductive capacity or indirect consequences to byproduct species or the wider environment. There is also the danger that once a market is established at a certain level of demand, it becomes difficult to resist going higher than the agreed off-take. In a poor year this may lead to long-term impacts on the population. As a result it is now more acceptable to aim for 'optimal sustainable yield' (OSY). This is the level of effort that maximizes the difference between total revenue and total cost. Or, where marginal revenue equals marginal cost. This level of effort maximizes the economic profit, or rent, of the resource being utilized. It usually corresponds to an effort level lower than that of maximum sustainable yield. In environmental science, optimum sustainable yield is the largest economical yield of a renewable resource achievable over a long time period without decreasing the ability of the population or its environment to support the continuation of this level of yield.

Calculating the level at which maximum economic yields can be maintained is one thing, but ensuring that a fishing community contain their harvest within that level is quite another. It is a classic case of the 'tragedy of the commons'. Many approaches are tried.

2.8.3 Agreed quotas

If the optimal yield can be measured or deduced from monitoring the resource population against off-take, then an overall quota of allowable harvest can be agreed by all stakeholders and shared in some agreeable way. This only works if all stakeholders agree with the quotas, if the stakeholders is itself a limited group and if there is some mechanism to check each shareholders take. This requires a high level of cooperation and supervision.

A reverse quota system may be easier to operate. This involves agreeing on an adequate survivorship by allowing a certain number of adults to breed without interruption or a safe number of eggs, young etc. to be reared each year. So long as this survivorship is guaranteed then the rest of the harvesting needs no further monitoring. Variants of such a reverse quota system can be seen in turtle hatcheries and in restocking schemes.

2.8.4 Closed seasons / no fishing areas

A closed season is one way of operating a closed reverse quota system. Harvesting of a specific resource is prohibited for a critical period, usually the breeding season, to allow enough recruitment for restocking the waterway.

2.8.5 Use of size limits

Having size limits is another way to ensure that only a certain proportion of the population is harvested each season and is gauged to guarantee that enough young individuals survive to become future breeders. This system is very commonly used to manage sport fisheries where the fun of catching smaller fish can still be enjoyed even if the undersized fish all have to be returned to the waters.

2.8.6 Net mesh size

Regulations regarding the type of nets that can be used in a given fishery especially the mesh size, can be very effective in ensuring low mortality of young or undersized fish or non-target species. Some nets, however, do prove to be very dangerous for rare and protected species such as diving birds, marine mammals and turtles. The wetland manager must be very careful in deciding what fishing methods and tackle can be allowed in a protected area.

2.8.7 Prohibited fishing methods

Fishing with poison, electricity, and explosives is prohibited by law in China but law enforcement remains weak in many areas and these methods still persist.

Two main types of poisoning are practiced:

1) With some variants, divers chase valuable food fish, octopus or ornamental fish into holes in coral reef then squirt in a cloud of potassium cyanide from a plastic bottle and nozzle. This stuns or immobilizes the target fish which can then be easily pulled out of hiding into the collectors bag or basket.

2) Fishermen beat or scrape roots, leaves or fruits of certain poisonous plants such as *Derris elliptica* or *Barringtonia asiatica* and then place the soapy fibres into pools or ponds or other slow moving water bodies. The poison spreads slowly through the water and then gills of fish causing them paralysis or breathing difficulty. Fish float to the surface for easy collection or drift downstream to be collected by waiting colleagues or in nets. This is a very efficient

method of catching fish and can be used in research to assess fish populations but also used as a selective method in fishery management to remove excess or unwanted predators, ill fish etc. However it is unsustainable, selfish and banned as a commercial fishing method in China and most countries.

Other unsustainable fishing practices lead to fast depletion of fish stocks or long term damage to the fishery environment or killing of non-target species and are banned in local regulations or specific sites. Such methods can include use of fine mesh nets, gill nets, drift nets, trawls etc.

2.8.8 Some traditional fishing methods in Asia

Ice fishing, use of cormorants, vegetable poisons, Cochin traps, marine traps, driving dolphins are all endemic methods used for fishing in East Asia. All can be done in a sustainable manner but all can be undertaken too intensively and cause overfishing.

The killing of whales and dolphins which are warm blooded mammals, much loved by large sectors of the public, remains controversial on animal rights grounds as well as endangered status of many of these species. Their use is governed by international treaties such as the International Whaling Commission.

Some traditional methods: expanding crab nets (top left), Cochin net trap (top right), ice fishing (middle left), driving of dolphins (bottom left) and simple bamboo fish taps (bottom right)

2.8.9 Signs of over-fishing

Classic signs of overfishing include overall reduction in yield, reduction of yield per unit of fishing effort, reduced mean size of caught fish, reduced proportion of high value fish in total catch, loss of key species, reduction of fish indicators (fish-eating birds etc.). Overfishing can lead to a dangerously accelerating and vicious cycle. As fish resources are reduced, the fishing community tries to compensate for this loss of revenue by increasing fishing effort. This is the exact reverse of what is needed to rectify overfishing and leads to even faster depletion of stocks.

2.9 Bio-geographic principles for protected area design

Studies of the birds that occupy different sized oceanic islands and the fish that occur on different sized coral reefs led to the realization among biologists that there are a range of geographic principles that govern how biologically rich a given unit of land or water can be. These principles of island biogeography became adapted to the design of nature reserve and nature reserve systems since isolated nature reserves lose or retain species in much the same way as other islands.

Large islands or land units contain more and retain more species than small land units and in-tact units of habitat retain more species than fragmented sites or habitats.

The underlying principle is that a given site achieves balance in species richness when its rate of local extinctions is equal to the rate at which new species colonize or recolonize the site.

Factors that favour low species extinction rates are large size, habitat diversity, proximity or connections to other sites and benign climates with adequate water.

Factors that create low colonization rates include small size, remote isolated location, low habitat diversity, harsh climates or periodic water shortages.

Small isolated islands or nature reserves tend to lose large animals, high trophic predators and terrestrial animals more that they lose small animals and flying or swimming animals with better dispersal. Retention of the highest trophic carnivores of a site is a good indicator of local ecological health.

The right scheme is often presented as a model to demonstrate application of bio-geographical principles for protected areas design. The wetlands manager usually has limited options for increasing the size of his protected area but he can often do a lot to improve connectivity of water courses and corridors of other habitats such a forest or grasslands.

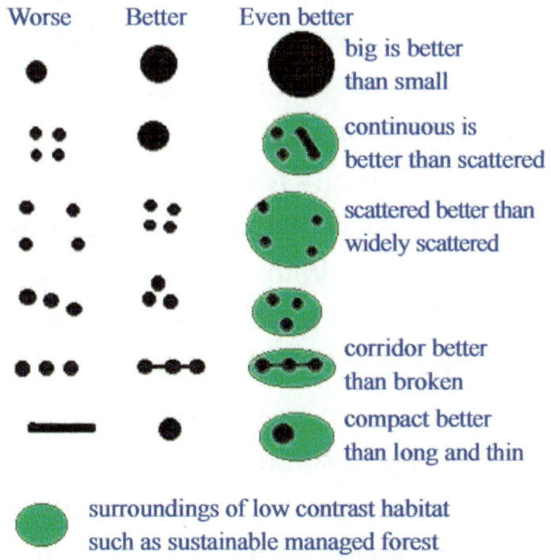

Biogeographical principles in reserve design
(after MacKinnon et al. 1987)

2.10 Mitigating climate change

Climate change is increasingly recognized as a threat to wetlands (see section 2.4.11 above). Man's continued burning of fossil fuels has changed our atmosphere, doubled its concentration of CO_2, added other 'greenhouse gases' such as methane and nitrous oxide and depleted the protective ozone layer. All these changes have significant impacts on global temperature and climate. Wetlands are being negatively affected in several ways.

2.10.1 Extreme weather events

Overall global warming causes the readjustment of vegetation zones and animal distributions. The wetlands manager must anticipate the arrival of new species, disappearance of others. Many fish for instance need to find cool streams to breed in. Such habitat may disappear in a given river catchment. Whilst the overall trend is towards global warming, the changing climate is becoming more extreme so there will be more cold spells, more heat-waves, more floods and more droughts. Coastal areas will experience more and bigger storms, cyclones and tidal surges.

2.10.2 Permafrost

The extreme northern parts of China and on high grounds of the Tibetan Plateau and some mountain ranges have permanently frozen water under the soil surface called permafrost. Impacts on the underground permafrost will have significant impacts on vegetation and the distribution and nature of wetlands in extreme boreal areas.

Relationships with permafrost are complicated and sometimes contradictory. Logging, thinning of forests, soils and vegetation cover loss all **reduce** thermal insulation allowing greater loss of earth's heat to space and have caused an **increase** in the total mass of underground ice and reduced the seasonal depths of unfrozen soil. Increased permafrost stunts the growth of trees, even of shallow rooted larch, blocks land drainage and therefore leads to **increased** area of marshy wetlands.

Changing climate on the other hand is causing the fast melting of much of the underground ice. This typically results in:

- faster tree colonization of some wetland meadows (ability to root in unfrozen soils)
- improved drainage and hence water level **loss** of some wetlands
- collapse of land profiles (ice has bigger volume than water which also flows away)
- release of frozen methane gas which acts as a greenhouse gas accelerating further climate change

Ironically in some terrain, as melting permafrost releases more water, but drainage remains poor, this can result in raised water levels and **increased** wetland area or **new** (neo-natal) wetlands appearing. Long frozen natural springs may suddenly flow again.

It becomes important to monitor changes in permafrost, wetland meadows and marshes and recolonisation by willow (*Salix* spp.), birch (*Betula* spp.) and larch trees (*Larix gmelinii*). There may be needs to interfere in the form of slowing or speeding up drainage to maintain a preferred balance between grassland, forest and wetlands.

As seas warm they expand and increase in energy. This causes rising sea levels and bigger tides eroding and flooding coastal areas and mangroves. As temperature continues to rise the great ice sheets of polar regions and Greenland are melting, giving rise to yet more sea level rise. Estimates of the speed of such rise vary in the long term we must expect several metres and will need to abandon many coastal cities such as Shanghai. But even within this century we are likely to see more than 1 meter sea level rise despite all efforts to contain global warming below 2℃ above pre-industrial levels.

2.10.3 Coral bleaching

As more CO_2 is released into the atmosphere, much is captured by the sea resulting in the seas becoming acid and preventing the normal growth of corals. Coral reefs are dying because of acid bleaching. Other organisms thrive in the changing conditions and we must now face increasing plagues of jelly fish and toxic plankton.

2.11 Recognizing main water-birds

Plovers
Terns
gulls
Waders
Pelecans
Divers
Painted snipe
Rails
Kingfishers
Thickknees
Oystercatcher
Ibisbill
Skuas
Murrelets
Tropicbirds
Boobies
Frigatebirds
Shearwaters
Osprey
Vultures
Eagles
Buteos
Raptors

2.12 Recognizing main mollusks

In China's coral reefs and intertidal waters we can find examples of all seven living classes of mollusks. The Cephalopods which include squids, octopus and nautilus are easily recognisable and the squids and octopuses are an important economic fishery. Many of the marine Bivalvia such as clams and oysters are also commercially important and some species such as the giant clams that live in the coral reefs are highly endangered due to over collection. Many sea snails or Gastropods are edible, others are valued for their beautiful shells. The other four classes are rare and less known though large chitons can be found in Chinese sea food restaurants. Only two classes Bivalvia and Gastropoda have colonised freshwater and terrestrial habitats.

Freshwater mussels and snails are important for healthy rivers, streams and ponds. They provide food for wildlife such as crocodiles, otters and many birds. Through their gills, mussels filter out small particles from the water and transform them into food for fish and other animals. Since mussels are filter-feeders, they clean water as they feed. This eating habit unfortunately makes mussels vulnerable to water pollution, which often kills them. Mussels, just like people, need clean water to survive.

Many mussel species are of conservation concern. Since most mussels stay in a single spot their entire lives, they need stable living conditions. Their most serious threat is river damming, which reduces or removes currents necessary to most species. In-stream sand and gravel mining also greatly disturbs stream channels and stream bottoms, where mussels live. Pollution from herbicides, pesticides, fertilizers, mining waste and residential and livestock sewage kill mussels and other aquatic life.

2.13 Recognizing main fish families

China is one of the countries with the most abundant freshwater fish richness with 920 species, which is higher than that in the U.S.A. (800 species), and much higher than Europe (233). There are 33 families of freshwater fish in China but half of all species come from the single family Cyprinidae with a total of 473 species of Cyprinidae. Balitoridae, Cobitidae and Sisoridae account for a further 25.11%, which is equal to the contribution of Cobitidae and Siluridae.

Most significant of the table fish are the various members of the carp family, whilst sturgeons, trout/salmons, catfish, eels are also familiar in Chinese kitchens.

It requires a specialist to accurately identify all local fish in a given wetland but it is useful if the manager has a basic knowledge of the main families involved.

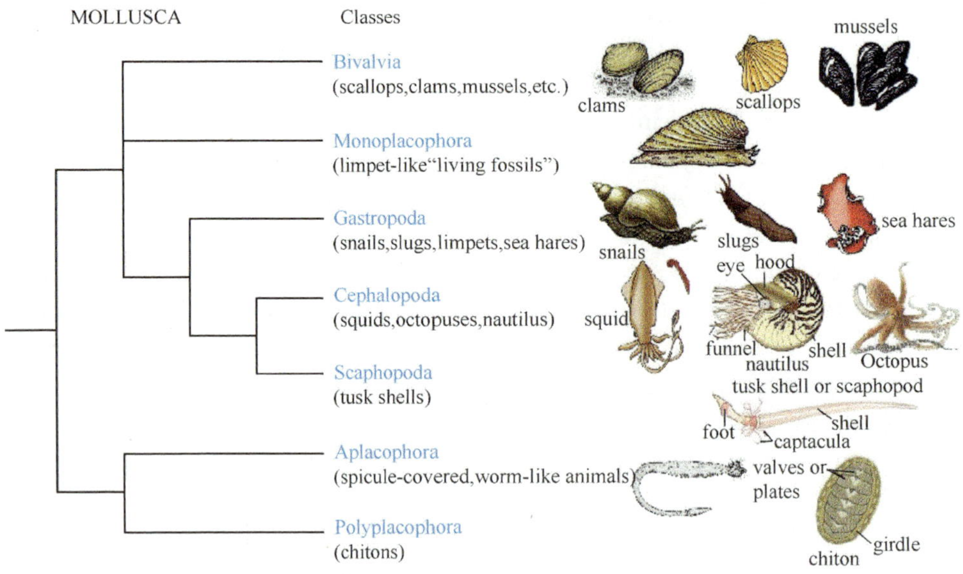

2.14 Recognizing amphibian families

The following family tree shows the range of amphibian families encountered in China and their respective relationships.

2.15 Recognizing main water insects

The following chart figures the main groups of water insects in China.

Insect group	Larvae	Adult
Water beetles		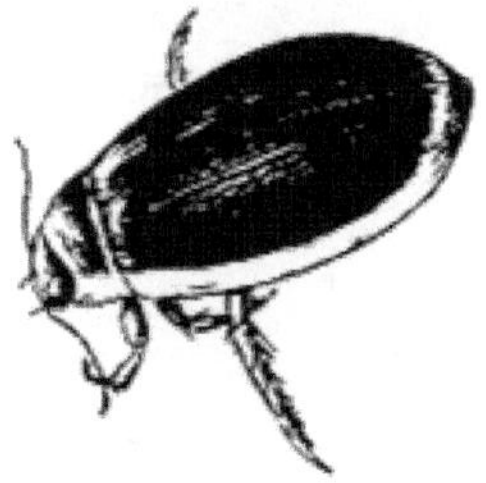
Pond skaters		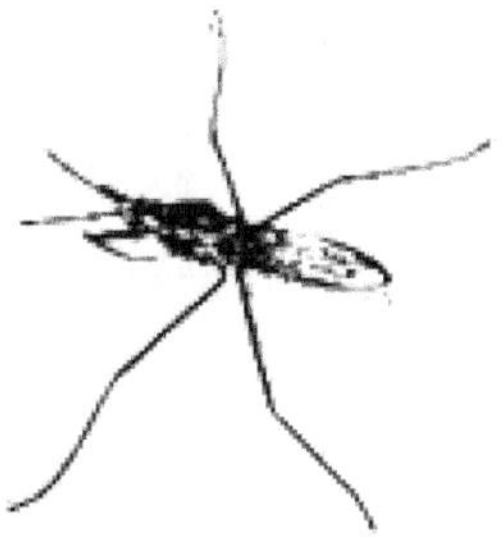
Whirligig beetles	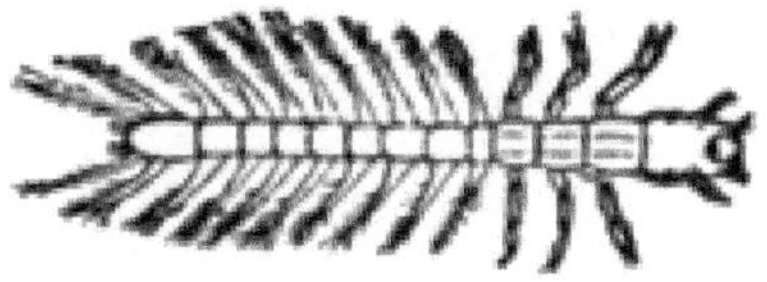	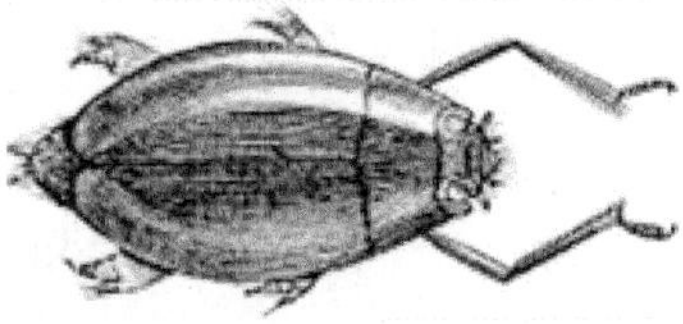
Water scorpions		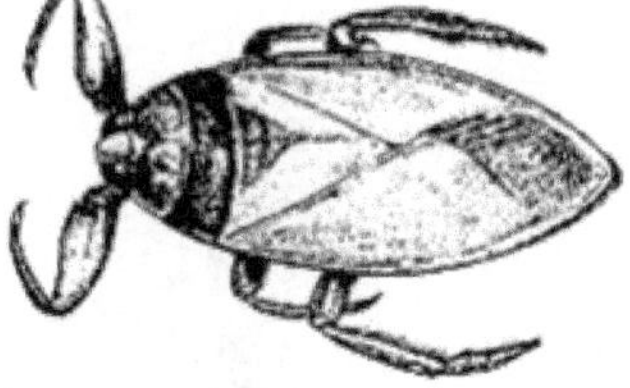
Dragonflies	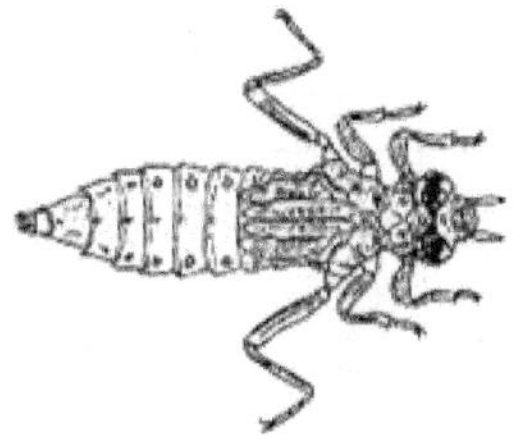	

Insect group	Larvae	Adult
Damsel flies	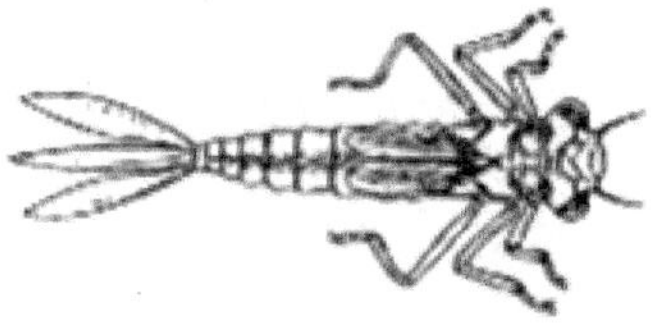	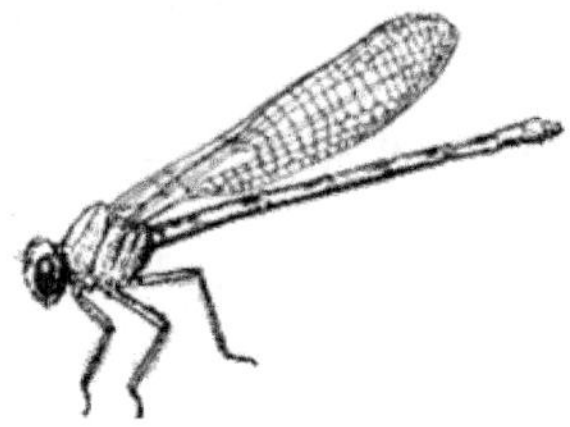
Stoneflies	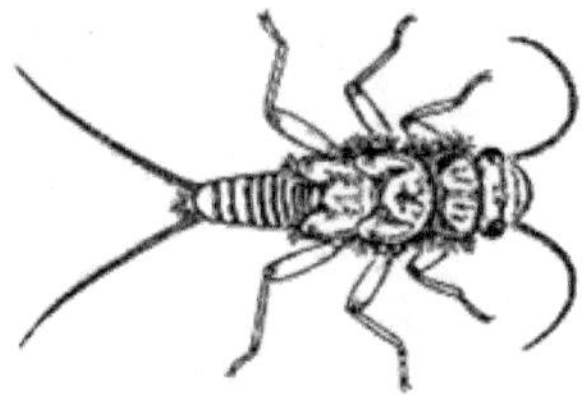	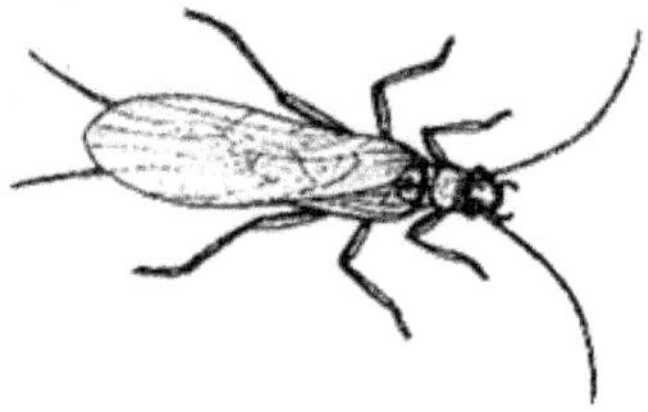
Caddis flies	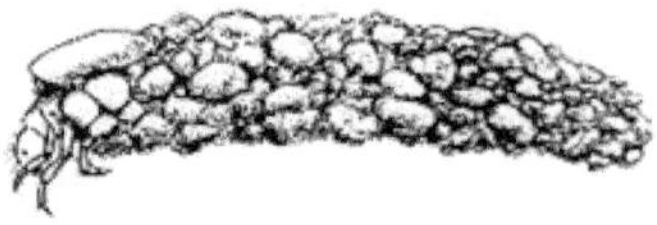	
Mayflies	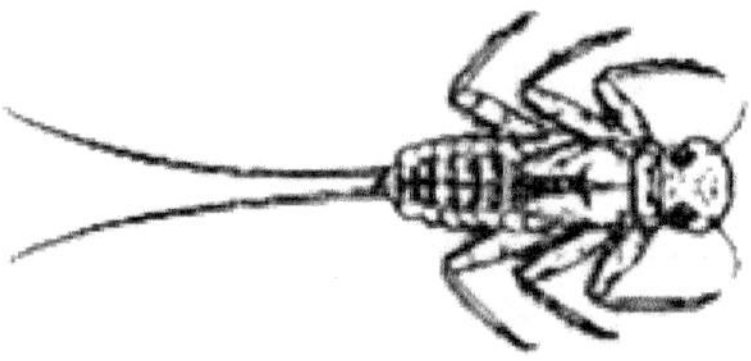	
Mosquitos	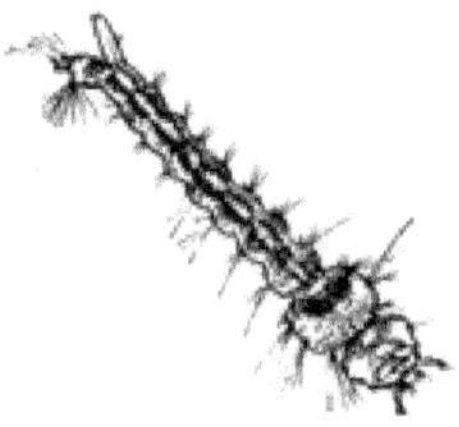	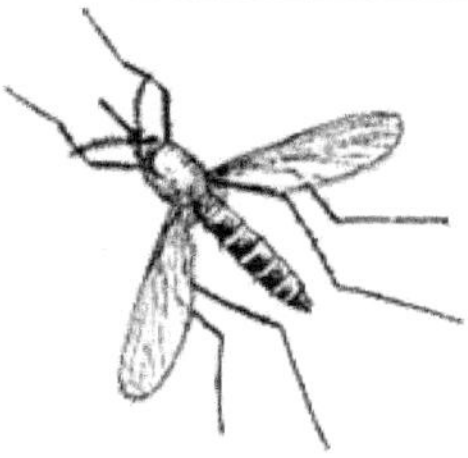
Midges		
Water boatmen		

2.16 Recognizing main wetland vegetation

Sedges—*Cyperus*

Sedges—*Carex*

Bullrushes—*Typha*

Duckweed—*Lemna*

Reeds—*Phragmites*

Reeds—*Arundo*

Water lilies

Irises

Lotus

Couchgrass—*Echinochloa*

Lovegrass—*Eragrostis*

Carpetgrass—*Arthraxon*

Elephant grass—*Miscanthus*

Finger grass—*Digitaria*

Goosegrass—*Eleusine*

Cheatgrass—*Bromus*

Meadowgrass—*Poa*

Amaranthus

2.17 Ecology of mangroves

Mangroves formerly occurred in the intertidal zone around most of the tropical coastline of Hainan, other cities of S China. Today more than 80% of these mangroves have been destroyed or converted to other land uses, notably salt pans, rice fields and fish ponds. This conversion causes loss of globally significant biodiversity, damage and weakening of the physical coastal environment and many direct and indirect losses to economic sectors dependent on the health of mangroves.

Estimates of the value of mangroves to mankind are constantly increasing as awareness grows of their role as nursery areas for many marine fish thus sustaining offshore fisheries, their function in trapping silt to create new lands whilst protecting adjacent corals from siltation, their role in supporting significant biodiversity, potential for eco-tourism, supply of sustainable products and coastal storm amelioration and protection of life and property. The Indonesian tsunami of 2004 illustrated clearly how those villages that had preserved mangroves suffered little damage whilst those that had cleared or converted their mangroves suffered enormous damage. Hundreds of villages were destroyed. China is in an earthquake zone where tsunami must be expected. Costanza et al. 1997 estimated the global average ecosystem service value of mangroves as $3,713 per ha/yr but had to greatly revise this figure up to $786,193 in the updated paper of Costanza et al. 2014.

The healthy mangrove biome consists not just of mangrove trees, but a mosaic of other plants, mudflats and shallow tidal waters. Health is assured if this ecosystem is diverse, resilient to damage and adaptive to changes in silt levels, coastal profiles and climate.

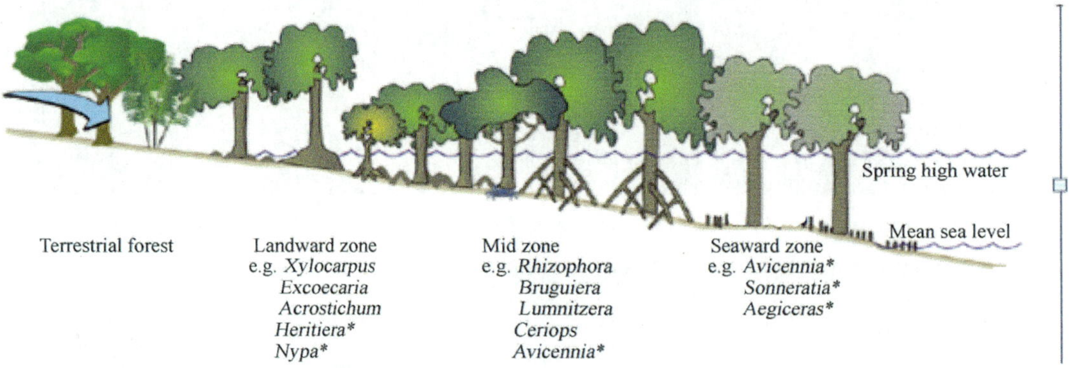

Mangroves form a complex ecosystem of plants, birds, fish and many invertebrates as well as the physical environment of intertidal water flows and deposition of mud, sand and other alluvia.

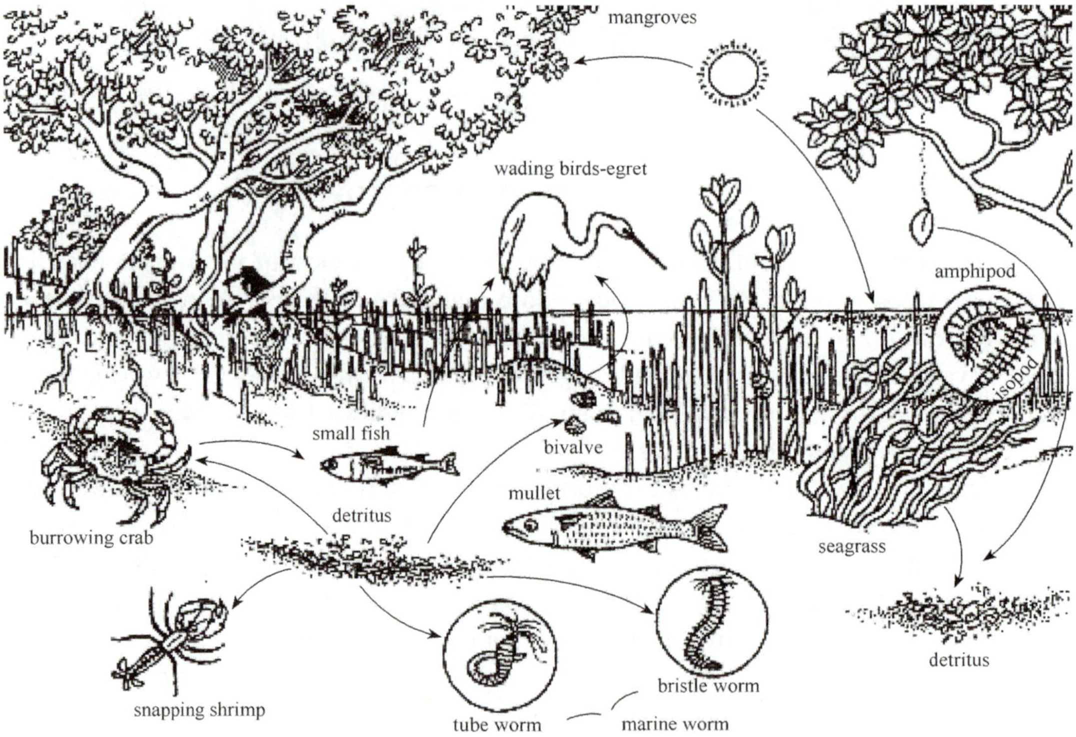

- The mangrove ecosystem is a mosaic of forests, herb areas, ponds, streams, sand banks, mud flats, open water and intertidal beaches. Different species require different habitat elements but a diverse and resilient ecosystem needs all its parts

Mangroves are organised along the high tide to low-tide profile with different species occurring at different places along this profile dependent on their relative ability to cope with water inundation and levels of salinity.

As a result we find mangroves form striations of vegetation in zones that parallel the coast and water courses. Different plant species dominate different zones.

The mangroves themselves form part of a broader mosaic of coastal ecosystems and are highly integrated with the other ecosystems around them. In front of the mangroves we have ever deeper mudflats and other shallow seas. These are often the important habitat for sea grass beds. Further out we may have coral reefs. Behind the mangroves we have inland forests and on the sandy areas with less soil and less water we find strand vegetation of coarse grasses or beach forests. Land birds and invertebrates move between the different habitats so these habitats to some extent depend on each other. Damage or disconnection of one habitat will indirectly affect and degrade the other ecosystems adjacent to it.

Bees, moths and other insects and some birds are important for pollinating the plants and trees of mangrove forests. Weaver ants, insectivores, spiders and dragonflies plus many birds are important in controlling the levels of insect pests that may damage the mangrove trees and other plants. Other birds act as seed dispersing agents spreading the seeds of figs and small mangroves and mangrove associate trees. Crabs and mud lobsters dig deep holes in sand and mud which helps drain and aerate the substrate, removing pollutants and allowing the roots of mangroves to absorb CO_2. Much overlooked is the role of bats. Bats are important for pollinating many of the brush-like flowers of the Rhizophoraceae family and genera such as *Eugenia* and *Barringtonia.*

Reforestation should involve the mangrove associate species; many of these are important for supporting and attracting some of the fauna required by the mangrove trees themselves for a healthy ecosystem. Moreover the zone of mangrove associates or forests adjacent to mangroves are highly suitable to a wide range of useful and economic plants and other crops. These include mangos, coconuts, papayas, lichees, bananas, cashew nuts and for keeping of bee hives for honey.

2.18 The ecology of fires and forests in NE China

Whilst forests are not themselves wetlands, they are often essential to the health of downstream wetlands in terms of acting as rain water catchments, soil formation sources and providing leaf litter and other nutrients into the upper catchment water systems that in turn feed rivers and wetlands further downstream. Many forests indeed contain small lakes, ponds or wetlands. Destruction of forests can have severe negative impacts on wetland ecosystems. They must be protected from unsustained logging and from natural or artificial fire.

The larch forests of NE China are one of the few places on earth where natural fires starting from lightning are common, especially in the springtime. Older trees often bear the scars of old lightning strikes. Such fires are usually of small scale and not too much heat. They have had the effect or periodically clearing the underbrush in forests, burning back the herbs and grasses in meadows and killing only young small trees. By periodically killing off the invader young trees at the edge of grasslands and meadows, fires have the effect of shaping clean boundaries between forest patches and grasslands and many of the admired scenic effects of Daxing'anling wetlands is a result of this process. Fires have also had an enriching effect by repeatedly creating spaces where new colonizer species can settle and in creating a patchy mosaic of forests in different stages of seral succession offering a wider diversity of microhabitats for fauna and flora to utilize.

We can conclude that this background pattern of natural fires is desirable and may be necessary to maintain the original rich diversity of the region and should be allowed to continue.

Part Three Planning for Wetland Protected Areas

The director or manager of a wetland does not own the land and therefore cannot run it like his/her own business. He is merely a custodian entrusted to look after it on behalf of the state and people. Just as the curator of a museum cannot sell or rent out the precious items he is entrusted to conserve. Seeking greater financial security or profits of the bureau may be a way to help reach objectives but not the real job which should be:

- Identify threats and their drivers
- Identify management barriers
- Design a series of actions/programs to diminish these barriers
- Monitor the delivery of these planned actions
- Monitor the impacts on the PA
- Revise management accordingly
- Err on the safe side—precautionary principle

Safest management options:

- Maximize size and connectivity
- Halt or control hunting and extraction
- Maximize habitat diversity
- Minimise alien invaders
- Zone sensitive places out of harm's way
- Use precautionary principle in allowing uses or developments
- Document all decisions
- Use expert panels

3.1 Systems planning and networks

Plans need to be made at different scales and over different time periods.

At the larger scale should be a systems plan at the national, regional or catchment scale in which the entire policy for protected area development is laid out spatially over the broad landscape. The protected area system is seen as a network of sites offering various degrees of protection to the full range of habitat, species features and delivery of vital ecological services expected.

The systems plan is based on an assessment of needs guided by stated national or regional policies. The plan is concerned with how representative is the PA system based on a gap analysis, spatial coverage, connectivity issues and outstanding sites coverage. The timeframe for systems plans can be medium to long term.

The systems plan deals with identification of areas that need to be added to the PA system, sites whose boundaries or zoning require adjustment, locations where connectivity issues of sources of pollution or physical threats are identified and compatibility of sites with adjacent land uses. A stand alone plan developed by one sector has little chance of acceptance. It needs to be mainstreamed and integrated with the overall development plans of other sectors and agencies across the whole region.

Such plans are developed by teams of experts at higher government levels but wetland managers should be involved in such planning to ensure their own issues of size, connectivity or other needs are incorporated.

Like all plans, the systems plan is not set in concrete. It provides a broad road map but new issues are ever emerging and some elements of the plan may not be achieved, fail or be no longer a priority. All plans need to part of a continuous planning cycle. Planning-implementing-monitoring impacts and review or periodic revision. As the China government operates via a cycle of 5-year plans, it is best if the systems plan is also revised to match this time frame.

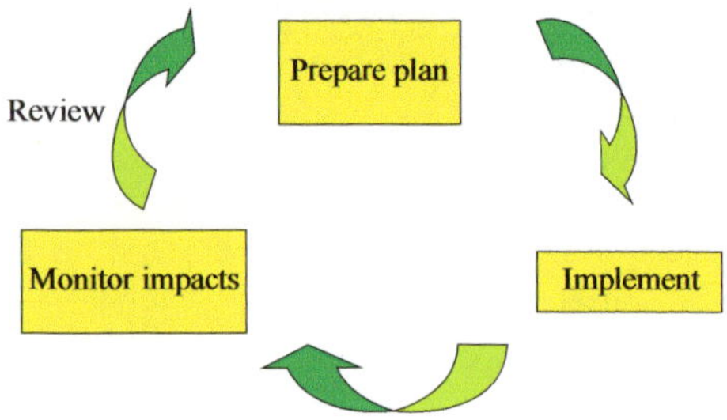

3.2 Management planning at site level

Here the planning focuses on the individual site. It should review the objectives of the site, suitability of boundaries and zones, identify key threats to be tackled and key features to be conserved, outline the actions necessary to accomplish those objectives into a series of programmes—law enforcement, species and habitat management, research needs, visitor access and educational functions, relations with local communities, infrastructure needs, staff capacity needs and finally budgetary requirements.

The management plan is not just spreadsheets of budget expectations or requests and not a constructions masterplan for adding new buildings. It should serve as a guiding tool, reducing

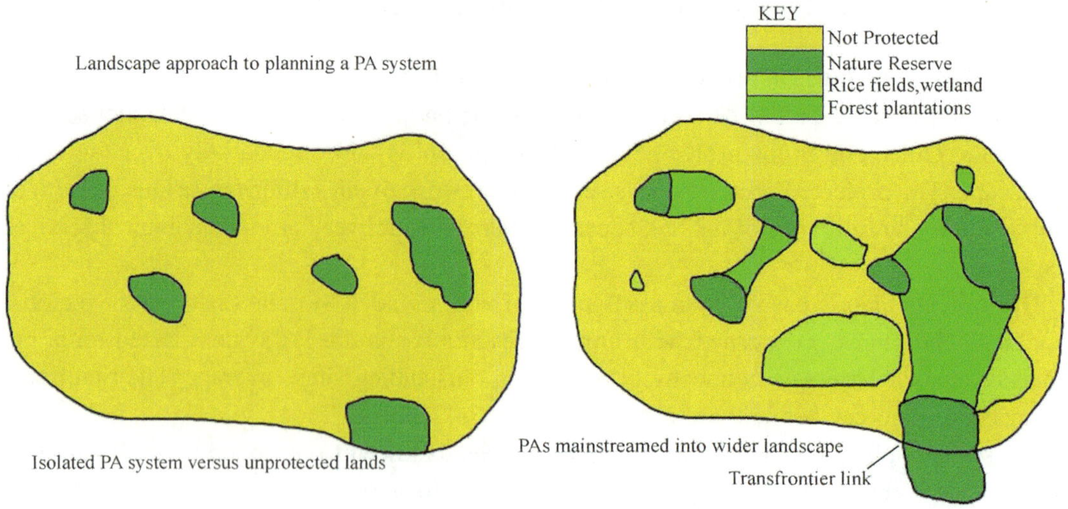

the responsibilities of the project manager in decision making over the coming years. The plan also acts as a communications tool, informing other agencies, sponsors, and stakeholders of what the PA management is trying to achieve.

Ideally the management plan is reviewed by an independent panel of experts who may ask for modifications, explanations or more justification for planned innovations or requested budgets. Funding of the plan should be controlled by its acceptance and approval.

Management should also be subject to independent supervision and performance appraisal to ensure that implementation remains on the approved track.

The development of management plans is still poorly developed in China. Managers often have no technical background in wetlands conservation. No system of expert review and approval is in place and there is little follow up supervision or performance appraisal. The government is aware of these weaknesses and much reform of the way PAs are managed and funded is in the pipeline.

Government policy of 13th 5-year plan involves much 'deep reform'. Reform implies that is not perfect and we must address the areas of imperfection. The following is a list of common management mistakes:

- access roads and fences for management
- core area in centre
- large office distant from site
- lack of clear objectives
- no baseline and no monitoring of biodiversity impacts
- perverse eco-compensation systems
- planting wrong trees in wrong places
- rescue/breeding centres and museums
- isolationism—lack of connectivity and mainstreaming
- masterplan for development rather than protection
- ignorance about climate change
- Netting e.g. airports

3.2.1 Boundary and zoning

The boundary is always a negotiated compromise. Many sectors and stakeholders compete for user rights of both wetlands and their immediate access and adjacent lands. The wetland manager must realise that protecting a wetland requires control of activities far beyond the mere water surface and as far back into the water catchment area as can possibly be justified.

In the competition for management authority, the wetland manager must use all arguments and all alliances he can muster. Importance for biodiversity is rarely enough argument to persuade local government to assign a wide mandate for wetland protection. The manager needs to show the economic value of ecological services, the high costs of repairs or alternative engineering that would be needed if the wetland health cannot be maintained, the additional values for health, recreation and security from floods, climate change etc.

In negotiating for the establishment of a wetland protected area, the manager needs to have a good sense of the key sites in highest need of protection and also the necessary connections that would be required to maintain the ecological health of the wetland.

Any application for establishment of a new protected area needs a strong justification with good maps, baseline data, feasibility study and cost/benefit analysis.

Land-use security must be immediately strengthened by boundary demarcation, boundary notice boards announcing the protection status of the site and prohibited activities therein. The manager should organise events to raise local awareness and support for the new status and should ensure that this status is fully registered on land-use cadasters of different planning bureaus, not least the Department of Lands and Resources.

Having secured a protected boundary the manager should establish a system of different management zones within the PA.

The current nature reserve regulation (1994) specifies only three main zones—core area, buffer zone and experimental zone along the lines advocated by the UNESCO Man and Biosphere Programme (MAB). Historically many of China's first nature reserves were MAB reserves established to study the interaction between human activities and the natural environment. But most newer protected areas were established with different objectives—protection of key species, key ecosystems, scenic areas or ecological functions. The old zoning system is inappropriate and that is one reason why so many new sites opt to be not listed as 'nature reserves' but adopt different titles and unclear status—forest parks, wetland parks, national parks etc.

Identify sites of high conservation values/importance/justification

- Key species (endangered, endemic, wild domestic relatives)
- Representative ecosystems
- Ecosystem services
- Cultural landscapes/intangible knowledge
- Recreation/ecotourism
- Sustainable use of renewable resources

Zoning: It is recommended that each registered natural protected area have a core area to protect the most sensitive or sectors but should otherwise have at least 80% of should be zoned to match the overall PA category of the site but the other 20% can be assigned to other zones such as:

- Wilderness zone—basically left wild but accessed for research, photography, monitoring etc. under special permission
- Management zone—area where habitat or species management is undertaken
- Visitor access zone—where visitors can have controlled access
- Restricted use zone—some uses allowed within management plan objectives
- Administration zone—areas needed for offices admin buildings, stores, public facilities, roads etc.
- External buffer zone—some limits on activities required beyond the boundary of the PA

The core area should not simply be positioned in the centre of the PA. This is often the top of a mountain or deepest part of a lake. these are the areas with lowest threat and lowest biodiversity. Rather the core area should protect the key features for which the PA is established. The core area should often extend to the edge of the PA and may link to adjacent corea area of an adjacent PA.

Core area should include:
Most bio-rich areas
Most fragile areas
Critical weak points
Fullest range of habitat types
Irreplaceable key breeding, roosting etc. sites

3.2.2 Buffers and red lines

In addition to internal buffer zones, all wetland sites require wide buffer zones outside and especially upstream of the protected area to ensure that external activities do not compromise the function and health of the wetland due to undue disturbance, pollution or sedimentation. Such buffers will extend way beyond the management mandate of the PA but it is important that the wetland manager is able to review, comment on and influence roader scale land use planning to ensure that his/her site is adequately mainstreamed into the full landscape and river catchment.

The wetland manager is helped in this task by new policies from Central Government. The 18th NPCCP made strong decision to implement a system of 'ecological red lines' to protect the delivery of key ecological services. Eco-compensation funds will be made available to compensate those who may have to abandon plans or potential activities that are determined to threaten the delivery of such services. Plans exist for designation of more that 30% of the terrestrial area of the country as ecological red lined—a major increase from the 15% currently protected as formal protected areas.

Content of the Management Plan

- Description of site
- Identify conservation values, key ecological services delivered and identify key threats leading to identification of priorities for protection or conservation
- Justify category of natural protected area registered
- Clarify protective management objectives in order of priority
- Describe boundaries and zones with any revisions proposed
- Outline sub-plans for: protection and monitoring , management oriented research needs, any habitat modification, control, restoration or species management, controls that may be appropriate (e.g. control of AIS), communication, educational and awareness responsibilities; visitor use and extension, information ; Contingency measures for fire fighting, combating desertification or meeting natural or artificial catastrophes
- Programme for involvement or benefit sharing with local communities
- Plans and justification for infrastructure investments or equipment purchases
- Details of staff development needs including training and other capacity needs
- Details of investment budget, staff costs, operational budget requested
- Proposal for use of any funds raised through user fees, compensation mechanisms etc.

3.3 Planning operations

All operations and major activities need their own plans. This helps to ensure that they happen with greater efficiency but also the plan needs to be approved by a superior manager to get the go ahead or to approve the necessary budget expenditure. The following content checklist may help in preparing such plans.

Content	Notes
Simple title for the operation or event	Add unique event ID
Dates , time and location	Complex operation may have several dates and locations
Objective	e.g. Remove illegal nets from north end of lake
Brief description of activity	e.g. Public celebration of 'Love Birds' Day. 500 children invited to PA
Planned schedule	Timing of the event or any pre-event preparations
List of individual activities	Each component activity should be listed with timing, responsibility and cost
Materials needed	List physical materials required, how sourced (purchased, borrowed, hired etc.) and cost
Responsibilities	Identify who is responsible for each activity or purchase
Promotion needs	How to advertise the event or action. Do you need to inform the media? etc.
Reporting	How will the action/event be reported
Summary budget to be approved	Totals required and from what project or PA budget lines

Part Four Habitat Management

5.58% of the land area of China is classed as wetlands. China's wetlands comprise many types—swamps, peatlands, wet meadows, lakes, rivers, floodplains, deltas, mudflats, mangroves, reservoirs, ponds, and tidal zones (see section 2.1 above). Many of the interior lakes of the Tibetan Plateau are saline to various degrees and even though there is little replenishment from rainfall, there is a constant trickle of water from the melting of glaciers.

China's wetlands provide great economic, ecological and social benefits, and so their conservation is very important (see section 1.2 above) . These services include flood mitigation, water storage, climate regulation, water purification, erosion control, land formation and creation of scenic and relaxing environment. They also provide homes for a wide array of plants and animal species including birds, fish, amphibians, crustaceans and many plants and animals that are used by humans for food, including some very valuable shrimp habitats.

It is a little-known fact that China's peatlands store more carbon that all the forests of China combined. Preserving such peatlands from drying or burning, both of which release the carbon into the atmosphere, is a great global contribution to the ameliorating climate change.

There are thousands of lakes scattered across the face of China. These are important for fisheries and many other aquatic species but also serve as natural reservoirs and hold back a lot of flood water during periods of high rainfall.

4.1 Wetlands habitat management challenges

Wetlands are seriously endangered (see section 2.4 above) but are also a very important case of habitat that requires special management and protection. They are vital for the survival of a great many important species, but they also deliver critical ecosystem services in acting as natural water storage reservoirs, carbon reservoirs, water purification and flood controls. This importance is underlined in the policy of the State Council—Wetlands conservation needs shall be routinely taken into account in all areas of government planning.

The management of wetlands remains complex and a great challenge wetland ecosystems are very open (water and species inflow and outflow); they are vulnerable to so many external influences beyond the control of the site manager and almost always they are used for so many different purposes by a great range of stakeholders. A further complication is that they are very dynamic in nature, constantly changing in character and even in position. This last point is well illustrated by the shifting distribution of wetlands shown by the historical maps of Dongting Lake at different dates.

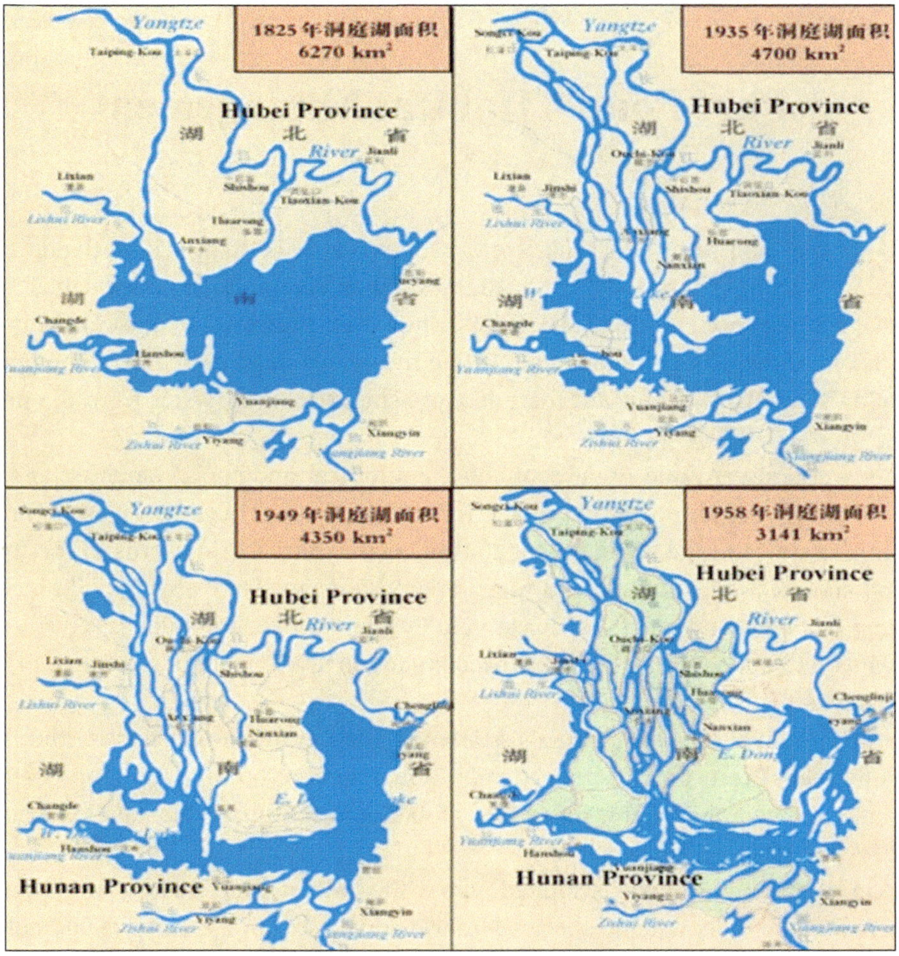

The dynamic nature Dongting Lake, Hunan

4.1.1 Defining the job—maintaining wetland ecosystem functions

The secret to maintaining ecosystem functions around wetlands includes:

- ensure that wetland ecosystems are protected and any harvesting from wetlands is done only on a sustainable basis
- protect water from pollutions, check siltation levels and avoid eutrophication
- maintain surrounding soil depth and health by reducing erosion through terracing, erosion engineering
- slowing down water flow by terracing and use of check dams
- reduce water wastage by good maintenance of irrigation channels, ditches and water pipes, bamboo gutters etc.
- avoiding fire (which kills many soil decomposers) by allowing agricultural waste to rot down into the soil as green manure

- reduce disease by maintaining high diversity
- maintain healthy natural and semi-natural ecosystems (woodlands, wetlands, grasslands), adjacent to wetland NRs
- do not cultivate slopes steeper than 25% and return steep farmland to grass and tree cover
- protect adjacent forest by no cutting, no burning, no grazing closure

4.1.2 Need for guards on the ground

However clever the plans, however sophisticated the monitoring equipment, protective management is only as effective as the rangers and guards deployed in the field. They are the eyes and ears of management. They are ones who check what is really happening. They are the visible protectors that need to establish the respect and cooperation of local people.

Guards and rangers are rather useless when they are sitting around a guard station. They have to get out and around the site. They have to be vigilant 7 days a week, day and night.

The secret of good protection is regular patrolling in unpredictable patterns and covering all parts of the protected area.

Management cannot act on what the guards see unless there is an effective system of reporting. All patrols should submit a report after each patrol. These reports should include sightings of interesting species but should also report on general conditions of the habitat and the activities of humans or domestic animals. More detailed monitoring roles can be attached to this system of basic patrolling. A simple minimal patrol report form is shown in Appendix 9.8 but each PA may wish to have a more elaborate form and collect additional information.

Expecting too much detail from patrols will reduce morale and motivation. Keep the reporting burden light and simple. Patrolling is best motivated when guards can see that their reports are taken seriously, acted upon and influence management decisions.

In sites that face serious poaching patrols can use the software system 'SMART' (Spatial Monitoring and Report Tool) (http://www.smartconservationsoftware.org/) developed and distributed by WWF. This software application used a GPS recording approach to monitor patrol routes, sightings of animals or their signs, evidence of poaching etc. and can be used to tighten up patrolling, identify where more patrolling is needed and focus on most urgent areas etc.

4.2 Protecting different habitats

4.2.1 Lakes

Lakes are larger bodies of water that form in land depressions. They are generally fed by rivers or streams and also discharge via other rivers or streams.

Lake in China are found from high altitudes to coastal lowlands and vary enormously in their character. Some plateau lakes are to varying levels salty.

Lakes support a great range of aquatic fauna and flora with different endemic species occurring in different river catchments. Many rare and endemic species have narrow distribution ranges in China and a great number of different lakes need to be protected if we are to hope to protect examples of most aquatic species.

4.2.2 Protecting reservoirs

Reservoirs are artificial lakes constructed for various hydrological reasons—water storage, flood control, irrigation etc. But many reservoirs turn out to be also very valuable sites for biodiversity and can often be used for multiple purposes.

Given that these are not natural, the manager has more scope to legitimately 'improve' their biodiversity function by some engineering or planting of special vegetation, introduction of various local wildlife etc.

4.2.3 Ponds and ditches

Wetlands are very dynamic. Artificial ponds and ditches clog up and become blocked with vegetation. They require regular maintenance and clearing to retain their original functions. Even natural ponds will change over time as they gradually fill up with sediment and weeds. The manager must decide whether the objectives of the wetland site are best served by allowing these natural and seral processes to continue to change the site or whether to arrest change at some desired state and manage the vegetation and water flow to maintain that condition.

4.2.4 Coastal habitats

China's long coastline forms a narrow but important habitat for many kinds of biodiversity. Mudflats, sand beaches, cliffs, gravel beaches, coastal marshes and fringing mangroves all contain very different but rich assemblages of mollusks, crustaceans, fishes and plants. They are important feeding areas of waders and other birds and serve as important stop over points for those birds that undertake annual migrations from their northern breeding areas and tropical wintering areas. Many species of waders nest along our coastline and even such marine mammals as seals breed along the coasts.

Coastal habitat yields many foods for human consumption and is the site of fish ponds for rearing fish and crustaceans. The coastal vegetation is very important in consolidating the alluvial mud and sand washed from China's interior by it large rivers. This alluvium is trapped by plants and builds up to form new fertile land for agriculture and even coastal cities such as Shanghai and Xiamen.

4.2.5 Marine habitats

China's seas range from the cold temperate seas of the north, the semi-enclosed Bohai Sea, the oceanic east coast and all the way down to the tropical coral reefs of the Xisha and Nansha archipelagos.

In the northern seas one can find marine mammals such as seals together with a wide range of sea gulls and other birds whilst below the water surface live great fisheries of tuna and mackerel, squid and bass.

In the south are colourful coral gardens teaming with colorful fishes, strange mollusks, dolphins, sharks and marine turtles.

Deep in the ocean giant whales sing their mournful songs and make their annual passage between polar waters.

China derives a great bounty from its seas in the form of fish, mollusks, crabs, shrimps and sea slugs. Many of the most treasured Chinese dishes depend on this sustained supply.

4.2.6 Additional marine issues

Development of protected areas in the China's seas lags behind that on land. This is partly a result of the difficulty of demarcating or patrolling marine protected areas. There is also an issue of disputed waters.

Other problems facing the sea is their heavy use by world shipping and fishing fleets and the nature of the threats of pollution and siltation which are way beyond the control of the Sate Oceanic Administration that is responsible for marine protected areas in China.

4.2.7 Protecting river catchments

Wetlands are dependent on their water sources—rainfall, stream and river flow, water tables, tides etc. The protection of these sources may extend hundreds of km upstream into the upper catchments of streams and rivers, but also includes allowing enough water to flow through artificial barriers such as dams, diversions, wells etc. China has several policies for protecting and improving the function of its water catchment areas. At a macro scale the Ministry of Environment has identified a number of key ecological zones, selected primarily for their role in protecting water sources. In addition the government is now embarking on a programme of establishing ecological red lined areas where the extent and type of development permitted will be limited in areas essential for maintaining ecosystem services, primarily water catchment. At the same time, large areas of the country are now protected by logging bans and the State Forestry Administration is busily engaged in protecting natural forests and planting new forests to improve water catchments.

Trees play a large role in improving water catchment functions. Firstly forests create soils and both the soil and deep penetrating roots of trees make the land surface more permeable to rainfall so that more water is stored in the catchment sponge and less water runs straight off to flood rivers. The leaf litter and understorey of forests protects the fragile soil from being broken up and washed away by rainfall. Thus upstream forests benefit wetlands in several ways.

By increasing soil penetration forests reduce flood run off and prolong the periods of the year when water still drains out of the land into water courses.

By creating and protecting soils, forests improve the quality and clearness of water supply so it is filtered, cleaner and healthier.

Breakdown of leaf material and other forest litter enriches the water in streams and lakes providing the food base for many foodchains.

Forests shelter the upper water courses from direct sunshine so that water is cooler and can contain more oxygen.

The manager of a wetland must know and understand his/her water supply, lobby for good forest protection in the upstream catchment and also consider threats of supply caused by physical damming, diverting or pumping of water out of the system and equally influx of polluted or sit-laden waters into the system.

Water is in short supply in many areas of water. Even in Hainan which is regarded as an evergreen wet forest ecosystem, the rate of human use of water exceeds supply and the level of reservoirs and water tables is decreasing. The wetland manager is only one in a queue of users eager to access and the same limited water. Local government must apply common sense and fairness to decide who gets what water but eh wetland manager needs his arguments to be strong in justifying his/her own demands for adequate water flow. The secret to ensuring such flow is integrated land-use planning normally undertaken by the Lands and Resources Department and local units of Reform and Development Commission.

4.2.8 Grasslands and fire management

Fire is a great controller of vegetation. Its capacity to destroy is well recognized, so the default opinion about fire is usually negative and the manager tries hard to prevent wildfires, put out any fires that do occur and help restore burned vegetation areas as soon as possible. But fire is not always negative for biodiversity and many landscapes have been shaped by regular burning. Fire suppresses of forest and shrubs and encourages annual grasses and herbs.

Wherever the wetland manager needs to retain or increase grasslands he may need to use fire as a management tool together with controlling trees by cutting, grazing or flooding.

Reed beds and various grasslands support their own fauna but many other species need short grass or open areas rather than dense swathes of tall species. In nature there would have been large grazers such as deer, wild cattle or even elephants that would maintain a balance between tall unbroken reed or grasses and opened up or grazed patches. In the absence of these original herbivores the wetlands manager can justify using controlled levels of reed or grass cutting or cattle grazing to open up some of the wetlands to provide a wider range of habitats and thus maintain a wider range of species.

Reed cutting and grazing

4.3 Habitat restoration

4.3.1 Restoring degraded ecosystems

China currently faces many problems of loss of ecosystem services resulting from decades of degradation of many vital ecosystems.

Forests have been logged, cleared, burned, converted to farmland and fragmented into small patches. Their natural recovery is slowed down by pressure of harvesting firewood, grazing by domestic animals and forest fires. Forests protect wetlands and their water sources. Their healthy maintenance is therefore directly relevant to protecting productive wetlands.

The following sections describe some ways in which wetland managers, foresters and farmers can assist the nation and improve their own ecological environment by taking care of and helping to restore the natural ecosystems around their landscape.

Saving the wild rice

Wang Xingsheng of Jin He village in Yunnan Province reported to local authorities on finding wild rice *Oryza officinalis* growing in a forest scheduled to be converted into a rubber plantation. The villagers have now made an agreement with Ministry of Agriculture to protect the site, use no chemical weed-killers nearby and cancelled the planting of tea or rubber. In return, the ministry's wild plant conservation programme is compensating the village with technical know-how on how to build a pig farm.

4.3.2 Returning steep farms to forest

Farming on steep slopes leads of great loss of the most precious resource of the country—its fertile soils. Soil loss in China amounts to billion of tons per year. Imagine what a loss that is. But with the loss of absorbent soil sponge and the silting up of river beds that act as drains in times of high rainfall, farming on steep slopes has been proved to add considerably to the damage caused by floods in China each year.

Example of steep farming in Sichuan in urgent need of regreening

For this reason the government has been implementing a special programme to encourage farmers to return steep farmlands to green cover. Farmers are helped with finding seedlings to plant and are paid a compensation (sometimes in the form of grain) for areas of farmland that are allowed to re-green.

There are other reasons why farmers may need to re-establish forest cover. Wooded areas also add to the attractiveness and living environment of a village, provide appreciated shade and wind shelter, improve soil and water holding conditions (reducing the need for watering) and encourage local pest controlling wildlife.

When deciding what trees to plant, avoid planting too much of the same species as monocultures are scenically less interesting, biologically less valuable, less attractive to wildlife, less efficient in providing environmental services and more prone to attack by pests, diseases and open fire.

Ways to reduce soil erosion on farms

- Construct terraces
- Construct physical or living (e.g. bamboo lines) bunds to reduce rate of water flow down hill
- Build check dams in erosion-prone gulleys
- Cover land scars with vegetation as quickly as possible
- Increase water penetration of soils by introducing trees (agro-forestry)
- Encourage vigorous green manure cover in fallow periods
- Plant wind breaks

Also avoid planting a stand of equal age trees. This results in a dense crown, which shades out the under-story which becomes bare, unsightly and difficult to maintain. Aim to establish a compatible species mix of local species with good age distribution so that the plot can be self-sustaining without further management inputs. This will involve planting suitable under-story shrubs, young saplings, as well as young trees.

In areas where broadleaf trees are deciduous for part of the year such as temperate woodlands or monsoon tropical forests, mix in a few local evergreen species such as conifers and hollies in the temperate zone or conifers in the tropical zone.

Wood plots can be given greater wildlife appeal by ensuring that they provide shelter in the form of hollow trunks and branches or nest boxes, attractive flowering trees that attract both nectar-feeding birds and insects (such as *Erythrina* and *Cassia* in tropical regions or Lilacs and Lime in temperate regions); and especially fruit species (figs in tropical regions and hawthorns and acorns in temperate regions).

Obtaining the plant resources to develop such wooded areas may require a bit more work. It may be necessary to develop a small nursery area for rearing your own trees and bushes based on seeds or seedlings taken from local natural habitat. Special permission may need to be obtained to collect such seed from public parks, woods or nature reserves.

4.3.3 Nursery establishment

Identify a place on a farm or in the corner of a village where the nursery could be established. The nursery should be partly shady to protect the establishing plants from direct heat of the sun. A small nursery is easier to maintain than a large one.

Identify plants that are native in the area. Choose from these plants which are to be propagated in the nursery. Some plants can be transplanted directly from source to a landscape bed.

Wild seedlings ready for out-planting

Propagate only the plants that are regularly needed for replacement on landscape beds. Otherwise, the nursery will be quickly full of overgrown plants.

When transferring seedlings or small trees from nursery to the open landscape, great care should be taken in preparing the ground. It is a tragedy if all the hard work of raising seedlings is lost due to taking shortcuts at this vital stage.

A few hints to increase survival rates when planting out.

- Prepare individual holes for each tree or seedling, add suitable soil mix and water well after planting
- Plant out at times of the year when plants face lowest stress from dessication (heat and wind) or cold. In temperate regions the winter period is a good time to plant when leaves have fallen
- Protect young trees from deer, pigs or other animals that may eat or disturb them
- Minimize root disturbance during transfer by suing planting bags or cutting root system out together with adhering soil (you can get special tools for this)
- Water vigorously for first weeks after planting till new roots get established, shape soil profile around plant base to catch and hold water like a cup

4.3.4 Steps in establishing wood plots

The figure on the page 237 shows that various tools and steps in the job of restoring tree cover.

4.4 Restoring and creating wetlands and ponds

The area of wetlands in China has been reduced over the past decades by drainage and impoundment and by over-use of ground water through use of boreholes for agricultural irrigation. Now we are suffering in the form of lowered water tables, increased erosion, loss of water holding capacity leading to increased floods and accelerating climate change. The government policy is now to halt and reverse the loss of wetlands and farmers are asked to contribute in this great task.

On some large lakes such as Dongting in Hunan, there are big programmes to return recently impounded farmland back to the lake. Here farmers allow their poldered fields to flood in summer and raise crabs, crayfish, fish and ducks as an alternate livelihood to planting crops.

But much can be done for many smaller lakes and wetlands all across China by simply blocking drainage ditches and allowing the dried wetlands to flood again. Damming the ditches with bags of sand is often sufficient to restore a small wetland and allow many of the valuable wetland biota and ecosystem functions to recover.

In addition it is often worth establishing new wetlands or ponds by lining holes with clay or a plastic liner and allowing them to fill up with surface water. By varying the depth and floor substrate (mud, gravel, rocks) you can create a wide range of habitat conditions that support a wider diversity of biodiversity. This can be done on an individual farm or done as a village pond by communal effort.

The advantage to farms is that the pond serves multiple functions as a reservoir for watering crops, drinking for domestic animals, rearing fish, improving local climate and supporting a range of insect eating species such as frogs, dragonflies and birds. If an island is left in the pond; this will greatly attract birds that feel safe from carnivores and use islands as roosts and breeding places.

The creation of small lakes in the form of check dams can reduce gulley erosion and form part of the local soil conservation effort.

4.4.1 Ensure healthy waterways

Clean, well-oxygenated water is more aesthetic and permits the co-existence of greater numbers of living organisms. Polluted, muddy or eutrophic water kills aquatic organisms, creates nasty smells, unsightly scum and a breeding ground for mosquitoes and dangerous bacteria. It also leaks out into wider ecosystems causing wider environmental damage and health risks outside the course area.

Choking of waterways by excessive water weeds leads to large management costs in clearance activities and disposal problems, reduced oxygen levels in the water and die off of oxygen hungry species such as some fish species.

The secret of waterway and wetland management must be on ensuring good natural or artificial filtration, avoiding pollution, and keeping good water flow for aeration. The more the number of organisms that live in the aquatic environment of the course the more stable the system will be and less prone to sudden die-off of different species, algal bloom, clogging by

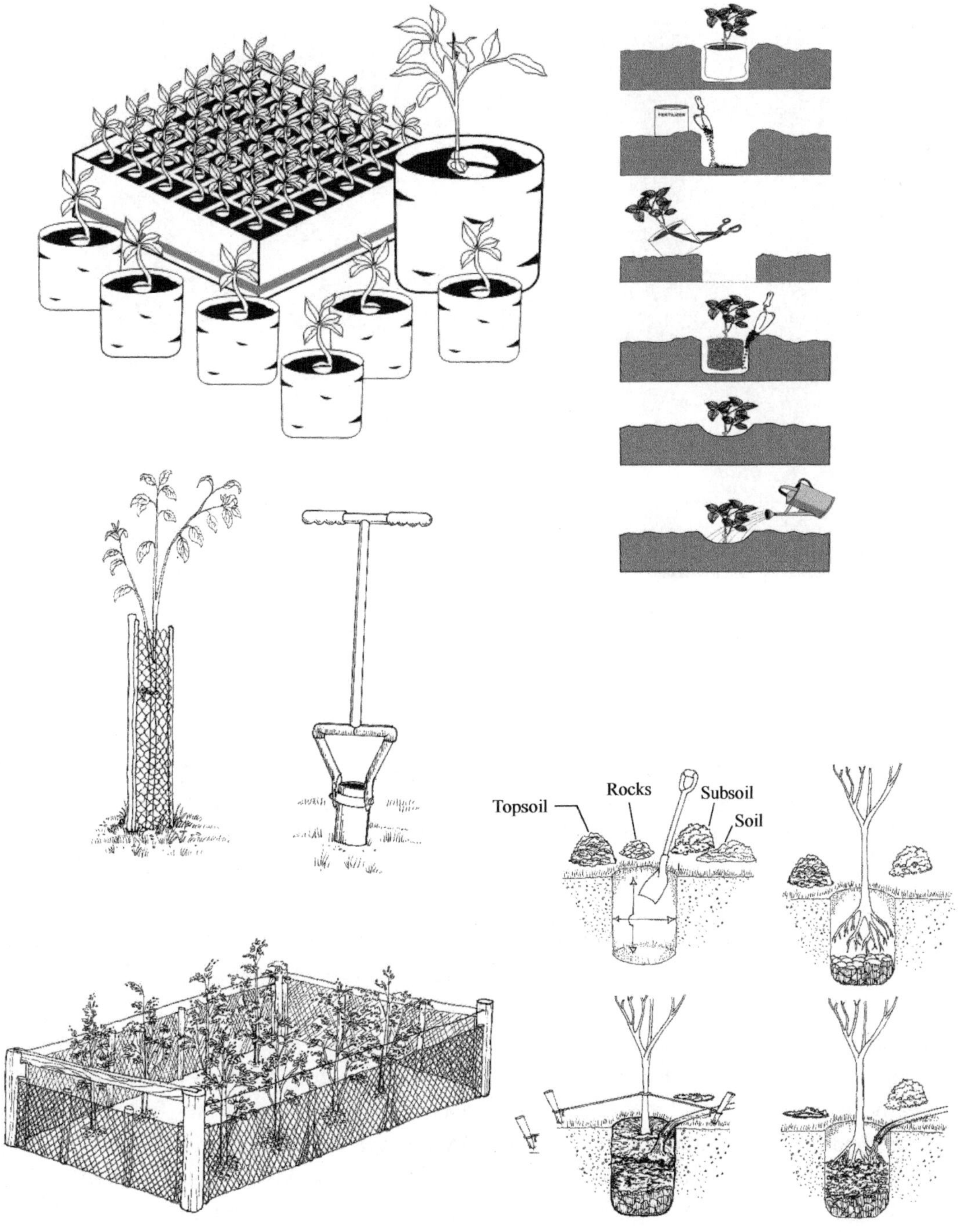

Various tools and steps in the job of restoring tree cover

Wetland outlet blocked by bags of sand by local villagers

water weeds, etc.

4.4.2 Creating a pond

Filtration of silt and pollutants can be achieved by passing water through gravel substrate streambeds and by encouraging filter feeding organisms such as bivalve mollusks. The latter require soft sand or mud substrate to live in. Mud substrate should only be used in areas where water turbulence is minimal and the mud does not get stirred up into the water. Mud can also be fixed by growth of appropriate water weeds.

Some water plants that at first might appear attractive can become serious weeds such as water hyacinth *Eichhornia crassipes* or water lettuce *Pistia stratiotes*. Try to eliminate these plants completely to avoid re-colonization.

It is often possible to leave some permanent or seasonally marshy areas in corners of a farming landscape. These can be very valuable miniature nature reserves in their own right. Globally, wetlands are under tremendous threats from development and few remain in a natural condition. Such sites can be colonized by many endangered marsh plants (including wild rice, water chestnuts and other species of high germplasm value) and wildlife—amphibians, turtles, dragonflies, and birds. Remember dragonflies are great controllers of insect pests on the farm.

Water birds and turtles may require safe perches to rest or hunt from. Turtles gather on emergent rocks or tree trunks on lakes. Egrets and cormorants congregate on tree stumps emerging from water or growing on small islets. These can become important night roosts or breeding colonies for many species. Other birds need dense covers of reed beds or scrub (rails, small herons and warblers).

Artificial pond with plastic liner can be rich for biodiversity and fish

In some countries, even the protection of crocodilians may be feasible in artificial wetlands and this is possible in East China for the Yangtze alligator.

4.5 Managing water levels

Wetlands management is largely a matter of controlling water levels to enable a desirable wetland habitat regime to continue. Where the natural hydrological regime of a wetland remains intact the best conservation action is to understand and safeguard the naturally functioning hydrology.

However, in many cases the original or natural hydrological regime has become man-modified as a result of water diversions, constructions, lowering of water tables, dam construction or other impacts and the manager of the wetland needs to use engineering or mechanics to raise or lower water levels as needed.

Recognising when more or less water is needed on a certain area of the wetland requires deep understanding of the local ecology and a clear plan of the desired outcome. Timings will depend on when migratory birds visit the site, which species the manager wishes to encourage and knowledge of the requirements of different species. Waders and shorebirds may require muddy and sandy banks and water depths of 5-10 mm, whilst ducks and geese may prefer water depths of 20-50 mm.

Timing of flooding or drainage will affect what types of plants thrive or fail, so again the manager needs to know what species his wildlife depend on and may have to provide suitable water for growing of those plants well in advance of the arrival of those species.

This type of knowledge and skill takes many years to acquire and is generally lacking in appointed Chinese PA managers. This is not knowledge that can be gleaned on a short training course. Which is why the supervisors of protected areas in developed countries are usually ecologists rather than administrators and they are driven by deep love and appreciation

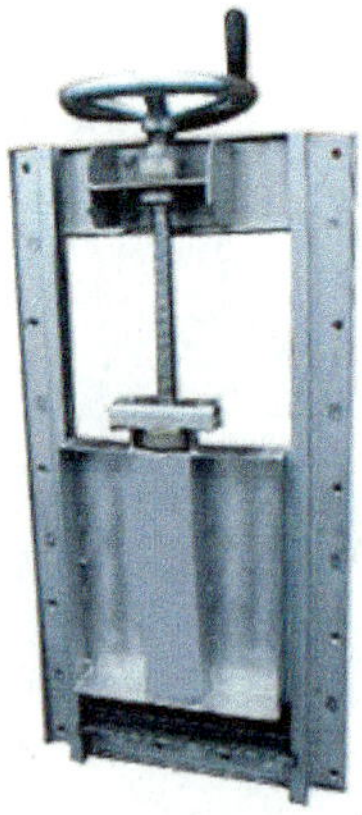
Simple water gate for control of smaller streams

for their wildlife as a vocation, not simply treat their appointments as a job.

We can divide water manipulations into two types—1) ways to lower water level by speeding drainage of an area or 2) ways to raise water levels by impeding or blocking drainage channels.

More rarely, the manager needs to make water go uphill by mechanical devices such as water wheels, windmills or pumps.

Most water control needs can be managed by a system of small simple sluice gates as figured. Larger lakes may be controlled by large gates operated by the Water Resources Department. Where drainage needs to be permanently blocked, outlets can be sealed by bags of sand. A great range of structures are available to wetland managers and UK's Royal Society for Protection of Birds (RSPB) have published technical guidelines for their application (https://www.rspb.org.uk/Images/Water_management_structures_tcm9-214636.pdf).

4.6 Fire prevention reporting and fighting

Fire can destroy an ecosystem very fast and almost all wild fires are caused by accidental or deliberate human activity. Lightning is usually accompanied by rain which makes it unlikely that a wild fire will occur, but there are occasions where the rain is formed high in the sky and evaporates before it reaches the ground. Under such storm conditions there may be 'dry' lighting and this can trigger a burst of wild fires.

Hot, dry and windy weather are the most dangerous time for wild fires, but almost all regions of China are susceptible to fire and sometimes these may get out of control of fire fighting services and burn great areas of forest, brush, farm land and even destroy houses and cause loss of life.

It becomes the duty and responsibility of all rural citizens to be careful not to cause fire, to be vigilant in spotting wild fires early and in assisting professional services in tackling fires should they occur.

Local media will give warning if dry conditions develop that are very dangerous for wild fires to occur. At such times extra vigilance is needed. Especially in China where so many men are smokers and often careless in the way they toss away their cigarette butts.

Most forested regions of China have already established networks of fire breaks, watch towers and procedures to follow in the event of fire.

If you live in an area where wild fires are frequent, you should make sure you have suitable clothes and a basic tool for fighting fire. Good boots are important to prevent damage to the feet when walking over burned ground. Have clothing of a bright colour so you are easily spotted if you are in trouble or need to be rescued. Clothes should cover as much of your skin as possible but not to too heavy to limit your movements. Face guard, hat and gloves are invaluable. Keep an axe or bush knife handy and have a broom or flat bladed fire tool for swatting out flames. A shovel is useful for clearing trenches, removing dry ground cover and also can be used to swat out flames.

The following basic guidelines in fighting wild fires are useful:

A bushfire is unpredictable and dangerous because it is out of control. Fighting bushfires is about bringing them under control and putting the fires out. This has to be done in two steps. The first is to stop the fire from spreading by **containing** it to some boundaries and the second step involves putting the fire out completely by **mopping up** all of the hot spots and continuously **patrolling the edges** to ensure that nothing is missed.

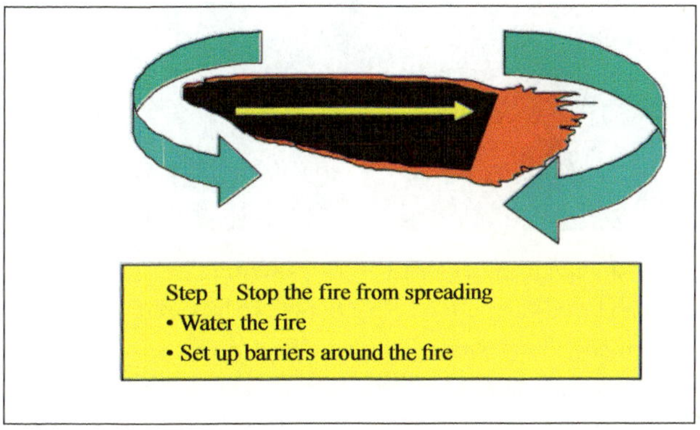

If it is safe, a direct attack can be made at the front or **head** of the fire. This is where the rate of spread is fastest but also where the heat, flames and smoke are worst. If it is unsafe to attack the head fire then an alternative is to work along the sides, called **flanks**, starting at the back and working towards the head fire.

A flank approach will generally be a safer tactic as the firefighters can always retreat to the burnt out ground. Alternatives to the direct attack.

When flames and smoke make a direct attack too dangerous, the crew leader may decide to step back from the fire edge for a distance of up to 20 metres. A line is then built **parallel** to the edge of the fire, **but always in sight of the flames**. The fuel between the line and the fire edge must be burnt out as the line is built.

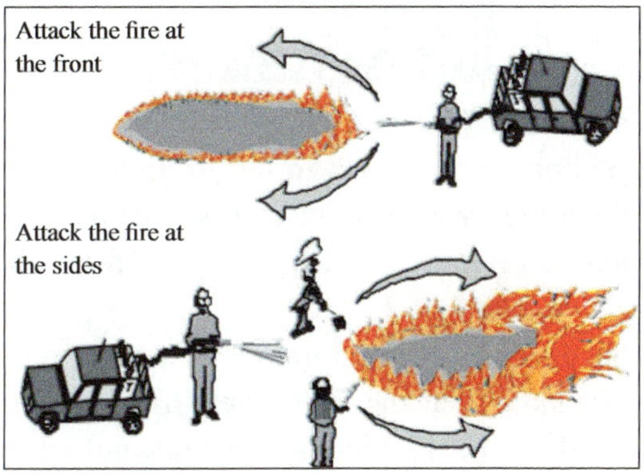

Safety issues when making a parallel attack on a bushfire.

If the space between the fire's edge and the line being built gets too wide, it's possible for the fire to flare up and run over the firefighters. A number of serious accidents have occurred because of this. This is why the remaining strip of fuel must be burnt out as the line advances

Fighting fire with fire— "backburning"
When bushfires are too active to put out by direct methods, "backburning" is sometimes possible. This can be a dangerous tactic and must always be supervised by experienced crew leaders

and the fire edge must be kept in sight at all times.

Safety issues when using the indirect method of firefighting.

"Backburning" is an indirect method of attack. A good fire boundary is chosen, which may be a long way ahead of the bushfire. In suitable conditions, fires are lit along the boundary and allowed to burn back towards the approaching bushfire. If conditions are not right, the new fires will just escape, causing a much worse situation and putting firefighters in danger. Sector commanders will give clear instructions if an indirect attack is to be made and these must be followed.

Firefighting usually involves a combination of all the methods described. The choice of tactics depends on the weather, fuels, slopes, fire behaviour, safety and the resources available.

Step Two involves careful patrol of every metre of the fire's edge, putting out all of the burning and smouldering fires. This is the hardest, dirtiest job in firefighting. It is also the most important because it only takes a single spark for the fire to escape and the whole job has to start again.

Methods for containing bushfires safely

The way that a bushfire is approached depends upon the fire's behaviour. Radiant heat is an important safety consideration along with the speed at which the fire is moving, the type of fuels it is burning and the general lie of the land ahead of the fire. If a fire is small, moving slowly and there is no apparent risk to the firefighters caused by either the weather or the slopes, then the fire can be attacked right on the edge of the flames. This is called a **direct attack**.

This method gets right onto the flames and keeps the fire small, but the firefighters have to work in heat and smoke. A direct attack must not be made if the firefighters cannot get close to the edge or conditions are unsafe, because of steep slopes or dense fuel.

Part Five Species Management

5.1 Recognising useful species

Many of the species we take for granted are highly beneficial to the ecosystem and wetland environment. Yet without attention and some protective measures we can easily lose these species and their benefits they provide freely to us. Useful species can be classed into a few common categories:

- Pest control species—raptors, insect feeders, snakes, amphibia, fish
- Pollinating species—bees, butterflies, moths, beetles, bats
- Protective and wind shelter species—tree belts, woodland plots, adjacent forest
- Soil forming organisms—earthworms, decomposers, fungi etc.
- Wild foods—ferns, mushrooms, fruits, edible leaves

Some useful species in wetlands—pollinators and insect feeders

- Medicinal species
- Plants with special uses—fibre, poles, bamboo, dyes, fish poison, leaves used for wrapping, cooking, thatch etc.
- Wood plot species—source of fuel wood, mushrooms, medicinal plants etc.
- Soil enriching species—nitrogen-fixing legumes etc.
- Wild relatives of domestic species—wheat, rice, fruits, ducks, jungle fowl

5.1.1 Water purification service

One of the most valuable ecological functions of wetlands is their role in water purification. The purifying agents are many bacteria and other microorganisms in the mud but also much cleansing is done by filter-feeding organisms such as hydra polyps, mollusks, worms and other creatures. Other mollusks and some fish also eat much of the algae that would otherwise smother and stagnate wetlands.

5.1.2 Pollinating agents

Sunbirds, spider hunters, bats, moths, butterflies, flies, beetles, and especially bees all serve as vital pollinators of most wetland herbaceous flowers. The diversity of plants adds to the resilience and ecological efficiency of wetlands and so preserving and maintaining adequate pollinators is an important management function.

Regular monitoring should establish that populations of pollinator species are maintained adequately but if species start to decline and disappear, it may be necessary to identify causes and reverse such trends.

In fact the widespread use of insecticides in agriculture is suspected of being the main reason for worldwide decreases in many of the these species.

Despite a great system of nature reserves in Europe, good habitat and healthy food plants for most pollinator species there has been reported worrying declines of bees, moths, butterflies and many other species. Destruction of caves and improved standards of roof construction on houses has contributed to big loss of bats.

It seems China is no more immune to these changes and bees in some regions are too scarce to fully pollinate agricultural vegetable and fruit crops and must be becoming similarly scarce in wild wetland meadows.

Methods to encourage pollinators include:

- Allowing bee hives in and around wetland sites
- Planting especially favoured nectar flowing bushes to boost wild populations
- Reaching agreements with local farmers to reduce dependence on insecticides
- Planting additional food-plants for specific butterflies and moths (see section 5.8.1)
- Leaving hollow trees, caves etc. for bats to roost in putting up bat boxes to encourage local populations

5.1.3 Seed dispers

All vegetation types need to use a variety of different seed dispersal mechanisms to distribute their seeds, colonize new available habitat and ensure ecosystem richness. Some plants including several mangrove species drop their seeds in the sea and the tidal action will move them around the intertidal zone. Some trees such as willows (*Salix*, *Populus*) use the wind to disperse their seeds. Many other plants have hooked seeds that attach to the fur of passing animals or human trousers, and thus get transported by unwitting agents. Trees that produce edible nuts like oaks (*Quercus*, *Lithocarpus*) or chestnuts (*Castanea*), take advantage of squirrels or jays to take their fruit away, bury, store or drop some fruits and thus help distribute the species. But trees and bushes with sweet fruits need to attract the attention of fruit eating birds or mammals to either transport the fruits to feeding perches or to eat the fruits and pass the seeds out in their faeces.

This mutual cooperation between fruit trees and fruit eating animals is part of nature's way of natural reforestation and gap filling but also greatly adds to the diversity and interest of wetlands in bringing in a larger array of plant species and fruit eating animals.

Fig trees (*Ficus*), haws (*Crataegus*), *Sorbus*, persimmons, wild plums, *Lantana* and many colonizer shrubs—*Rubus*, *Sambuccus* etc. are maintained in wetlands in this way but would quickly disappear if the fruit eating doves, starlings, barbets, mynahs, flocks of little white-eyes (*Zosterops*) and various squirrels.

One interesting group of birds is the flowerpeckers of gens *Dicaeum*. These colourful little birds feed on the sticky fruits of parasitic mistletoes (*Loranthus*). Some seeds get stuck on their beaks and it is the habits of whipping extra sticky seeds from bill to tree branch that transfers the seeds to new rooting places. The flowers of the mistletoes also attract the beautiful little sunbirds that brighten anyone's mood who cares to watch them flitting actively about.

Since they are mostly active at night, bats role in maintaining floral diversity is often overlooked but bats are important both in pollinating many of the feathery flowers of mangroves, *Barringtonia* and other trees. Larger bats are fruit eaters. They pluck large fruits, carry them off to a convenient perch and then slowly eat them. Mangos, *Terminalia*, some large figs, Papayas are almost entirely distributed by bats. Plants that depend heavily on bats tend to have pendulous hanging fruits that can be more easily accessed by the upside down feeding animals!

Some fruit trees that overhang waterways such as the wild figs depend on catfish to eat the falling fruit and distribute seeds back upstream to new bank sites.

5.2 Control of herbivores

Density and type of herbivores present determines height and density of herbs and grasses. Try to maintain optimal natural levels. Where too many herbivores are degrading grassland, the manager needs to find ways to reduce grazing pressure. Often where natural herbivores have been hunted to low density, grasslands become invaded by bushes and trees and the manager may wish to increase herbivore levels by allowing grazing by some domestic animals.

5.3 Rescue and rehabilitation

5.3.1 Usually no need of rescue centres

Many Chinese wetland nature reserves have invested in special facilities to hold and care for rescued or injured birds or other animals. Such facilities are rarely worthwhile. They cost a lot of money to construct, equip, maintain and staff. The need for rescue is tiny compared to the numbers of wild healthy birds that need better protection.

Given that several million birds a year are killed in mist nets in China and that most wetland nature reserve protect many thousands of individual birds, it is evident that the need to have a special centre for rescuing a few individuals or tens of individuals of injured birds or mammals is clearly a very low priority.

Most PAs do not need a rescue centre and would serve conservation much better by concentrating its resources on better protection of the wild populations in situ. Only in the case of extremely rare and precious species is there justification for undertaking the expense of buildings, staff, veterinary care, feeding and animal care that would be needed to maintain such a centre. The centre also drains attention and staff from the primary conservation role.

5.3.2 Rescue and treatment of oiled birds

Whenever there is a large oil spill from a tanker or facility the floating slick poses a huge hazard to water birds and those that get covered in oil face a painful death unless rescued and treated quickly. Cormorants, grebes, shearwaters and waterfowl are most vulnerable. In desperately trying to clean the oil from their own feathers these birds eat oil which is very toxic.

Most birds contaminated by oil die. Even those that do not die quickly may suffer long term health problems, lower life expectancy and lowered reproductive success. Contaminants can also be passed on to eggs and offspring and to predators that may eat the contaminated bird.

The numbers of oiled birds that get washed up on beaches represents only a small proportion of those that die at sea. Some species such as shearwaters, cormorants, grebes and diving ducks have very low survival even with best rescue and treatment techniques. But some birds such as swans, geese and other waterfowl may respond well and can be released after a few days. The National Wildlife Research Center (NWRC) of USDA have published a paper on how to treat such birds: http://www.nwrc.usgs.gov/wdb/pub/wmh/13_2_8.pdf.

Successful oiled bird rehabilitation involves six basic procedures:

- prompt intervention and retrieval of contaminated birds
- stabilizing the bird
- removing oil from the bird's feathers
- removing the cleaning agent from the feathers
- restoring waterproofing; and
- acclimating the bird for release

It is often more important to prevent more wild birds getting into the oil spill than saving those that have. This involves various measures such as scaring birds away from dangerous spills by loud explosives or other devices. Meanwhile efforts need to be made to clean up the spills themselves. Oil spills can result in decades of slow recovery for wetlands.

5.4 Captive breeding and reintroduction

China continues to lose some important and valuable wetland species. The Yangtze dolphin was recently declared functionally extinct as was the giant soft-shell turtle *Rafetus swinhoei*. Probably extinct in the wild is the Yangtze paddle fish and critically endangered candidates include finless porpoise, several sturgeons and many more.

Whilst the bulk of wetland species and all wetland ecosystem service functions are best conserved by *in-situ* protection and restoring habitats to good health, there are some species that have become extinct or almost extinct in the wild that may only be saved through programmes of captive breeding and reintroduction. But there are also examples of over-investment and over-reliance on maintaining captive populations.

Some species of fish now have no way to get from their adult feeding areas and their traditional upstream breeding rivers leading to local extinction. Skills in captive breeding and reintroduction need to be maintained and improved.

Captive breeding and reintroduction—the last resort—three Chinese examples

Once widespread in China, Japan and Korea, this attractive bird became extinct over most of its range. The last small population of 7 birds remaining in Shaanxi province was found in 1981. In 1986 6 fledglings were taken from the wild to form the basis of a captive breeding colony in Beijing Zoo and a successful breeding and reintroduction programme has continued ever since. Some eggs are hatched naturally but more by using incubators. This has been one of the most successful projects at captive breeding and reintroduction. There are now several hundred of these birds kept in various captive breeding stations and a few hundred more released into the wild including reintroduction in Japan. The total population new exceeds 500.

Crested Ibis—*Nipponia nippon*

Formerly widespread in the lakes and waterways of the lower Yangtze but virtually exterminated by conversion of wetlands to rice-fields, pollution, eating of poisoned rats and other human factors despite their small size and the fact that they never attack humans. A captive breeding reserve was established in Anhui and Zhejiang since 1979 and has been very successful. A total of over 10,000 captive alligators are now held in the two facilities. These should form the basis of a good reintroduction programme into

Chinese Alligator—*Alligator sinensis*

Père David deer—*Elaphurus davidianus* or 'milu' reintroduction

the many protected wetland lakes within their range. But conservation interests have become confused due to introduction of commercial interests in the farming business.

Formerly distributed over much of E China, the species was already hunted close to extinction in the 19th century with a captive herd held at the imperial hunting park of Nan Yuan at Nanhaizi near Beijing. By early 20th century the species was extinct in China but a few animals taken to Europe by Pere Armand David continued to thrive in captive collections. In 1985, 22 animals were brought back to China, among which 20 animals were released in Nanhaizi, and a further 18 animals were also released in Dafeng marshes of Zhejiang in 1987. Both these colonies have thrived under close protection and good veterinary care such that there are now several hundred of these splendid large deer in China. Herds have been released into new nature reserves in Hubei and in the Yangtze River basin. Also incidental escapes have resulted in establishment of some free-ranging wild populations. Total population in China is now estimated to be more than 2000.

5.4.1 Rules for resorting to captive breeding

1. Captive breeding should be undertaken as a last resort only when attempts to save a species in the wild seem likely to fail and the wild population has dwindled to the last few individuals.

2. Captive breeding programmes of critically endangered species should be entirely conservation driven and not undertaken to show off rare animals or for commercial purposes.

3. The decision to remove some of the last wild individuals to form a breeding group should not be taken unilaterally by managers or even national authorities but on the advice and agreement of the appropriate international Species Survival group under IUCN and preferably in line with an approved species survival plan.

4. Such programmes should be undertaken in secure, healthy facilities under supervision of vets and biologists.

5. Stud books should be maintained of all individuals animals in the programme recording their history, origin and parental background and all animals should to tagged for individual identification.

6. Great care should be taken in selecting parents to ensure that subspecies are not mixed and to ensure that genetic diversity is maintained by allowing several adult males to contribute to the next generation.

7. Care should be taken in selecting release sites to ensure that these areas are suitable, within the correct geographical range and safe from immediate hunting. Released animals may need some half way training or acclimatisation prior to full release. Released animals

should be monitored closely to evaluate success and learn from experience. This may involve radio tagging some individuals.

5.5 Reintroductions

With wetlands being very dynamic they face fast turnover of resident species. They lose species when they change nature, extreme weather events or water diversions. In nature species could quickly relocate to suitable habitats or recolonize new or changed habitats. This is no longer reliable since most wetlands in China are fragmented and enjoy poor connectivity to the overall water system. This places a management responsibility for artificially assisting sites to retain or gain their most diverse and suitable biota through a policy of monitoring, control and where necessary reintroduction.

Reintroduction for fish, plants, some insects is relatively easy but may be very difficult with some delicate species of birds and mammals.

The Species Survival Commission of IUCN has a specialist group devoted to the subject of reintroductions and periodically publishes new state of the art guidelines on methods and safe procedures (http://www.issg.org/pdf/publications/RSG_ISSG-Reintroduction-Guidelines-2013.pdf).

Reintroduction is a greatly misused excuse for seeking budgets for more buildings, staff and facilities. It is usually quite unnecessary and not the job of wetland protected areas. Where reintroduction is justifies it should be undertaken by a specialised dedicated centre and not a tag on activity of a nature reserve or wetland park.

Two forms of reintroduction are relevant. Translocation involves capture of individuals from one site and release in another. This operation may be needed to restore connectivity between populations that have become artificially separated or to restock a local population that has done extinct.

Wetlands that were formerly all connected are now separated by dams, weirs, or stretches of now polluted water unsuited to many species.

In the case of translocation, it is important to keep the birds or mammals for as short a time as possible. They need only a minimal check that they are healthy and able to look after themselves in their new home. Reduce stress by keeping animals in boxes or even loose cloth bags so that they do not damage themselves struggling and they are unable to see frightening humans, dogs, cars etc.

The second main form of reintroduction is when animals have been either bred in captivity or by using injured rescued wildlife that has been kept in captivity to recover fitness before being released. Unless part of a deliberate operation to reintroduce the species to a new location, rescued birds should more normally be released as near as possible to where they were found.

5.5.1 Re-stocking

Recreational and commercial values from many wetlands can be increased by artificial fish stocking. The numbers of commercial or favoured sport fish can be boosted by release of artificially reared fry from a fish breeding facility. Naturally occuring species that have become scarce or locally extinct can be reintroduced in this way. This may be essential if the natural passageway of adult fish to reach upper stream breeding sites has been artificially blocked by downstream dams or other developments.

Care must be taken not to introduce species that may be harmful to the indigenous wetland biota. It is usually possible to select suitable species for breeding and stocking from the list of locally occuring native species.

5.6 Managing wild animals

5.6.1 Managing top predators

Trophic cascades can occur when predators in a food chain suppress the abundance of their prey, thereby releasing the next lower trophic level from predation (or herbivory if the intermediate trophic level is an herbivore). For example, if the abundance of large piscivorous fish is increased in a lake, the abundance of their prey, zooplankton-eating fish, should decrease, large zooplankton abundance should increase, and phytoplankton biomass should decrease. This theory has stimulated new research in many areas of ecology. Trophic cascades may also be important for understanding the effects of removing top predators from food webs, as humans have done in many places through hunting and fishing activities.

In lakes, large predatory (piscivorous) fish can dramatically reduce populations of smaller zooplanktivorous fish, zooplanktivorous fish can dramatically alter freshwater zooplankton communities, and zooplankton grazing can in turn have large impacts on phytoplankton communities. Removal of piscivorous fish can change lake water from clear to green by allowing phytoplankton to flourish. Similarly removal of large and medium predatory fish can result in overabundance of algae and midges.

5.6.2 Reducing alien or undesirable predatory fish

With almost all waterways highly disturbed, their nature much changed and connections broken, and compounded by many alien exotic species established in different parts of the country the fish fauna in many lakes and other wetlands may be seriously out of natural balance. Man may need to take control of the management and depending on what the manager wants to achieve. If you want more game fish such as salmonids, it might be necessary to remove some of predatory fish such as pike that feed on the young ones. One way that this can be done is by use of electric fishing. The operator runs and small generator carried in backpack or on a boat and puts both positive and negative terminals into the water about half a metre apart. Fish are attracted to the negative terminal which can be shaped as a wire mesh net for scooping the fish out of the water. The operator can then select which fish to remove and which to put back.

5.7 Encouraging birds

Birds can be encouraged by providing suitable habitat, planting trees that provide fruit, flowers, perches, shelter, nest sites etc. Birds respond to protection. Simply keeping and area free of hunting, netting or loud disturbance will attract birds to use the area more. In wetlands with lakes, birds recognise the added security of islands and will use small islands as resting areas and nesting areas.

5.7.1 Use of nest boxes

Nest box preferences by different species

Small boxes with holes	Large boxes with holes	Open fronted boxes	Very big boxes
Tits	Swifts	Robin	Kestrel
Sparrows	Starlings	Pied Wagtail	Tawny Owl
Nuthatch	Woodpeckers	Spotted Flycatcher	Stock Dove
Redstart	Little Owl	Mandarin Duck	Jackdaw

Construction

Before you start building your nest box study the below nest box illustrations, dimensions and advice on the materials to use and the hole sizes. Make sure that you put your box up as soon as it has been built—the longer it's up, the more chance it has of attracting visitors.

Materials

It is important that the inside of the box doesn't get too cold or warm and that the box is durable.

- Nest boxes should be **made from wood**. Metal and plastic are unsuitable materials as they may cause the contents of the nest to overheat or allow condensation to build up inside the box, wetting eggs and chicks.
- The type of wood used is not critical but **hardwoods**, such as oak and beech, will outlive soft wood, such as pine.
- Rather more critical is the **thickness of the wood**, which should be at least 15mm to provide sufficient insulation and to prevent warping.
- **Nail boxes together** rather than gluing (use galvanized/stainless steel nails to stop rust) as this allows water to drain.
- **Drill a couple of holes in the base** of each box to ensure that any rain that does get in can drain out quickly.
- **Do not fir a perch on the front** of the nest box as this is not necessary and may aid access to a predator!

Hole sizes

Different size holes are suitable for different species:

- 25mm or larger for Blue, Coal and Marsh Tit;
- 28mm or larger for Great Tit and Tree Sparrow;
- 32mm for House Sparrow.

Access for inspection and cleaning

Nest boxes should have a means of easy access for both inspection and cleaning.

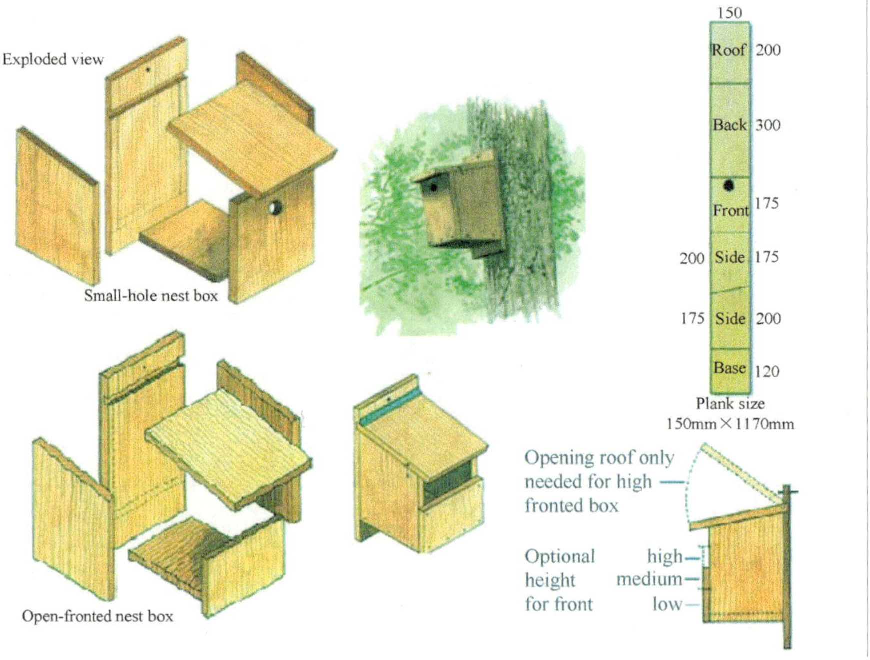

Examples of the structure of two types of nest box

Attach a waterproof hinge to the roof of the box, so that it can be lifted easily but won't fall off. Car inner tubes or Butyl rubber are ideal waterproof materials to use. Cut the rubber to the width of the box, and then nail the rubber along the back of the box and to the roof.

Some very large birds can be encouraged by providing platforms on which to nest. Chinese examples include the platforms for rodent-eating raptors across the High plateau grasslands and the use of wooden platforms for breeding of White storks in NE China.

5.7.2 Bird feeders

Providing birds and some mammals additional food in the lean seasons or water during the dry seasons may help them get through the bad times but will certainly make them more visible to visitors or researchers for observation. For that purpose it is often a nice idea to place feeders near to visitor centres, cafeteria etc.

Different species need different types of food and a wide variety of feeder designs can be employed, constructed, viewed over the internet or ordered from specialist suppliers.

It is generally best to present food and water well off the ground where visiting birds would be very vulnerable to predation by cats or natural predators.

If squirrels become a nuisance raiding feeders designed for birds, the feeders may need placing inside squirrel proof cages.

The following diagrams indicate some standard bird feeders.

Bird feeders and water bath

5.8 Encouraging insects

5.8.1 Encouraging butterflies

Butterflies are beneficial on farmlands and wetlands. They are good pollinators of fruits and beans and add to the beauty and peaceful atmosphere of a healthy countryside. Their continued well-being is a sign that all is well with the environment. Their absence is a sign that too much insecticide is being used on the land. Butterflies lay eggs on many wild plants in wetlands and open country but they can be encouraged by ensuring that suitable food plants are available for rare or specially attractive species. Adults can also be attracted to feed

Mating birdwing butterflies

at particularly attractive flowers. Rearing butterflies is a matter of growing the correct food plants for the species to be encouraged. Most species are rather specific on what they lay their eggs on, but such information is generally well known. A lot of details can be downloaded from the web following a search for butterfly food plants.

The table below lists major genera of food plants and the main butterflies that eat them.

Tropical areas:

Citrus—several swallowtail butterflies

Aristolochia vines—many birdwing butterflies *Troides* and *Atrophaneura* spp.

Cassia—*Passiflora*

Palms—many hesperids

Oleander family—several hawk moths

Temperate areas:

Urtica—many nymphalids

Thistles—some nymphalids

Wild Carrots—Common Swallowtail

Salix—*Apatura* spp.

Vetches—many lycaenids

Prunus—many species

Privet, lime bedstraw—hawk moths

Attracting adult butterflies to flowers and other baits

In addition to attracting and encouraging butterflies by planting their larval food plants, adult butterflies can be attracted to flowers and other baits. Certain flowers are more attractive than others to attract showy butterflies to ornamental areas and areas where people spend most time.

Attending vines planed for rearing of birdwing butterflies

Examples of very attractive plants are *Buddleia davidi* for temperate regions and Hibiscus and Lantana for tropical regions. Lantana must be well controlled as it can become an invasive alien weed (see section 5.9).

Some spectacular butterflies can be attracted by other types of bait. Purple emperors *Apatura* and *Charaxes* butterflies are attracted to animal (civet) dung or decaying carrion. Other *Charaxes*, nymphalids and swallowtails are attracted to animal urine or decaying fruits (bananas). Many tropical butterflies can be attracted to puddles or ground saturated with mineral rich water. Carpets of pierids, swordtail and *Graphium* swallowtails and even the spectacular birdswing butterflies *Troides* regularly visit such sites.

5.8.2 Encouraging dragonflies

Dragonflies lay eggs on vegetation overhanging water and most of their life cycle is spent as an aquatic carnivore. The final nymph stage climbs out of the water to allow the emergence of the adult to fly around over wetlands feeding on small insects that it catches in mid air. The entire life cycle depends on the health of water and abundance of suitable prey. Dragonflies are a good indicator of wetland health and diversity is ensured by retaining a variety of wetland habitats—fast water, slow water, deep water, shallow water and different types of water vegetation.

Butterfly attracted to Buddleia flowers

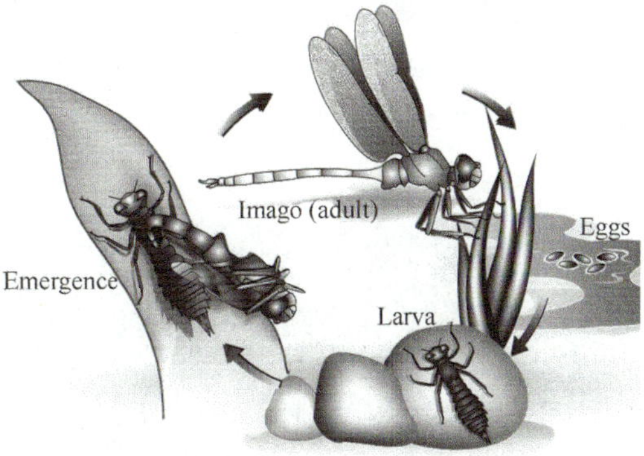

Life cycle of dragonfly

5.9 Control of AIS

Actions local manager can take to help:

- Watch out for the arrival and spread of new species
- Report potentially damaging species to local
- Do not experiment with introduction of exotic seeds, crops or ornamentals unless these have been okayed by local authorities
- Cooperate with other land managers in early control measures if there is a
- Seek technical assistance from local institutes or universities to tackle serious problem species

Recognising the biggest AIS threats in Chinese wetlands

The following responses are needed:

- National and provincial laws and regulations controlling the introduction and release of alien species
- National database of distribution and experience with such species and how to recognize them
- Improved awareness about the problem and the AIS species already recognized in each region
- Site level monitoring of problem AIS
- Control or eradication programmes for serious problem species

Golden apple snail invades rice fields. Watch out for pink egg masses. Can be eaten.

African giant snail invades southern orchards and fields. Eats many plants.

Black-eared toad—a good pest control animal but also destroys local endangered amphibians! It is poisonous. Indigenous only in southernmost parts of China, alien elsewhere.

Lantana camara. Attractive to butterflies and forms a living fence but inedible to animals and invades pastures. Very hard to remove.

Louisiana crayfish introduced from America. Good to eat but displaces native species and burrows into dykes, weakening flood defenses and irrigation systems.

Tilapia introduced from Africa. Good to eat but replacing native fish species.

Water hyacinth. Attractive flowers but chokes up waterways, blocks boat ways and reduces oxygen levels so that fish die. It is very hard work to remove.

Water lettuce. Can spread very fast blocking waterways and reducing oxygen levels at expense of fish and crustaceans.

Eupatorium. Several species now invade fields and open spaces in southern 2/3rds of China. The tiny wind-blown seeds spread aggressively. Some farmers are happy to plough this species back into the ground as a green manure.

Spartina cordgrass has invaded China from USA and now spreads fiercely along the coast outcompeting local beach vegetation and mangroves. Very difficult and expensive to control.

Warning: Biological controls are less destructive to non-target species that any use of toxins, but biological controls can involve introduction of further AIS!

A valuable resource is the Global Invasive Species Database of the Global Invasive Species Programme (GISP). This database needs translation to Chinese (http://www.issg.org/database/species/ecology.asp?si=1890&fr=1&sts=&lang=EN).

A search on country China and habitat wetlands generates a lists of major recognized alien species with individual profiles and details of biology, threats and methods of control.

5.10 Encouraging natural pest control

The International Rice Research Institute (IRRI) actively focuses attention on maintaining biodiversity in the rice landscape, especially on restoring the diversity of predators and parasitoids to enhance ecosystem services related to pest invasions and regulation.

Rice is a short term crop and thus the herbivores that do well in these habitats are usually 'r' strategists or species that have short life cycles, small size, move a lot and lay a lot of eggs. Thus invading females when they come into a new rice field will lay a lot of eggs, and without the predator and parasitoid biodiversity, the survival of the eggs may be 90% to 100% and eventually leads to great damages. But if the rice field has rich biodiversity, most of the eggs of the invading pests don't survive, usually less than 5%; and thus damage is low and crops compensate and no loss is expected. The factors that destroy the biodiversity are pesticides, extreme mono cropping and low genetic variations.

IRRI rice plant-hopper project restores local biodiversity by encouraging farmers to increase floral diversity to bring about increase in resources for the predators, such as nectar and shelter.

A pair of barn owls with five chicks will eat at least 3,000 rodents in one breeding season. They're not territorial birds, and it is possible to have many nest-boxes in action, with dozens or scores of owls working in one area—a whole colony. In one study 48 nest-boxes accounted for at least 17,000 rodents eaten in eight weeks. Add the thousands more rodents that would have been born in those weeks without the owl control, an entire plague can be avoided.

Some practical ways to help maintain natural pest control:

- Leave natural cover and erect nest boxes for key raptor species
- Protect bat roosts in caves and house roofs
- Welcome swallows and swifts to nest in barns and house roofs
- Look on snakes as agricultural friends not enemies (most are not poisonous and none attack unless provoked)
- Protect wetland habitat for dragonflies, frogs etc.
- Place convenient perches for shrikes, raptors, starlings, drongos, etc.
- Limit use of insecticides on the farmlands

5.11 Aquaculture issues

Biodiversity in semi-wild aquaculture systems is typically higher than in dedicated fish ponds, far less susceptible to disease and as it requires less additional food, creates less eutrophication and pollution. Disadvantages are the great risk of introducing alien species, hybridization with wild species or transferring new diseases to wild biota.

Freshwater cultured species

About 50 commercially important species are cultured in China. The most commonly farmed species are carps, Chinese bream and blunt-snout bream. Since the 1980's, with increasing domestic and international market demand, various species have been developed or introduced from abroad for commercial cultivation in China such as Japanese eel, mandarin fish, sturgeon, soft-shelled turtle, Chinese river crab, loach, snakehead, crawfish, giant river prawn, tilapia, rainbow trout, paddlefish, catfish, frog and european eel. In 2003 China produced a total of 17 782 734 tonnes of freshwater aquaculture products. The approximate share of cultured species of the total freshwater aquaculture production was as follows:

Species	Percentage
Silver carp and bighead carp	30.10%
Grass carp	20.20%
Common carp	13.20%
Crucian carp	10.00%
Black carp	1.30%
Chinese bream and blunt-snout bream	3.30%
Tilapia	4.20%
Chinese river crab, giant prawn and soft-shelled turtle	3.40%
Eel	1.0%
Others	13.30%

Marine and brackish waters culture species

About 40 commercially important species are farmed. Traditional marine culture is largely limited to four groups of molluscs: oysters, clams, blood cockles and Manila clams. Scallop and abalone were developed in the 1980s. Seaweed production was developed in the 1950's. The shrimp culture industry was the major cash industry in the 1980s. *Penaeus chinensis, Penaeus japonicus, Penaeus monodon, Penaeus vannamei, Penaeus merguiensis, Penaeus penicillatus, Metapenaeus ensis* are the major species farmed in China, of which *Penaeus vannamei* is now becoming the dominant species in terms of production. Marine fish began massive production in 1990s. Sea bream, milkfish, sea perch, Japanese flounder, mullet, yellow croaker, grouper and puffer fish are the major fish species cultured. Species introduced from abroad such as seabass, large mouth bass, turbot, redfish have also been successfully farmed. Of the total marine culture production (12 533 061 tonnes in 2003), the proportion of species groups is as follows:

Species	Percentage
Marine fishes	4.10%
Crustaceans	5.30%
Molluscs	78.60%
Seaweeds	11%

The graph below shows total aquaculture production in China according to FAO statistics.

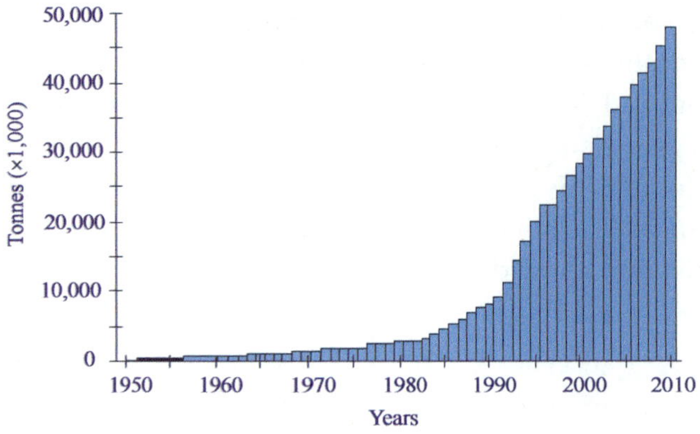

Part Six Mainstreaming and Communication

6.1 Need for cross-sector mainstreaming

Although the State Council has assigned SFA to be lead agency responsible for wetlands conservation in China, SFA cannot undertake this massive task alone. Wetlands are used by or deliver services to so many other sectors that it is imperative that SFA should coordinate its acitivites with the needs of those other stakeholders and also call upon their assistance in regulating many types of activity well outside SFA's control. We call this cooperation 'mainstreaming'.

Achieving harmonious integration of the plans, activities and needs of so many sectors and agencies involves establishing some cross-sectoral committees and the full integration of wetlands plans and strategies into the broader development context.

6.2 Involvement of local communities

Biodiversity conservation cannot simply be ordered or even commissioned by government. It requires the compliance and support of local communities whose activities ultimately determine whether biodiversity will be destroyed or conserved. Local communities need to be more aware of biodiversity issues, and more involved in planning and decision-making concerning actions designed to help protect biodiversity and restore ecosystems. Local communities need to be properly trained and motivated so they can participate more fully in conservation, ecosystem protection and monitoring activities.

Incentives to better motivate local community involvement could include eco-compensation schemes, payment for conservation activity, encouraging reporting biodiversity news and information, land-use security, better benefit sharing, protection of indigenous knowledge rights, etc. These aspects should be kept in mind during the process of review and revision of more streamlined biodiversity legislation.

Such efforts can also help to alleviate rural poverty. Most of China's poor people live in rural areas, depending directly on local ecosystem services for their livelihoods and well-being, including food production, freshwater availability, and protection from hazards, among other services. Investments in ecosystem service maintenance and restoration can enhance rural livelihoods and be a stepping stone out of poverty.

The story of Apo Reef

Apo Reef in the Philippines is like many other coral reefs where fishermen live from what they can harvest from the sea. Overharvesting had resulted in declining catches.

Prof. Angel Alcala, a marine biologist from the local university, was able to persuade the villagers to undertake an experiment. They declared 20% of their fishing area as a 'no fishing' protection zone. Within two years the fish in the protected zone recovered enough to produce 'extra' young fish that started to repopulate the surrounding reefs that were still being fished. The total harvests have continued to rise for 25 straight years! The success of Apo allowed Prof. Alcala to persuade many more villages in the Philippines to set up their own 'no fishing areas'.

6.2.1 The responsible fisherman

A fisher is a person who depends upon or enjoys the exploitation of a fishery resource. A responsible fisherman is one who exploits such a resources in fair and legal fashion at a level that can be sustained and causes no irreversible damage to the fishery environment and fishery resource population. The responsible fisher is respectful of rights of other persons to the same resource base and does not waste or destroy resources that might also be needed by others. The responsible fisher respects the rights of the animals he harvests, takes only what he needs, kills in a humane way and tries to cause no unnecessary pain or cruelty to the harvested creatures.

6.2.2 Requiring farmers to reduce use of agricultural chemicals

China faces growing problems as a result of pollution.

Fertilizers and pesticides (insecticides, herbicides, fungicides) are the main types of chemicals used on farms. All are hazardous to the environment. These chemicals could pose threats to humans, birds and wildlife and plants. Water quality in streams draining from farmland can also decline due to contamination. Minimizing the need for such treatment is both good for the environment and good economy.

Fertilizers are expensive, they add to weeding costs and they can cause algal bloom in waterways or eutrophication of ponds and lakes if not controlled. Sudden death of fish is commonly associated with a new application of fertilizer.

Ways to reduce the need to use fertilizers:

- Selection of hardy local crop varieties
- Planting at natural growth periods of the year
- Leaving weeds to rot on ground rather than pile to burn
- Use of natural fertilizers (peat, compost, green manure) rather than chemicals
- Fence off growing areas to reduce stress from feet, vehicles and animals
- Employ cultural practices to improve soil structure and health and turf growth
- Conduct soil tests to determine exact nutrients needed by plants

Ways to limit use or impact of pesticides:

- Manual weeding
- Use of biodegradable chemicals only
- Use of target specific agents
- Avoid using chemicals close to waterways or during rainy weather

- Use of mulch on landscape beds
- Use biological controls i.e. introduce weeds as edible plants backed-up by research
- Ensure rich biota with natural enemies of target pest species
- Erect suitable fences
- Set threshold levels for pests to treat area only when necessary
- Learn health hazards, pest identification, and proper handling of chemicals

6.2.3 Using green fertilizers

Green manures are generally under-used, yet they are easy, cheap and have a number of benefits.

- First, they are a great way to hold nutrients in the soil that would wash out over winter. Nutrients that would leach from the soil are held onto for the spring crops. With light sandy soils, where there would be little benefit to winter digging, a green manure is ideal.
- Secondly, they suppress weed growth. Two plants cannot grow in the same place and a green manure will prevent weeds from getting a hold.
- The third benefit is that green manures improve the soil structure and add humus. The roots keep the soil friable as well as drawing up minerals and nutrients that would be otherwise wasted.

As well as holding nutrients in the soil, some green manures have the ability to fix nitrogen from the air. This ability to fix nitrogen means that the plants are actually providing you with fertiliser.

Whilst purists may condemn the various species of Eupatorium as being alien invaders, they are certainly good green manure species and look upon favourably by farmers in southern China. They grow dense and tall, shading out more difficult weeds; they die when cut close to the ground and their break down quickly when left to rot or ploughed into the soil before the next planting.

Aquatic *Azolla* ferns form a useful nitrogen fixing green manure in irrigated fields as do the semi-aquatic legumes *Sesbania*, *Astragalus* and *Aeschynomene*.

Leguminous crops such as beans, clover, alfa alfa have the ability to fix nitrogen (N) biologically from the atmosphere. This can benefit not only the legumes themselves but also any intercropped or subsequent crops, thus reducing or removing the need to apply N fertilizers. Improved quantification of legume biological nitrogen fixation (BNF) will provide better guidance for farmers on managing N to optimise productivity and reduce harmful losses to the environment. Many of these plants can triple up as good fodder plants for feeding to domestic animals, flowers for honey production and also good green manure for encouraging on fallow lands before plowing in before a new planting.

6.3 Co-management

Co-management involves the sharing of management responsibilities between government and non-government parties. Co-management means a much deeper cooperation than simply cooperating with the local community. It means sharing some of the management responsibilities in terms of planning, monitoring, reporting and workload.

Co-management does not mean duplication of functions but the forming of a team where each player's best strengths can be best used. The table below suggests roles for which government and local community may be best suited to perform separate roles.

Government bureau	Public/community
Planning/analysis	Surveys/consultation
Patrolling/law enforcement	Monitoring
Reporting (formal)	Reporting (informal)
Zoning control	Guides
Visitor centre	Cultural entertainment
Protective management	Additional labour
Supervision/control	Homestead tourism
Maintain roads, docks etc.	Transport (boats, horses etc.)
Regulations	Buffer activities
Visitor centre shop	Handicraft items
Collect/share revenues	Shareholders
Contracting	Concessions

6.3.1 Forest responsibility system

Proper protection of catchment forests is often an essential requirement of maintaining wetland health and richness.

Under recent legislation the government is allocating tenure of forest areas of former collective forests to local villages or individual households to protect and manage. Tenure varies from a few years to up to 70 years and sometimes with the guarantee of further renewal if management has been satisfactory. Results have been mixed. There are certainly a lot of great success stories, but there are also failures and in some areas local authorities have been disappointed in the level of management delivered by the tenants and have had to revoke tenure agreements to bring lands back under government management.

In regions that fall under the state program for protection of natural forests, tenants have little freedom of choice in management, harvesting or replanting but in most ex-collective owned forests the tenants do have free choice in the management approach.

Poor production and poor ecological conditions are caused by:

- Inadequate duration of tenure
- Inefficiency due to fragmentation of household forest tenure holdings
- Lack of forestry skill and capacity
- Lack of access to best seed stock
- Inadequate compensation for environmental services

Based on experiences so far, government is trying to improve the first two causes through giving longer tenure periods, affording renewal guarantees and allowing the free sale and transfer of tenure. The government is also trying to improve the reward for ecosystem services by experimenting with a variety of eco-compensation mechanisms whereby downstream beneficiaries for good upstream water catchment protection would be taxed to pay for those services.

6.3.2 Homestead eco-tourism

Outdoor tourism is a fast growing industry in China and can generate a significant economic into economically poor regions of China. But tourists bring a lot of negative impacts on natural sites and can easily kill the 'golden goose' whilst the benefits often bypass the economy of poor local communities and flow to 'external' investors. Eco-tourism is by definition a less damaging form of tourism in which net benefits are left in terms of environment protection as well as contribution to local communities. In this regard it is a good policy to push for community based eco-tourism for which there is enormous undeveloped potential in China.

Two models of ecotourism development. Traditional model a) bulk of tourism going through large private companies or preferred model b) where bulk of tourism channeled through community based tourism. (red arrows indicate flow of tourism revenues, green indicate need for eco-compensation).

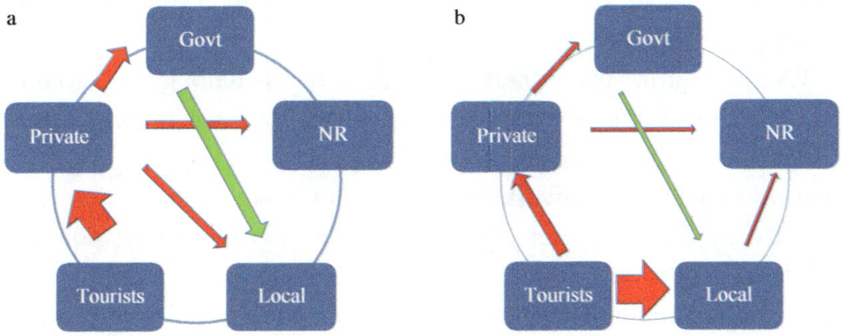

Two international examples where this model works very well are N. Thailand and Sabah, Malaysia.

Types of community based tourism that could be offered by local communities in China:

- Homestead experience
- Authentic local cuisine
- Boating, Horse riding, guided mountain walks
- Personal wildlife photo safari/fly fishing
- Winter cross country skiing
- Canoeing/rafting/rock climbing
- Mountain bikes/hang gliding
- Ethnic cultural entertainment (music, falconry etc.)
- Handicrafts

A group of households can coordinate to form a small association to operate a website to advertise their facilities and recreational options, take bookings and agree among the different households as to which bookings will be housed where. This type of tourism offers a much more personal and local experience to the visitor as well as providing a better livelihood to local people who then have a greater interest in helping to protect the locality, scenery and wildlife.

Caveat: There are risks from eco-tourism if facilities not developed to cope with the numbers of tourists to be catered for.

6.4 Police and law enforcement

Law enforcement in Chinese wetlands remains weak because:

- the laws and regulations are themselves weak
- the mandate of protection overlaps other mandates for use of the site
- Wetland PA staff usually have no police powers, forest police have little motivation to help
- courts have little understanding of the importance of wetlands and generally sympathize with poor local people
- patrolling, reporting and enforcement is slack
- hard evidence is usually lacking
- staff familiarity with criminal procedures is poor

All these factors must be dealt with. National and provincial authorities are in the process of strengthening and revising regulations.

It is up to the PA manager to ensure there is good awareness among local communities and local police and courts about what the regulations are, why they are important. He/she should cultivate good relations with local police and court officers.

The manager should establish a network of trusted public informants.

At least one guard for each protection station should be given forest police status with powers of arrest and confiscation.

The system of patrolling and reporting needs to be strengthened and report forms completed for all patrols to include information on any illegal human activity. Reports of illegal activity should be supported by photographic evidence e.g. by cell phone. A simple sample report form is given in Appendix 9.8.

Patrol routes and timing must be varied to make it difficult to avoid detection and should also cover night and holiday periods when many illegal activities take place. Where serious or consistent illegal activities occur, it may be necessary to launch special law enforcement operations. In such cases the details and timing of such operations should be kept secret for as long as possible to avoid staff leaking warnings to their friends or relatives. Major operations should also be accompanied by police.

6.5 The role of NGOs

Since the 1980s various International Conservation organizations have been working in China, assisting and sponsoring many conservation projects in cooperation with the government at various levels or sites—WWF, TNC, CI, FFI, WI, IUCN etc. Only since the 1990s have a diversity of indigenous NGOs emerged in China but there are now a few hundred environmental NGOs working with the government, with the media and with local communities. These national NGOs can be classed into several types: Registered NGOs, Unregistered voluntary groups, Non-profit enterprises, Student environmental associations, Bird-watching societies, Web-based groups, etc. But all can be useful to bring additional expertise, energy, manpower, motivation and support to the wetlands manager and help conserve important sites and species. NGOs form a valuable bridge between the public, the media and the government and can operate long-term at a more grass roots level that most government agencies are able to do.

There is a need to amend existing laws such as the nature reserve regulation of 1994 and new drafting laws for national parks and other categories of protected area to facilitate the important co-management involvement of NGOs.

6.6 Meet the Bird Force!

If you go strolling in the hills around Dalian of a weekend you may run into a fierce looking patrol of fit young people in military fatigues. You may think this is a crack special unit of the armed forces. But look again. Some of those smiling faces are a bit too cute to be the PLA! This is Bird Force!

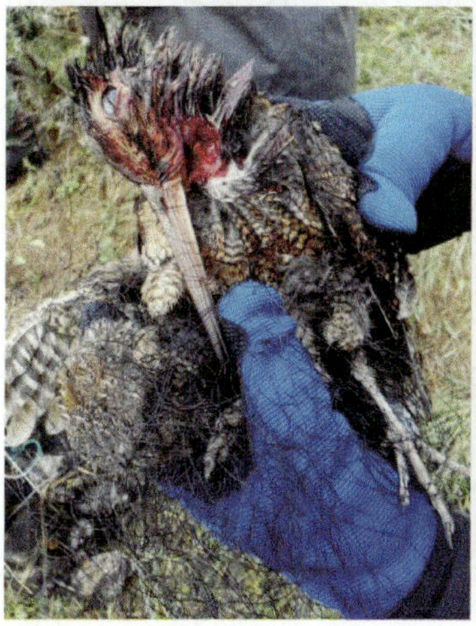

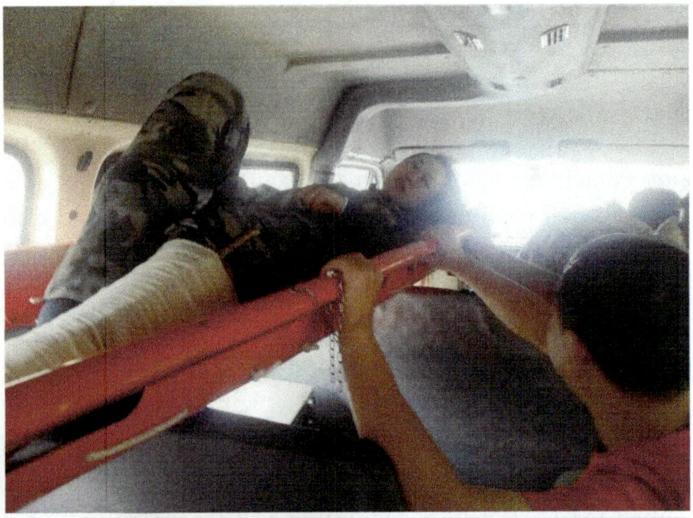

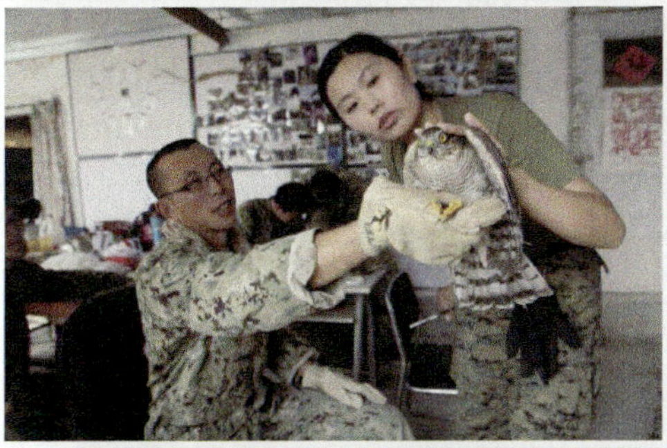

As Grace Xiao explains—"We are all volunteers, there are no funds to support us so we meet all costs ourselves to combat illegal hunting. Our leader was a military-fan who is now manager of a decoration company. He founded our team to protect birds on Laotieshan Mountains in DaLian City. He provides his van, saws, knives, binoculars, many military equipments that he has collected. Every weekend, our team members come to Laotieshan Mountains. It means long but interesting patrol days. We often walking six or more hours in the mountains. If we find nets they must be destroyed and we record the GPS location. In addition, if someone inform us there is an injured bird, we often give public notice in our We Chat group. Anyone who is free will come to get it and transfer the bird to the unique treatment centre for birds in Dalian City."

But wow are these young people fit and dedicated? They train like the army—climbing, hanging, abseiling, lifting great weights and running up and down rugged mountains. Accidents happen and one girl is whisked away in an ambulance with a broken leg.

Bird Force scour the hills and forest for illegal bird nets and traps. Sometimes they are too late but can still take down the offending nets and dispose of the grizzly haul of dead birds. But the rewards are when they can save the birds and release them back into the wild. Small birds are relatively easy but Dalian is on a raptor migratory flyway and many of the birds are larger hawks and owls. Sometimes these need examination back at the Dalian treatment centre and may need care, feeding up or even minor surgery before they are considered fit for release.

What a great bunch of people. If only China had more groups like this showing their love and appreciation for wild birds and helping reduce the ugly netting that is reducing our native bird populations so dramatically over the past three decades. We also owe a debt to all the other birdwatcher groups, their dedication, expertise and willingness to invest time and money.

6.7 Communications

6.7.1 Development of a communications strategy

Good communications is an essential component of good wetland PA management. A serious barrier to achieving the aims of good protection of critical wetlands is the low level of awareness of the value of those wetlands. Throughout China, at central and local levels, there is a widespread misperception of an inherent conflict between economic development and biodiversity. At all levels of government, the links between healthy wetlands, biodiversity and quality of life, successful development and well being is poorly understood.

One way to better explain the value of wetlands, the needs to protect them, inform about the threats and problems they face and stimulate interest in and support for the wetland PA system or a particular wetland site is through the development of a good Communications Strategy.

The strategy can address communication needs to a variety of different audiences. Three main target groups are immediately recognized:

1. Government Decision makers and local agencies and officials especially around WPA sites.

2. General public.

- **Youth** through formal education and student debate groups, bird watching society etc.
- **Wider public** through mass media and web site
- **Local communities** around WPAs

3. Intermediaries and partners that help broadcast environmental messages to the first two target groups but need some capacity enhancement.

- **Journalists of key media agencies** (press and TV broadcasters)
- **Teachers** to deliver new materials on wetland importance

The strategy should identify the roles of the main players in wetlands and biodiversity issues in the given area and different messages should be designed to suit different the target groups.

Increased commitment by the first target group is expected to result in increased resource allocation for wetlands and biodiversity conservation and improved integration of biodiversity concerns into provincial dialogues, programmes, plans and policies. An increased interest of Chinese government officials in wetland issues should also result in more frequent news about wetlands and biodiversity through official media coverage.

The second target group is necessary for the public to understand new regulations, cooperate with or even lobby government agencies into action as well as to take a greater interest and participation in wetland and biodiversity conservation and help reduce the environmental impacts of personal and public actions.

The third audience is expected to act as an intermediary for both the first two target groups.

The approach proposed to reach the target audiences, in order to achieve their specific desired outcomes, is presented under 5 lines of action:

- **Awareness** through mass media, publications, newsletter, web and events
- **Education** through introduction of new courses into school training
- **Training** to raise capacity of staff, partners and intermediaries
- **Information** to allow better decision making and to strengthen awareness
- **Promotion** of site activities, events and attractions

These communications are very important and deserve a serious investment of effort. It is recommended that each WPA hire a full-time professional communications officer but because the budgets are usually tight and staff limited the Communications Strategy should identify its approaches cleverly to have maximum impact. The following approaches can be followed.

- Harness more partners and resources rather than act alone
- Use mass media—TV and newspapers to reach maximum audiences
- Tag wetlands and biodiversity issues onto familiar faces e.g. celebrities
- Link wetland biodiversity issues to brand images
- Invest in approaches with snow-ball or high replication potential for increasing impact e.g. intermediaries to relay the messages
- Time activities to hit moments of heightened attention
- Be flexible and opportunistic to take advantages as new doors open in such a dynamic society
- Be responsive. Have materials ready in advance, ready for quick reaction
- Build on existing attitudes rather than create totally new ones—Taoist, Confucian, socialism, respect for balance and holistic approaches, some traditional beliefs, etc.

- Use Government and UN channels to reach sectors not easily accessed by NGOs—senior government officials, SFA, MEP, etc.

Impact will be increased by building up a partnership of aligned agencies, institutes and NGOs. Obvious strong government partners are Dept of Water Resources, Bureau of Environmental Protection, other divisions of Forests Department and Dept of Fisheries under Ministry of Agriculture and in coastal areas the department of Oceanography Administration. The PA manager should try to become associated with provincial plans for re-greening, biodiversity strategy and action plan, tourism plans and ecological red lining plans. Media Actions should be timed to coincide with moments of media attention during such events as annual environment and biodiversity days, bird-loving week etc. All elements of the project visibility should carry a recognizable brand image that will raise recognition and assist in monitoring impacts on the target audiences. It is worthwhile undertaking a survey of Knowledge, Attitudes and Practices (KAP) as the basis for determining information and awareness needs as a basis for developing the strategy.

Key messages should be developed based on concrete examples which are close to people's daily life and experiences. The key messages to be disseminated to the target audiences must clearly show that biodiversity conservation has economic and social benefit, such as (1) evidence that healthy wetland ecosystems are more resilient to climate change (2) evidence that wetland biodiversity has significant importance for human health: high quality wetlands result in positive water quality, prevents flooding etc.

Each target audience must be reached differently, yet it is important that the answers to the following questions must be embedded in all key messages:

- What is biodiversity exactly?
- How does biodiversity relate to my daily life?
- What is the value of biodiversity and why does it have to be conserved?
- What is the relationship between biodiversity and socio-economic development?
- What is the relationship between biodiversity and natural hazards?
- What are the species in danger?
- What are the consequences of biodiversity loss?
- What activities are damaging to biodiversity?
- What can people do to conserve biodiversity?
- What does the government do to conserve biodiversity?

6.8 Ways of raising awareness

Under the guidance of the Communications Strategy, the Protected Area should organise a programme of awareness activities. Possibilities are endless and internet offers a superb new tool to help disseminate awareness better. Plan activities to meet the needs of different audiences. A guest lecture might interest the adults but there needs to be something else offered a amuse the children. The following list suggests some of the ways that a PA might expect to raise its profile and raise local awareness and support.

- Publications—books, brochures, leaflets, posters, guidelines

- Internet—info sites, websites, blogs
- Education—teaching or development of teaching materials, courses, modules
- Creation of news—TV, films, articles
- Dialogue—engage all sectors especially local communities who are the interface with nature
- Events—use well chosen moments to focus attention. Attract attention by attracting participation of celebrities or local VIPs

6.8.1 Bird races / photographic competitions

Bird Races are increasingly popular in wetland sites. They serve as great visibility, raise awareness, can provide useful data and can even be organised as fund raising events.

Typically a bird race is organised over a single day or 24 hour period. Teams of birdwatchers register. Usually 2-4 persons per team. The rules are to record as many different birds as possible on that day. Each record must be witnessed by at least two members of the team (three if the team has four members. The team must stay together. Organisers can decide whether to allow identified birds heard as being acceptable records or not. Other rules usually include no luring with tape recorders, no lamping at night, stay within prescribed boundary of the competition, no access to private lands unless accessible to all teams.

Judges are on hand to check submitted lists. Some prizes are offered to the winning teams and local enterprises can be approached for sponsorship and also gain green praise for their support of conservation.

Photographic competitions can also be a good way to raise interest and awareness in your site.

6.9 Photographing birds and wildlife

It is important to be able to take good photographs of birds and other features of wetland protected areas. Photographs are excelled records for documentation, evidence for identification, ways to help record and monitor change. Photographs are useful for awareness materials, publicity, displays and news. The public is bombarded with dazzling photos in all walks of life. If we are to capture their attention, our own photos must also be of highest quality.

Bird Photography is a challenging and rewarding activity in its own right. Birds are small, timid and often only seen at moderate distance. We need a good camera with powerful lenses to get the detail of a small bird and we need good light of high speed to freeze their fast movements. The following tips are useful to ensure better quality of bird photos.

6.9.1 Photographic cameras

Ardent bird photographers help the Japanese economy and keep the boys at Canon and Nikon in their luxury cars. Big lenses cost big money and can be easily damaged if dropped or get wet. For management purposes much cheaper and more convenient cameras are available. The new range of bridge cameras offer zoom from wide angle to very long telephoto. The

cameras are small and relatively cheap. Quality challenges the artillery of the big boys!

Use of long lenses used to necessitate use of sturdy tripods or sand bag but digital cards are now so fast that you can apply a huge rating of several thousand ISO. which allows you to take fast pictures without the need to carry a heavy tripod.

Long lenses are very heavy; tiring to hand hold and carry up mountains etc. You need to find a balance between optics and convenience. Most people find they can easily manage a 400m lens without need of a tripod. Longer lenses really can only be used with a sturdy tripod but use of a ×1.4 or ×2 adapter is a possible compromise between weight and magnification. If you can get closer but use a shorter lens you will get a better result.

Digital cameras get better and better. You can now take high quality photos on a cell phone or ipad and these are fine for scenery, events, flowers and dead birds! But they are not really up to photographing wild birds. For that you need a bridge camera (which now can have tele lens zooming up to ×50) or a larger camera able to accept a range of telephoto lenses. The two leading competitors for this market are also Canon and Nikon.

Spotting scope. This instrument is almost essential to identify shorebirds and birds on large bodies of water. Many birdwatchers get quite attached to their scopes and use them also in forest and other locations. They have the advantage of being used with a steady tripod and having very high magnification. It is possible to take quite nice photos through most spotting scopes and adaptors are available to link to small digital cameras such as Nikon Coolpix camera etc.

Tripods. The longer the lens the lower the light, longer exposure time needed and more camera or subject movement. It becomes necessary to keep the camera steady with a tripod and a long heavy lens will need a sturdy tripod and lens cradle.

6.9.2 Tips for better photos

Use of hides. Most birds show little fear of hides, even when they can see miscellaneous movements or hear noises inside. Formal nature reserves may have well positioned hides for public access. Hides or blinds can be made at specific sites to conceal the photographer or small portable hides can be used. Camouflage netting is available from internet suppliers to make a quick light alternative. A car can act as a moving hide to approach roadside birds.

Good spots for photography. Walking around with a camera can result in a lot of mediocre

wildlife photographs good enough for identification or documentation but rarely good enough for publication. You will be able to take a much higher proportion of higher quality pictures by taking more care and finding good sites for viewing wildlife such as waterholes, feeding areas, tide lines, nesting sites, roosts, courtship sites and the best feeding spots of the season (fruit trees, grain crops, fishing sites, insect emergence etc.).

Time of day. 'The early bird catches the worm' (old saying). You need to be in place before the birds are active. Birds are generally sluggish through the middle of the day. Go home and have a big lunch!

Decoys and bait. You can attract birds to a selected 'shooting ground' like the old hunters by putting out attractive food bait or decoys (effective for waterfowl and pigeons etc.). Even an old stuffed or model owl may attract a lot of attention from small mobbing birds.

Sound playback. Some birders use playback of recorded songs or previously recorded or downloaded library sound files to attract territorial birds closer for viewing and photography. Do not overdo this technique. It can seriously disturb and confuse breeding birds. Playing owl calls may attract small mobbing passerines to come close.

Touching up. Un the old days of real film photography, touching up was regarded as part of the art. Specks and scratches could be removed, irrelevant sectors of the picture could be 'softened'. Exposure and contrast could be increased or reduced. With digital images the effects possible have become endless. But to the purists, too much photo-shopping is regarded a bit as cheating. Try to keep to the minimum.

Filing. With digital photography it is now possible to record thousands of images—each a large file. You have to be ruthlessly selective. Get rid of all marginal pictures and unnecessary duplicates, keep only great shots or shots you think you will need. File pictures intelligently by subject and date or you will waste vast amounts of time in the future looking for pictures that you know you have somewhere!

Video. Most stills cameras can now also take high quality video footage also. It is useful for the Protected Area to keep a collection of good video clips that can be cut into other programs at a later date. All the same tips for photography apply to video. However, video really has to be smooth or steady, so use of a good tripod is very important. Make sure you also capture suitable natural sound as background, especially if video shots have noisy machinery, music or people talking on the original sound track.

Automatic cameras. These cameras are increasingly cheap, increasing quality and being increasingly used to photo-trap shy wildlife. They can be set near the ground looking over frequented wildlife trails, feeding or watering spots etc. These cameras use infra-red beams to detect movement within the scene covered, can shoot B/W at night or colour in the day and can shoot stills or video. Inspecting the camera can be very exciting to discover what you have 'captured'. Conceal cameras well. You do not want to lose them to curious passers by or rogue elephants! More detailed advice on use of automatic cameras is given in section 8.3.5 below.

Using flash. Flash can be useful to fill in shadow or for taking birds at night but flask creates strange eye reflections that make photos look unnatural and flash from too close a distance can seriously dazzle and disturb your subjects. Only use a flash with responsibility.

Macro photography. Whilst bids may require much use of long lenses, you will need macro techniques to capture insects, flowers, wide angle shots etc. Depth of focus becomes of high importance, so it is common to use flash or ring flash in professional macro-photography.

Rules for photographers

- Do not injure or hurt birds in obtaining pictures
- Do not disturb or open up nesting sites
- Do not stress birds by coming too close, or flushing them away
- Do not confuse or interrupt breeding by playing territorial song calls
- Do not destroy vegetation to clear a better view!
- Do not endanger birds by luring them to bait in dangerous locations
- Keep quiet!

Back lighting and low angles adds to the allure of close up photography.

6.10 Visitor management and education

Visitor use of wetland parks and other PAs helps to spread awareness, helps educated the public, provides a healthy valuable recreation, earns useful revenue for the PA and economic opportunities for the local community; but excessive visitor use can "kill the goose that lays the golden eggs" and seriously damage the wetland and disturb its wildlife. A sensible balance must be achieved.

Visitors are welcome and encouraged to use and enjoy the nation's beautiful wetlands. It is important to generate wide public support for the maintenance of wetlands and to increase awareness of their economic and functional importance for environmental quality and safety. The level of use and interest shown by the public justifies the establishment of the protected area and encourages support and finance from local government.

In many cases, parks or protected areas need earnings from entrance fees to help cover basic operating costs. The danger is that desire to increase visitor revenues drives management in directions that conflict with needs for conservation.

Three ways to ensure that visitor use is kept within safe limits are:

1. Undertake careful assessment of visitor carrying capacity as the basis for determining strict quotas and controls on visitor numbers in each given areas or seasons.

2. Zone the site so that visitors are confined to less sensitive areas where their impacts are less serious. Key sites for protection should be zoned inside the core area where regular visitor use is forbidden.

3. Use a variety of methods to minimize negative impacts. Such methods include low impact boardwalks or pathways, screens and hides to minimize disturbance to wildlife, use of low impact toilets and efficient waste disposal and collection systems, notice boards instructing visitors how to behave—keep quiet, no littering, no fire, no dangerous climbing, keep with the roped areas, no collecting plants, no disturbing animals, no swimming, no throwing stones, no graffiti etc. Ideally visitors to the site have already passed through an information or education centre which also prepares them with advice on where to go, what to wear, and how to behave.

6.11 Designing a visitor centre

The visitor centre of a protected area is the main reception area where management can

interact with and provide advice and guidance to its visitors.

The visitor centre can be combined with ticketing office, education centre and even restaurant, toilets, souvenir shop and/or viewing hide as suitable.

The size of the centre will depend on what functions it contains and the numbers of visitors it is designed to accommodate.

The following hints should be kept in mind when designing or operating a visitor centre.

The visitor centre should reflect the style and environmentally friendly policy of the protected area.

The structure should be attractive but in harmony with the natural setting, neither too high or too garish and constructed with local natural materials such as stone and wood rather than concrete and steel.

Different audiences have different interests and needs. The centre should cater for a wide range of educational, informative and interactive purposes. It is useful to have a discrete children's corner where children who may be bored with looking at information displays can play happily with colouring kits, sand pits or interactive toys.

Plan for all seasons and all weather. Children can enjoy outdoor play adventure play area whilst parents take in more viewing or educational materials. The centre must have enough indoor space and interests to entertain visitors when it rains.

It is increasingly required by law to provide facilities for handicapped persons, so ramps for wheelchairs, wide doors, special toilet facilities etc. must be provided. The handicapped are limited in what they can do so have few chances to get out and see nature. Provide that chance and allow them to appreciate much of the scenery and wildlife second hand through good exhibits.

Displays need to be high quality and interesting. Dismal collections of badly stuffed birds and animals should be a thing of the past. Exhibits should be bright, full of life, stunning photos or videos and interesting captions or commentary.

Basic information—maps, guidebooks, identification materials and brochure advising visitors on how to behave should be available at the ticket area. A visitor shop may offer a wider range of books, souvenirs, bird watching equipment etc.

The centre should have a direction and flow to it so people can dally to take in information or move quickly on to the next item rather than be crossing hither and thither with cross and counter flows.

It is useful to include a class room or meeting room into the centre where videos can be shown or to be used for meetings events, guest lectures etc.

It is great if the centre can serve as the hub of a nature lovers group or be used to entertain and educate classes of school children etc.

6.12 Information for visitors

Basic information should be provided to enable visitors to get more out of their visit, select the best routes and activities for the season, their timing and interests, advise on appropriate behaviour in a protected area and provide some basic education about the wetland, its ecology, species and special importance.

Information may be delivered in many ways such as through trained guides, via the web-site for the PA, maps, brochures, booklets etc. available at the visitor centre, museum or information centre and through signs and notice boards scattered around the site. It is useful to allow visitors to regularly see where they are on the PA map, suggest route alternatives and show what features of landform, vegetation or species to look out for at different locations within the site and how to identify them.

Useful and up to date information can be provided on sighting boards or news updates. Birdwatchers also report re ent sightings on a variety of webs, blogs and telephone apps.

Outdoor information panels need to be well printed and protected in water-proof frames or under a rain roof. They should be angled and fixed at the right height for easy reading. Printing ink must of high quality or it will fade with exposure to sunlight.

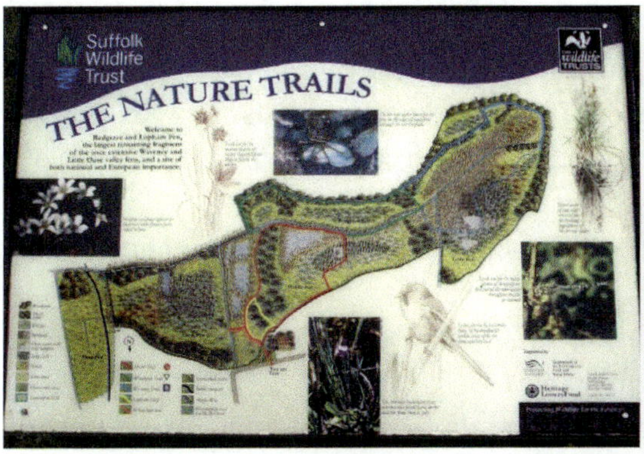

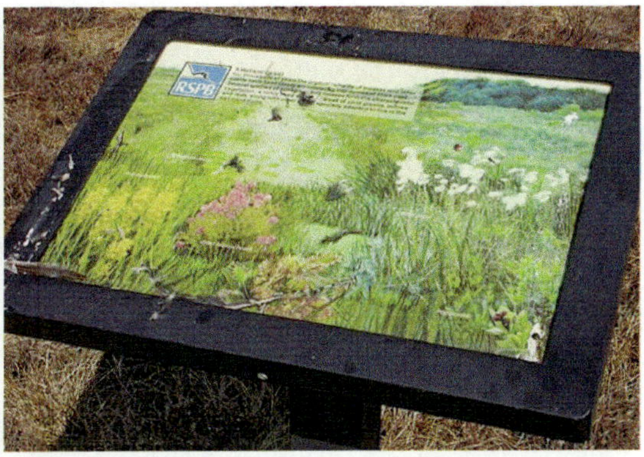

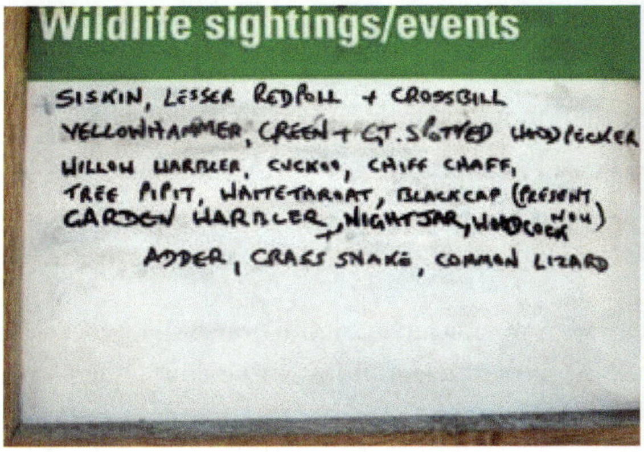

The following figures show various types of outdoor information displays.

6.13 Maintaining a museum

Many wetland nature reserves and parks feel the need to maintain a museum. The protected area should be primarily for protection of wildlife not the grizzly display of slaughtered animals! The reserve is also not a zoo to display captive animals. Please limit wasting time, money and staff on irrelevant display.

Small poorly maintained museums and zoos look really sad. Zoo animals are miserable and in poor condition. Stuffed animals are poorly presented, poorly preserved, start to fall to bits and gather dust and mould. The capture of animals for display, whether alive or as stuffed specimens constitutes a big loss of wild creatures that is totally inconsistent with the role of conservation.

6.14 Education programme

Many wetland nature reserves and wetland parks have developed successful education programmes accepting classes of school children, mixing education about wetlands with a fun day out of the classroom. The protected area may organise its own education programme or may simply offer its facilities to local schools or NGO who then organise the children's exploration and learning activities.

The benefits to the wetland site are that the programme fosters a greater audience of public support and may nurture individual nature lovers and future conservationists.

The specifics of the programme will depend on the nature and safety of the wetland site, the numbers and types of children or students to participate.

Children can undertake some helpful activities like helping to clear litter, monitor various species, plant seedlings etc.

If an education programme is to be encouraged, it is best if the management can offer a classroom with white boards, project facilities, toilet etc.

There is an enormous amount of educational materials on the internet for interesting children in wetlands. This can be downloaded and modified or translated to make it suited to local conditions.

Children are fascinated by life and love the opportunity to explore, discover things, draw things and talk about their experiences. Whilst their mothers' may be horrified, children like to get down to ground level, grubbier and muddier the better!

6.15 Newsletter or website?

Many nature reserves, parks and conservation projects have tried to raise awareness by means of regular publication of a newsletter. Some attractive and informative newsletters have been maintained but in general this approach is found to be too expensive too demanding of staff time, disappointing in results and reaches a very small audience.

A cheaper and easier solution is to host a website. But again unless the website is done well, it is usually difficult to generate large number of regular hits. The following advise may reduce wasted effort and disappointment and help the wetland manager reach a wider audience.

Part Seven　Operational

7.1　Structures, facilities and their maintenance

One way to limit impacts of visitors on wetlands, keep them within strict visitor areas and also provide safe convenient pathways across or adjacent to wet areas is to build boardwalks or walkways.

Typical designs of paths and board walks

Such structures are costly to build and require regular maintenance. Care must be taken in design and the following should be kept in mind.

Surfaces get wet and slippery, especially in cold weather. Ways to reduce slipping include placing wire mesh over sloping sections, adding strips of rough non-slip sheeting, and keeping surfaces as dry as possible. One way to speed drying is to leave gaps between boards of a boardwalk and this also allows light to reach the ground beneath, rather than creating a dark damp passage.

Materials for construction should ideally be wood or stone coloured to blent in more harmoniously with the natural background but there is no need to creat fake wood effect for railings. It is acceptable that rails, safety chains etc. are seen as they are. Pay attention that no pollution comes from the materials (e.g. toxic paints, preservatives etc.) i.e. screen all materials used in the construction of board walks.

Gravel roadways are noisy both for human footsteps or for vehicles. These should not be used close to sensitive wildlife. Roads through wetland sites should use speed notices or sleeping policemen humps to reduce speeds.

The routing of all roads or paths must be planned with respect to minimising disturbance to wildlife. Keep routes away from breeding areas and feeding areas which can be admired at a suitable distance from hides or towers. Also aesthetic enjoyment of the site from a distance has to be considered.

7.1.1 Construction of bridges and tunnels

Water flow and connectivity are essential for the health of the wetland so great care must be made to ensure these are not blocked by construction of roads or paths. Adequate tunnels or bridges need to be constructed to allow natural flows to continue and need to be regularly maintained and cleared.

7.1.2 Construction of shelters

The manager should provide adequate shelter for use of visitors as rain shelters, resting points etc. These should be provided with litter bins and may also have containers for leaflets, maps, information board or a sightings board.

7.1.3 Low impact toilets

Wetlands are themselves natural sewage treatment systems but overload of human waste will create problems of eutrophication.

If large volume of tourist use of toilets is envisaged they should be equipped with their own treatment units, septic tanks, separate drainage channels away from the wetland area or use a total removal system. Do not use a chemical treatment system as the disinfectant chemicals used in such toilets are very harmful to wetland biota and ecosystem.

7.1.4 Litter collection

The extent of litter in wetlands is increasing yearly. Litter gets dumped in rivers, ponds and the ocean. It is insightly but also dangerous. Plastics become ingested by wildlife, netting, fishing lines, lead fishing weights, sharp cans and other debris kill large numbers of wildlife. Much litter does not break down and remains a hazard to the environment for many years.

Three actions need to be taken by the wetlands manager:

1. Clean up beaches, river banks or water bodies in the protected wetland. This is costly in

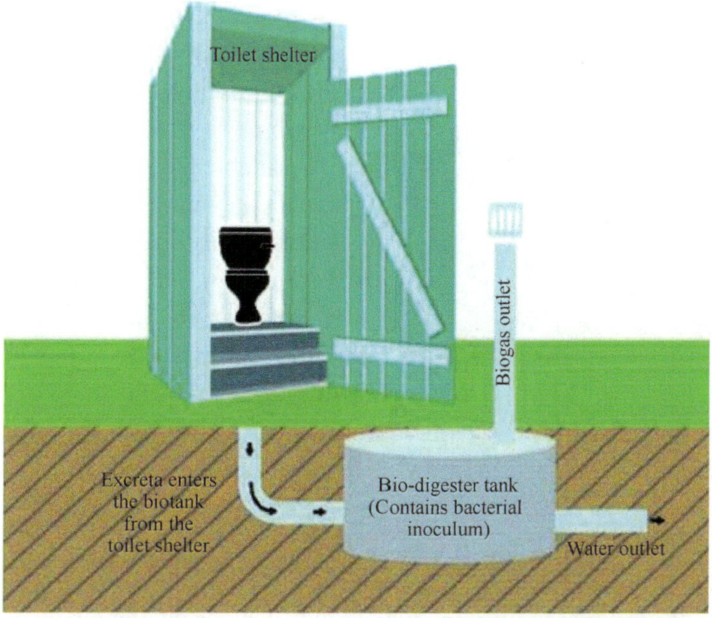

terms of staff time but may be reduced if the manager can enroll a volunteer force of school children, youth or adult helpers.

2. Promote greater awareness among local communities and farmers about safe and proper disposal of litter before it can enter the wetland site.

3. Ensure that there is adequate litter disposal facilities and collection frequncy to minimize litter impact within the wetland protected area.

7.1.5 Observation hides, blinds and towers

Wetlands are excellent places to see and photograph wild birds and other wildlife. But if visitors and photographers are to get good views of birds without disturbing them, it is important encourage and provide suitable hides, screens and towers that allow humans to remain unrecognised as the threats most wildlife see them as.

Many wetland reserved construct special watch towers or hides for this purpose. In other

cases they construct blinds so that people can walk close to wildlife without being seen.

Photographers may construct special hides from which to photograph a special site or may take a portable hide to be put up for concealment wherever they may need it.

A great deal of research has been undertaken on hide design. There is no need to re-invent the wheel. Follow well established models that have proved successful in other places. Get local businesses to sponsor hides and place a simple plaque inside the hide to acknowledge their support.

The main principle of a hide is to allow observers from within the hide to get a good view outwards but allow birds outside the hide to get very little view inwards. The best designs have horizontal slit windows so that it is darker inside the hide that outside, the observer has a wide horizontal view but wildlife cannot see more than a narrow dark slit.

The following illustrations show some typical hide designs and layouts.

Smaller hides or camouflage netting for photographers are also readily supplied online or through bird-watching societies or hunting, camping suppliers.

Pathways can be screened to allow access to hides without visitors being seen walking to and from them. These can be simple screens of fabric, planted hedges or constructed out of thatch etc.

Pathways leading close to wildlife should be designed to be quiet. The plonk plonk of people walking on wooden boards to the noise of feet on stony gravel can disturb nearby birds. Visitors should be warned to keep quiet at all times when close to wildlife.

Towers can be made overlooking good water bodies and can be useful as monitoring points. Birds are not very concerned about people well off the ground! However, towers are costly constructions, possible hazards unless well maintained. Cannot be moved and have limited view. In most cases in Chinese wetlands, they are not really a worthwhile investment.

7.1.6 Boats, moorings and boathouses

If water level is relatively constant a simple jetty fixed to pole supports will suffice for management purposes. If water level rises and lowers, the design can be based on a floating pontoon type jetty with a hinged ramp. For moorings to be used by visitors, it is important that safety rails are attached.

7.1.7 Creating fish ladders

Many fish species mate and lay their eggs in colder and better aerated waters of upper streams

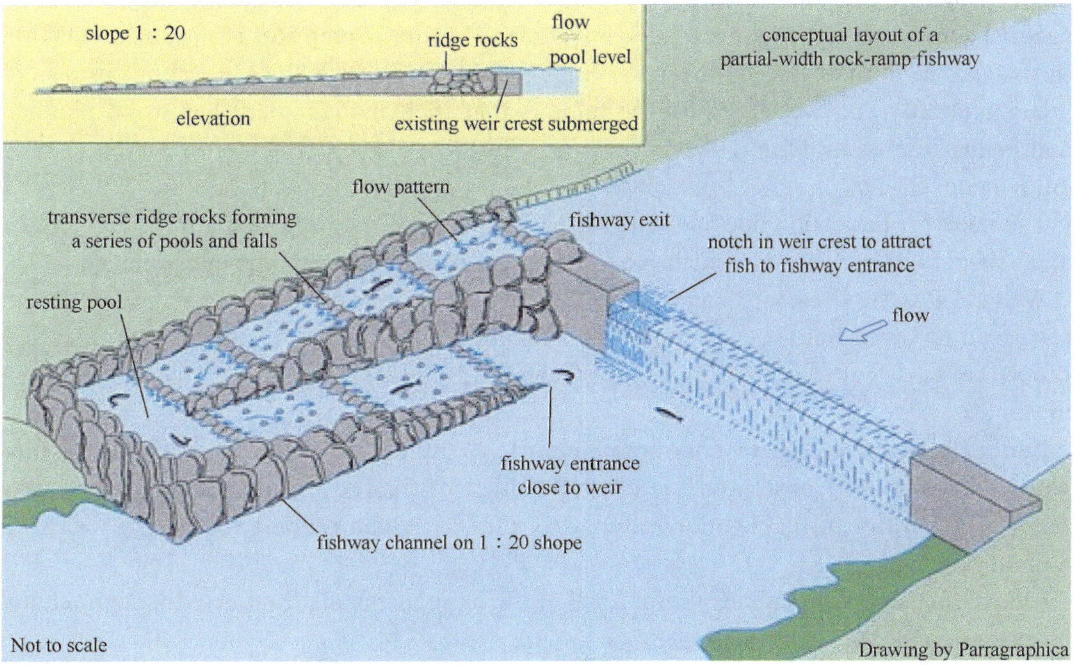

but then live as adults in lower reaches of rivers or even in the sea. They thus need to make breeding migrations up and down rivers. Construction of dams, wiers and other structures can make such routes impossible unless we artificially create separate gentle passage ways that fish can negotiate. We call these structures fish ladders and they can be a very effective way of maintaining connectivity along rivers. Many designs can be reviewed on the internet and can be incorporated into the structural plans and designs of all new and existing aritificial barriers. Be ready to insist on these on dammed rivers that feed your own protected area.

7.1.8 To fence or not to fence?

Fences are costly, require constant maintenance and cause large damage to wildlife. They should be used at the absolute minimum.

Some wetland managers see a need to put a fence around each wetland nature reserve to keep people and domestic animals out and keep wildlife in. There may be a few cases where this might be necessary but generally this perception is an out of fashion throwback to the isolated nature reserve concept. The current thinking about protected areas generally and wetlands in particular is that their management has to be planned and mainstreamed into the wider landscape. Wildlife need to permeate the landscape and enjoy connectivity.

The need for protection does not end at the boundary of the protected area and protected areas should **not** be viewed as exclusive closed areas by their surrounding communities.

Wetland PAs serve as safe harbours where many birds, other creatures and fish etc. can rest, feed or breed but they are mostly not confined within the PA boundary but venture out widely to feed over neighbouring farmlands, grasslands, other wetlands, even urban areas.

Fences may be useful to encourage people or domestic animals not to stray into sensitive parts of a protected area but the idea of excluding people by means of a fence is generally futile. The most expensive fence can be rendered useless by sawing a single rod!

Electric fences are sometimes useful for limiting large mammals from invading agricultural areas outside the PA.

7.2 How to tag trees properly

Whether you run a research plot for monitoring vegetation or you merely want to provide visitor with information about the names of the trees they are looking at, you may want to attach labels or signs to trees. Unfortunately this rather simple task is often poorly done resulting in damage to the tree or at least quick damage and falling off of the label. The point is that **trees grow**. You cannot just attach a label by hammering in a nail. Also some nails are toxic to trees.

There are many notices and quite a lot of tree identification labels displayed along the boardwalks. Botanical gardens have been labelling trees for more than 100 years now and know quite well how to do this so there is no excuse for the very poor labelling done at both sites. Labels have to be 1) legible from up to 5 m distance, 2) preferably in colours that are in harmony with the natural setting and 3) attached in ways that will last several years and not damage the trees. Again the managers seem to totally ignore that fact that trees grow and move! By nailing boards and labels tight to trees they are guaranteed to break off within 2 years as the trees swell and bend and snap off the labels. Choose materials wisely.

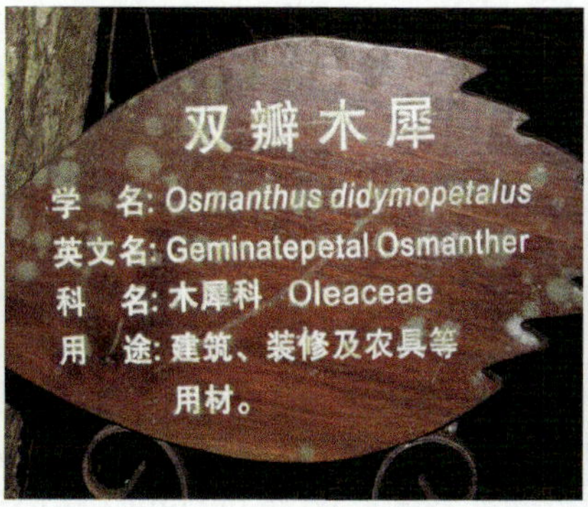

Rules of labelling

- Avoid attaching notices to living trees where possible by attaching to dead wood structures of boardwalk or independent posts.
- Make sure writing is large enough to read from distances up to 5m. and notices not too

Good message but garish

Good free-standing label

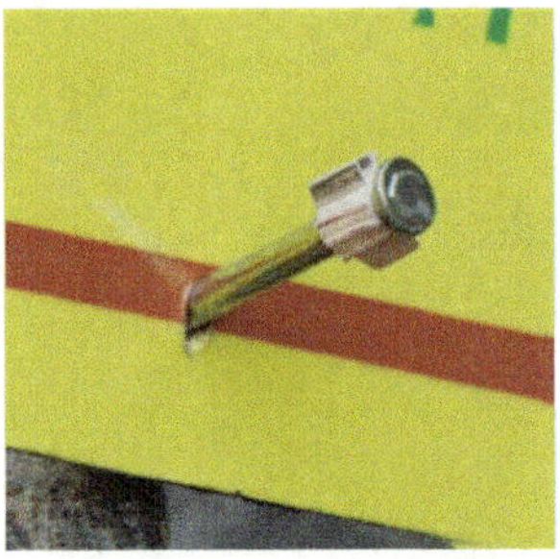

Label OK on flexible line

Font too small

Nails much too tight to tree

Better length for growth

glaring colours.

- Where labels need to be attached to trees, use an expanding necklace or if using nails these should be of aluminium or stainless steel. Iron is bad for trees and copper is toxic. Nails should extend long enough beyond the label to allow several years of tree growth and labels should be loose so they can slide along the nail and move in the wind etc. Each label should have only one nail.

7.3 Map skills and orienteering

Understanding the distribution of plants and animals across the extent of a PA needs good spatial awareness. All field staff need to know the terrain and lay out of their PA. They should be equipped with a clear map and be familiar with how to use it. If the terrain has any significant altitude range, the map should show topography. Rangers should be able to record sightings in terms of map grid reference or GPS coordinates.

7.3.1 Map skills

Maps are produced at different scales. Be familiar with the scale and projection of the map being used. Standard map grids are expressed in decimal degrees or in metres.

Contour lines reveal lines of the same height and hence the shape of the land. They can show the shallowest or steepest path between two points. Contour lines closer together represent steep terrain. The farther apart the lines, the gentler the slope, and the flatter the terrain. Valleys, ravines and gullies are represented by a series of V-shaped lines pointing toward higher ground (greater elevations). Contour lines denoting ridges or spurs are shaped like a series of V's or U's pointing toward lower ground (lower elevations). Another easy way to interpret terrain is to look for a stream running down the middle or side of the V's. You can tell if there is a pass or a saddle in a ridge by looking for an hourglass shape with higher contour lines on each side. A peak is depicted by the innermost ring of a near-concentric pattern of contour lines. It is often marked with an X or Δ and its elevation. Colours, or shading added t a map may distinguish between forested areas, meadows or alpine tundra. Other lines on the map may indicate roads, trails or land use boundaries. This may be useful information when looking for a campsite, a water source, good view, or a good place to view wildlife.

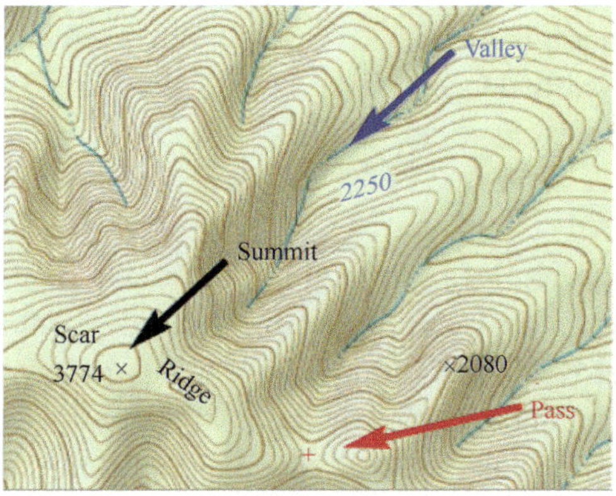

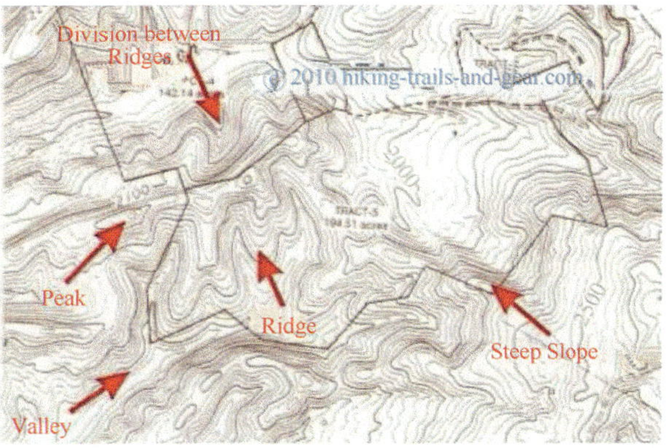

With the aid of a compass, you can take a bearing on any prominent peak or feature that you can recognise on the map to fix the direction for a travel route. By taking a bearing on two different points you can triangulate on the map to deduce your current position. You will need to align your compass with magnetic north. This may be slightly different from due north on most maps.

Compass skills are essential and are not replaced!

A more accurate compass uses a mirror or prism to give precise bearings. An altimeter is also useful to check which contour line you are at. Many modern watched have compass and altimeter built into them.

With good map and compass skills, rangers will be able to navigate the landscape with ease. But when conditions change unexpectedly, a thick fog or a snowstorm can obscure a landscape commonly used as a visual checkpoint. A GPS (Global Positioning System) receiver quickly pinpoints a location on the earth, providing the opportunity to navigate with a map without landmark visibility. A GPS can also be used to plot a route ahead of time, and determine the distance between way points along that route.

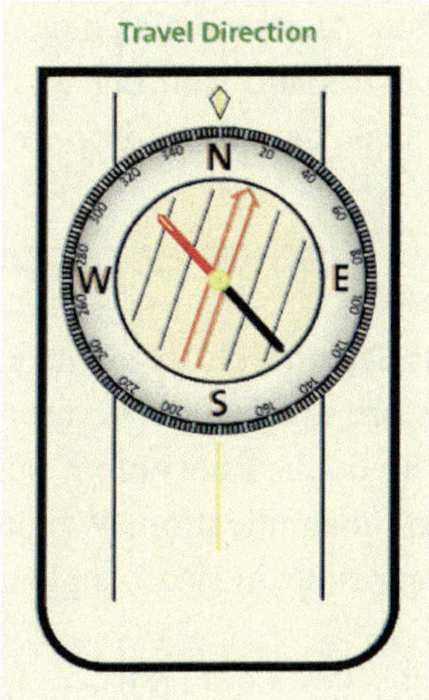

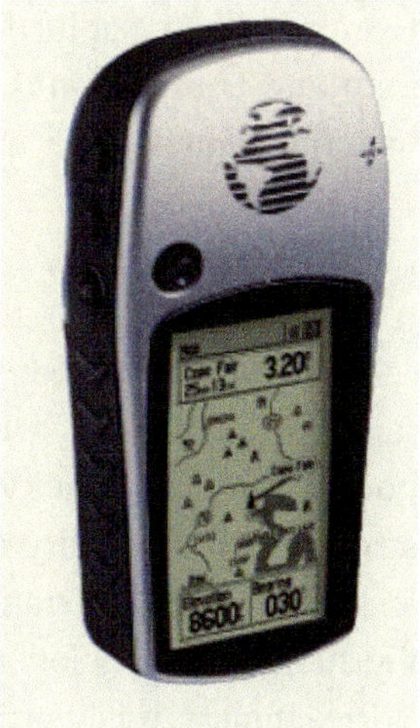

7.3.2 Telling the time by compass

If sun is visible you can tell the time of day with a simple compass. Align compass on east west access. Keeping left edge of compass level, raise right edge till the plane of the compass is pointing at sun. At this angle it will have no shadow! Move a pencil point around the edge of compass to find the position where its shadow falls at the centre point of the compass. This point can be read as the face of a clock in relation to the centre of the compass.

Conversely if you have a watch but no compass you can align your watch to find a compass bearing. This can only be done accurately if the watch has an hour hand (i.e. not digital display). Your time zone may be offset in relation to sun time but for most zones the following.

Point the 12 o' clock point towards the sun, then north is found half way between where the hour hand is pointing and that 12 o' clock position on the longest of the two possible arcs. In the southern hemisphere you will find south at that half way point.

7.3.3 Estimating distance with compass

If you have no map or GPS but can see a distant prominent point you can estimate your distance using a compass. Take an accurate bearing on the point, then move laterally 17.4m at right angles from the direct line and take another bearing. A difference of 1° computes to a distance of 1km. If the difference is less than 1°, continue moving laterally until you do get a difference of exactly 1°. The point is then n km away where n=total lateral distance moved to achieve 1° shift in bearing/17.4. (17.4 is 1/360 × circumference of circle radius 1km). You need a very accurate compass to gauge a 1° shift.

7.3.4 Use of GPS applications

If the PA has complete cell phone coverage, rangers can take advantage of various GPS apps available commercially for car drivers or off-road hikers. Dedicated hand held GPS unit is also available. This can give them accurate location at any point and the better apps will also provide altitude information. Some survey software such as WWF promoted 'Smart' programme can use GPS in recording field sightings of wildlife and signs of human activity. Even with these sophisticated technical aids rangers should be equipped with back up physical map and compass and have some training in map reading and orienteering. Cell phone batteries can fail, GPS unit can get wet or dropped etc.

7.4 Use of appropriate equipment

Binoculars have been the main field equipment of ornithology for many years and a wide range of makes and sizes are available. The user must make a balance between weight and portability. Large bulky binoculars become a burden to carry on long outings but do offer advantages of stability and magnification. Weight is related to inertia so the image remains less wobbly. Magnification results in small angle of view and consequent difficulty in locating the subject for viewing.

In summary it is worth having large powerful binoculars if they are to primarily used from a fixed hide or viewpoint but smaller, lighter binoculars with wide angle of view if they are to be used primarily whilst walking.

There are very light narrow tube binoculars available which are so light and convenient to carry around but cheaper brands are very delicate and can easily be knocked out of alignment, they have narrow lenses so not much light for bright clear view and usually very narrow field of view. Most birdwatchers select binoculars with a magnification of 8 and an angle of view of 30°.

It is worth having high quality, strong binoculars with waterproof casing. It is also important to keep your binoculars dry and protected from knocks. Once out of alignment or with lenses full of humidity and fungus, the binoculars are virtually useless and may even cause strain damage to the users eyesight.

Spotting scope. In birdwatchers carry a spotting scope and a light tripod. This provides the viewer with much greater magnification and steadier view that possible with binoculars. This is almost essential if you want to count or identify birds at long distance on water bodies and shorelines. There are now small cameras that can be put onto the eyepiece allowing the spotting scope to double up as a large telephoto lens for photography. The spotting scope does mean carrying quite a weight of long and cumbersome equipment so means a compromise on what else you can carry in your pack.

7.4.1 Care of equipment

Optical and electric equipment costs a lot of money and is difficult to mend or replace so it is important that items are given due care and maintenance and staff assigned to use such equipment should be duly trained in such care and proper use. The following rules are appropriate:

- Maintain register of all equipment including details of where kept, who is responsible and current condition
- Nature Reserve equipment is for serious work purposes not for frivolous personal use.
- Keep equipment dry, store in dry place, protect from rain and spray, dab off moisture if equipment gets wet
- A good tip to remove moisture from equipment is to place equipment in warm sunshine to dry out
- Best way to ensure equipment remains dry is to seal in an airtight container with a bag of silica gel crystals. Silica gel is blue when dry but goes pink as it absorbs moisture. Pink silica gel can be revitalized by cooking in an oil-free wok over a heat source until blue again.
- Avoid exposure to dust and sand and salt
- Avoid extremes of temperature
- Avoid knocks, take care against dropping
- Wrap carefully when transporting

7.5 Training staff

Management of protected areas and especially wetland protected areas is a complex subject requiring a wide range of skills and knowledge—ecology, conservation, geography, hydrology, administration, GIS, data management, planning, law enforcement, communications, engineering, boating, social relations etc.

Recruitment of suitably trained staff is difficult. PA staff have often been transferred from different forestry units with no specific PA management skills. Recruitment of new staff is limited by strict quotas by local government and funds for hiring extra specialists are limited. The best solution is in-service training.

The booklet Competence Standards for Protected Areas Staff in China presents details of skills or knowledge (competencies) needed by staff at different levels and provides a framework for identifying the training needs among staff. The training needs assessment can then form the basis for organising appropriate training for relevant officers. This can help in selecting suitable officials for different posts in the management organisation and also identifying where additional training is needed.

A variety of self guided online training modules are being developed and a large number of useful training materials, tools and guidelines can also be downloaded from the internet.

However, there is no substitute for hands on training and the wetlands manager should try hard to ensure that external training can be brought to his staff from neighbouring universities, institutes, NGOs, visiting experts, projects or international programmes. Staff exchanges with other wetland protected areas is another way to spread experience and lessons learned.

The manager should also track training courses that may be being organised by other agencies or programmes and seek to get staff placed on these.

WWF have been offering regular wetlands training programmes at their Mai Po centre in Hong Kong since 1990 and are often able to cover the costs of training of wetland officers from mainland of China.

WWF　for a living planet®

世界自然基金會
香港分會

WWF-Hong Kong

香港新界元朗擔竿洲米埔沼澤
野生生物教育中心及自然保護區
Mai Po Marshes Wildlife
Education Centre and Nature
Reserve,Tam Kon Chau,
Yuen Long, N. T., Hong Kong

電话 Tel:+852 2526 1011
傳真 Fax:+852 2482 0369
maipo@wwf.org.hk
wwf.org.hk

Wetland Management Training Programme

Use of competency standards should go hand in hand with use of a national certification system to ensure that staff hired have the necessary skills to do their job. This applies to site managers who are often transferred from other disciplines with inadequate ecological understanding to manage a complex ecosystem.

7.6 How to apply for further support

Money is always a limiting factor determining what level of management can be applied or protection afforded. Government investments in wetlands protection continue to increase but are still far from adequate in most provincial or county level sites. So long as the value of healthy wetlands remains under-recognised, wetlands management will remain under-funded. So where to find the extra money?

Here are many ways:

- Applications to national and provincial sources
- Applications to international programmes
- Research applications for specific research needs
- Applications for eco-compensation funds
- Seeking corporate sponsorship
- Development of legitimate economic ventures (controlled tourism in carefully selected zones, within sustainable impact levels—carrying capacity)

Everyone else also wants any free money available in funding sources so application is a competition. Your case will be much stronger if you present a clear need and justification plus assurance that money will be used in responsible and transparent way.

Needs homework:

- Undertake a gap analysis of staff and budget needs to show that your site genuinely needs more funds and maybe more staff plus capacity.
- Undertake a preliminary economic valuation of the ecological service values and other benefits or income generation of your site to show that further investment is worthwhile.
- Tailor your application to the specific objectives and requirements of the funding source agency. A research project needs to tackle a scientifically significant or conservation urgent question, whilst application to an animal conservation fund needs to align against appropriate species and an application to a philanthropic fund needs to focus of addressing poverty, human rights, gender issues etc.
- Get high level endorsement of your request before application.
- Make the application very clear with transparent logic, targets, timeframe and indicators of success.
- Raise profile of your site by aspiring to achieve greater status and recognition. The higher up the national or international status ladder you can get the higher your justification for and likelihood of getting additional funding.

Justification and demonstrating your value:

- The site makes money
- The site delivers valuable ecological services
- The site is loved, used and supported by public
- The site offers valuable recreational and health outlets
- Protection is part of China's national and international obligations

Any application should contain the following content but make sure it fits with the templates or TOR of the individual target funding source. Keep the applications short, clear. Do not lose the wood for the trees by adding too much irrelevant detail or information. Reviewers of applications do not want to waste time on long, boring, poorly argues or badly translated applications.

Section	Notes
Title	Keep it brief and punchy. The title should reflect the objective and ring a bell with the mission statement of the funding agency. Put their own buzz words into it.
Objective	Very clearly state what is the purpose of the application.
Background information	Basic background details about the site. Where, how big, status, what its main function.
Importance of site	Added detail boosting the importance of the site—rare, endangered species, unique features or details about economic services delivered to wider society.
Threats to site	Ranked assessment of threats being faced. Do not make this so heavy as to frighten away donor as a no hope site!
Barriers to be tackled	Bundle the main threats you hope to reduce in terms of underlying barriers or causes behind those threats. One source of funds cannot tackle all the threats but select those most relevant to the funders mission.
Current funding status	Lay out your annual expenditure against estimated needs showing where there are gaps.
Justification for application	Show how the application meets the terms and conditions of the fund source and its mission.
Details of activities to be funded	Present the activities that you intend to undertake with the additional funds bundled into a series of outcomes.
Logical framework	A spreadsheet that clarifies the logic of activities arranged under their desired outcomes and/or outputs, themselves arranged under the identified barriers to be tackled.
Indicators	Usually inserted into the logical framework as verifiable achievements that indicate if targets are being reached.
Timeframe of activities	Spreadsheet of activities over the timeframe of the project with specific milestones identified.
Reporting schedule	Schedule of how often and what sort of reporting you can deliver to the funding agency or as demanded by the agency.
Detailed budget	Costs of the various activities or purchases listed. In the case of a cooperative project include the inputs of any other collaborating parties.
Endorsements	Letters of support, commendation and agreements with any collaborating agencies or partners.

Part Eight Monitoring and Reporting

To know whether our management is being effective or adequate, and to identify where we have problems , we need to monitor the conditions of both the health of their environment. This is a big and important job but the manager only has limited budget and manpower to invest in monitoring so we need to be very efficient in selecting meaningful elements to monitor and cost effective methods to do this. Biologists have developed literally hundreds of different methods for field surveys, tailored to specific problems posed by different species and different natural conditions. There are many entire books devoted to this subject but most methods are variants of some classic standard approaches. To start with we need to be clear what questions we want to address.

What? Wetlands manager needs to know what he is dealing with—what ecosystems, what species, what problems. He/she probably needs a basic qualitative inventory or stock taking survey.

How many? Counting takes a lot of time and is difficult. The manager needs to select which species or aspects he really needs quantified assessments for.

Where? Biota is not distributed evenly across a site. Surveys must establish which spots or habitat types are important for different species.

Patterns of change? Monitoring the population trend of a species is often more important than knowing the exact number. This is the only way to assess impacts on status over time.

Why? The wetlands manager often needs to know why changes are occurring but identifying the causal factors of natural processes is quite complex and often indirect. This may require professional analysis and specialized research.

Different types of monitoring are needed for different purposes.

Type of monitoring	Area of improvement
Climate	Improved resilience
Pollution	Identify and close pollution sources
Vegetation cover	Better planning, zoning and habitat management
Vegetation condition, phenology	Better understanding of faunal responses
Target species (numbers, breeding success, migrating species)	Improve survival and immigration rates and reduce mortality factors
Indicator species	Early warning of shifts in ecology
Alien invasive species	Better control of harmful species
Migrating species	Improved protection at global scales

8.1 Baseline inventory

Wetlands ecosystems are complex and dynamic. It is usually impossible to exactly predict how status of all species will respond to each change. Management depends heavily on regular monitoring. It is necessary to watch and record what happens and only prescribe intervention management when key species or habitats are becoming threatened by change. Each manager must deploy enough staff and time resources to patrolling and reporting on the conditions of key species, indicator species and problem species as well as physical or climatic conditions and levels of human activity.

It is not enough to merely record changes. It is important to understand what and why conditions are changing. It is advisable for each wetland site to have some ecological experience to help interpret monitoring results and translate these into recommendations to the management.

Monitoring of wetland biodiversity is not an isolated activity. Ramsar therefore stress the need to see monitoring as part of a larger framework including inventory, assessment, monitoring and management:

- establishing the location and ecological characteristics of wetlands (baseline inventory);
- assessing the status, trends and threats to wetlands (assessment);
- monitoring the status and trends, including the identification of reductions in existing threats and the appearance of new threats (monitoring); and
- taking actions (both *in situ* and *ex situ*) to redress any such changes causing or likely to cause damaging change in ecological character (management).

8.1.1 SFAs 2nd National Inventory

SFA have established a national inventory of wetlands. To ensure that the inventory can also serve international and Ramsar reporting requirements, the following core fields should be harmonised.

Revised core wetland inventory fields
(Harmonized with Ramsar ecological character description sheet)

Site name:

Official name of site and catchment/other identifier(s) (e.g., reference number)

Area, boundary and dimensions:

Site shape (cross-section and plan view), boundaries, area, area of water/wet area (seasonal max/min where relevant), length, width, depth (seasonal max/min where relevant)

Location:

Projection system, map coordinates, map centroid, elevation

Geomorphic setting:

Setting in the landscape/catchment/river basin—including altitude, upper/lower zone of catchment, distance to coast where relevant, etc.

Biogeographical region.

Climate:

Overview of prevailing climate type, zone and major features (precipitation, temperature, wind)

Soil:

Geology, soils and substrates; and soil biology

Water regime:

Water source (surface and groundwater), inflow/outflow, evaporation, flooding frequency, seasonality and duration; magnitude of flow and/or tidal regime, links with groundwater

Water chemistry:

Temperature; turbidity; pH; colour; salinity; dissolved gases; dissolved or suspended nutrients; dissolved organic carbon; conductivity

Biota:

Plant communities, vegetation zones and structure (including comments on particular rarity, etc.);

Animal communities (including comments on particular rarity, etc.);

Main species present (including comments on particular rare/endangered species, etc.); population size and proportion where known, seasonality of occurrence, and approximate position in distribution range (e.g., whether near centre or edge of range)

Land use:

Local, and in the river basin and/or coastal zone

Pressures and trends:

Concerning any of the features listed above, and/or concerning ecosystem integrity

Land tenure and administrative authority:

For the wetland, and for critical parts of the river basin and/or coastal zone

Conservation and management status of the wetland:

Including legal instruments and social or cultural traditions that influence the management of the wetland; and including protected area categories according to the IUCN system and/or any national system

Ecosystem services:

(for a list of relevant ecosystem services, see the Ramsar ecological character description sheet)

Management plans and monitoring programs:

In place and planned within the wetland and in the river basin and/or coastal zone

8.2 Selecting appropriate survey methods

Select a method that is suited for the species that need to be surveyed and is manageable within constraints of time, manpower available and urgency.

8.2.1 Line transects

Transect sampling for birds.

Use same defined transects on each survey. Again try to keep to similar time of day within main bird active period. Teams should be of two observers once looking left of survey line and one to right. Progress slowly and smoothly. Do not depart from transect line. Keep quiet. Record all birds of target species observed. Record angle from transect line and distance and number of birds when first seen. This can be calculated into right angle distances for analysis of results later. Record very accurately the total distance surveyed. Attach notes to report form on weather conditions and visibility. Visibility may change through the year with degree of leaf on trees, height of herb vegetation, fog, water level etc.

Note that birds are not randomly distributed, they aggregate in flocks or disperse rather evenly on territories. They defy most statistical assumptions. Also in line transects they do not match expected distributions from transect line. There is generally a lack of records in band strips close to transect line because the birds have fled to a safer distance away from you, or where someone already walked before you or where they know people regularly walk (e.g. dyke). This means some correction of data is required and in fact every species covered needs to be calculated separately as they have different behaviour and visibility. It is generally ok to lump all ducks or all geese etc.

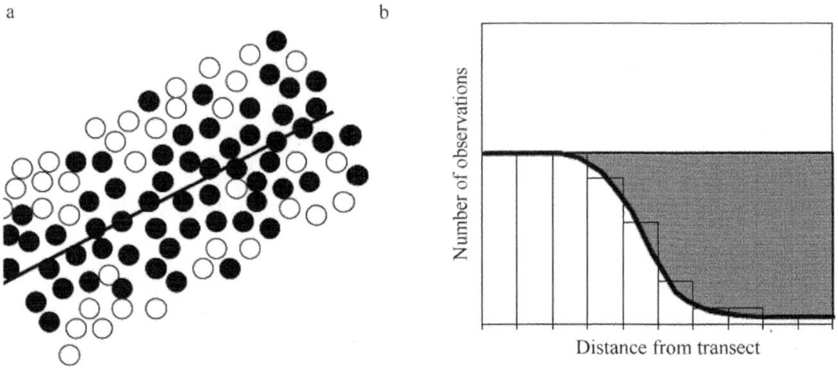

- Proceed along line transect (or existing trail) for measured length. Slow walking, from vehicle or airplane. Can be one side, both sides or predetermined strip width.
- Note all animals of target species sighted with estimate of perpendicular distance from transect line at moment of discovery (animal may have detected observer first and moved away from transect line before observer sees it).

Calculating density and population

- Density = number of individuals observed (sightings × mean group size)/(length of transect × effective strip width × how many sides surveyed)
- Population estimate for given habitat = density × area of habitat type available (this calculation assumes the sampled area is typical of the entire habitat)

The effective strip width is the mean distance of all sightings from the survey trail (see graph above). This is the point where as many sightings were missed within the strip width as were added beyond the strip width.

8.2.2 Spot counts

Point sampling for birds

Use same fixed point on all surveys. Undertake counts only in 0800-1030 or 1500-1700 time of day and for each station keep repeated surveys within one hour of each other. i.e. one station may logistically always be an 0800-0900 visit whilst another further away may be always a 0930-1030 visit. Use a tripod mounted spotting scope. Undertake 180 or 360° visual sweep counting and recording all birds of target species within 1km radius. Note visibility if less than this distance is sampled.

Roost counts

Position well in advance of dusk (1700hrs) on good vantage point to see all angles of sky approach to roost site. Record all flocks of arrival target birds (egrets) record time and number of each arrival. Add notes on weather conditions and visibility. Cease recording when light too dim for identification.

8.2.3 Sample quadrats

You do not have time to count over entire area so you can use a sample of quadrats either selected uniformly over entire area or selected randomly. Now count birds or whatever targets within the selected sample quadrats and scale up to get estimate for whole area. In the example below, 6 birds are found in a 20% sample intensity giving estimate total of 30. In fact the true total is 33. There is a small underestimate because the sample plots missed the concentration area in SE corner! If you know a species uses different habitats differently in overall density or degree of clumping, you can separately sample each habitat with higher (weighted) sample intensity in preferred habitat type. Add the sub-totals for each habitat to estimate total population.

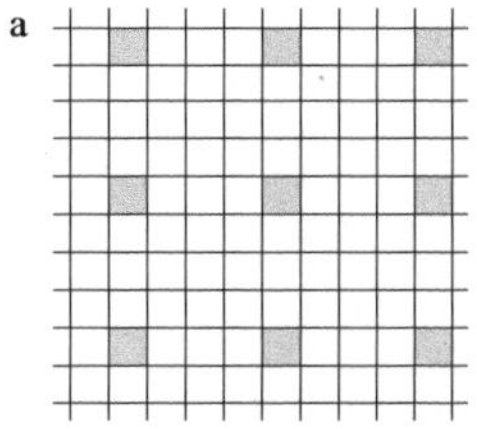

(a) Systematic sampling

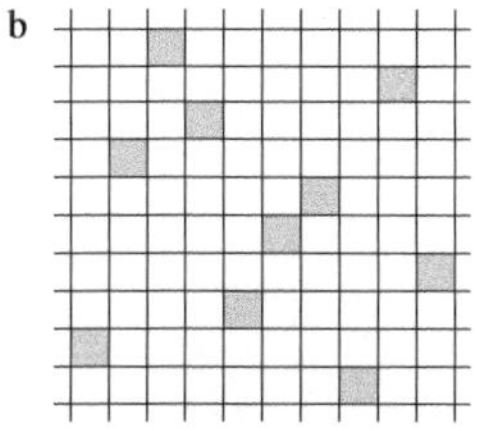

(b) Random sampling-Latin squares. The shadow squares are the sample quadrats

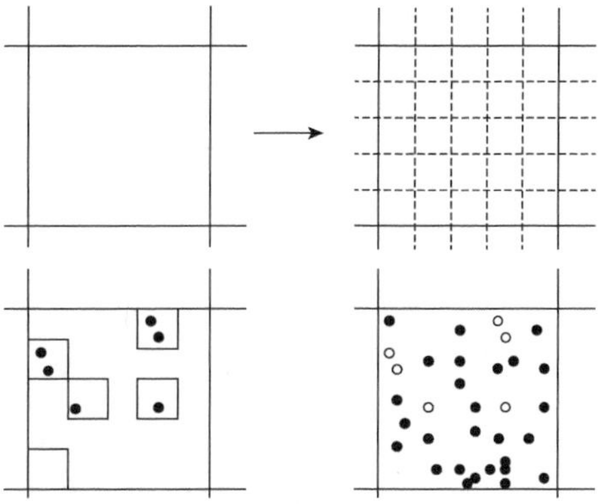

8.2.4 Approaches to monitor other species/factors

- Terrestrial mammals—field evidence (footprints, dung, feeding sign) or incidence in local markets.
- Common birds—regular counts from representative sampling points or transects
- Rare birds—total counts at peak season in their known favourite localities
- Insects—sweep samples or specialized traps e.g. moth trap (attracts flies, mayflies and other insects)
- Fish—fish catch, market availability and prices
- Aquatic mammals—sighting frequencies from boats patrolling known waters. Accumulated

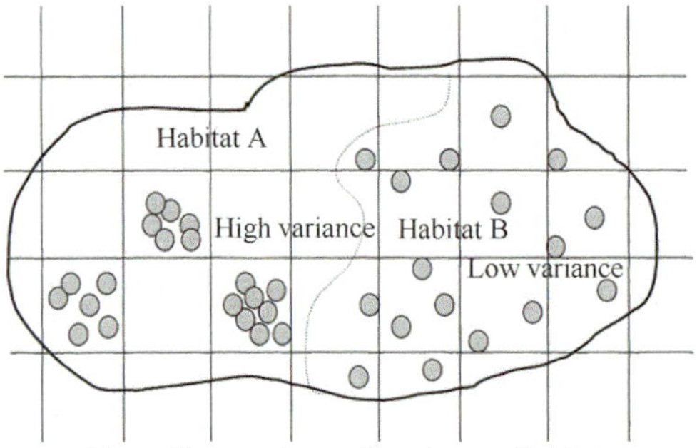

Nonuniform variance of two types of habitats. The black spots in the figure indicate birds

incidental observations.

- Invasive vegetation such as *Salvinia* and *Eicchornea* crassipes—monthly map of surface cover in selected sample plots combined with water measures of eutrophication (oxygen, nitrogen).
- Invasive crayfish—*Procambarus clarkia*—weekly monitoring of supply and price in market
- Habitat extent—Annual mapping from remote sensed images at lowest water level period.
- Habitat quality—annual biomass estimate for trees, monthly biomass estimate and height measurement for herbs in small permanent sample plots. Camera technique is available to measure canopy cover of trees in full leaf.
- Water quality—take samples monthly at selected points and test for oxygen content, nitrogen, phosphorous and coliform bacteria. Take samples in all cases of reported fish die off. Select points in ponds, shallow water, rivers and mid lake. Establish closer links to monitoring by provincial EPB.

8.2.5 MacKinnon lists and discover curves

This is a useful quick and robust method for assessing the species richness at a given site. A recorder documents the first n species he/she sees (hears) then starts a new list. For most moderate to rich sites it is recommended to use n=20 species on each list. In species poor localities it may be necessary to reduce the length of lists to 12 or 10 species.

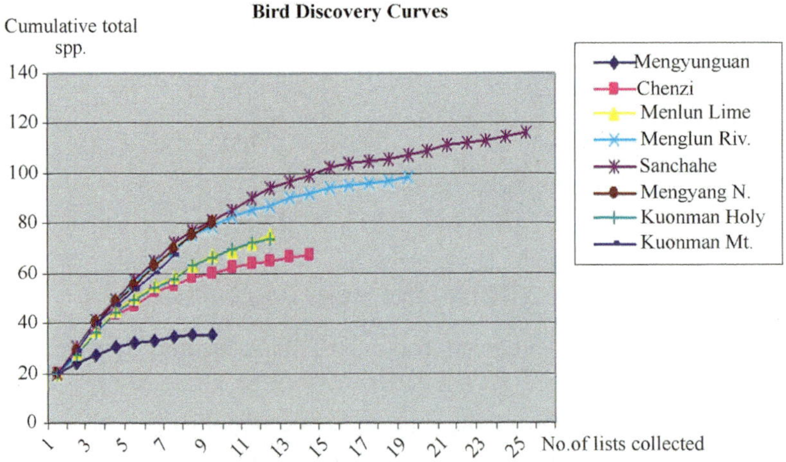

As the recorder progresses around the site to be surveyed more and more such lists are completed. It is then possible to plot the total number of species discovered against the total number of lists collected. The steeper the curve produced indicates higher species richness.

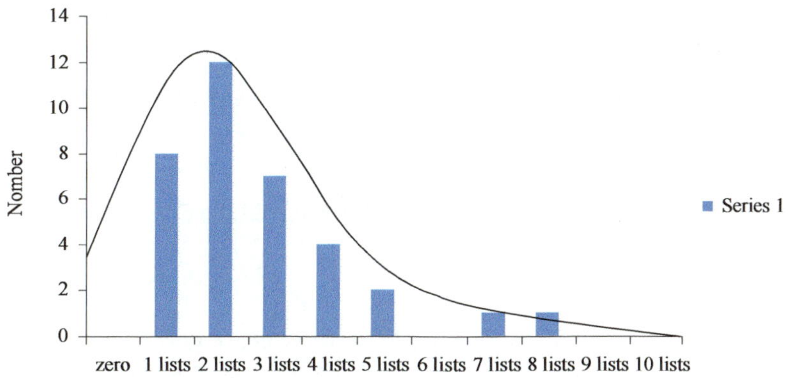

By plotting the data in another way as a histogram of how many species were recorded on 1,2,3 etc. lists respectively, one can graphically or mathematically (Poisson distribution) estimate the zero class of species still to be discovered. The sum of the zero class plus total already found is the best estimate of the number of species at that site at that time.

The method allows comparison between sites, between habitat types, between seasons etc. and more detailed analysis of the results obtained can allows comparison of the quality and composition of bird communities in addition to pure richness.

With minor modification the method can be used for assessing local richness of other taxa (insects, corals and even trees).

8.2.6 Making field sketches

When staff are familiar with the regular birds and other animals they will notice when they see something new and unfamiliar. If they have a camera or cellphone they should take a picture for later identification. Try to include an object of known size (coin, binocular etc. in the picture to enable the size of the unfamiliar species to be calculated).

Examples of how to draw field sketches

If the creature moves around too quickly or is too far away to take a photo then it is important to make a field sketch. Staff should practice making field sketches of familiar birds to develop some skill in getting shape proportions etc. correct and then should add notes about colour, patterns and even behaviour to the sketch.

8.3 Some standard field survey tools

8.3.1 Use of mist nets

Mist nets are an effective way of catching birds for research, ringing or monitoring purposes but unfortunately are also effectively used by hunters to catch birds for food, trade or protect crops.

Fine nets of various sizes are suspended between bamboo poles. Flying birds failing to see the nets get caught by falling into pockets where the loose nets form folds or get entangled in the mesh of the netting. Birds caught in such nets are very vulnerable to dehydration, starvation and cold so it is important to inspect nets at regular intervals to remove captured birds as quickly as possible. Birds can also be easily damaged by careless handling during extraction from nets especially if they get very tangled up. Extracted birds should be placed in soft cloth bags to prevent further struggling and reduce stress. They should be processed (weighed, measured, ringed, photographed or whatever and released as soon as possible.

In most countries the use of mist nets is strictly controlled and only certified experts are authorised to set such nets and remove and handle wild birds. Unfortunately in China there is no control on the use of these nets which are readily available on internet and other suppliers. Misuse of mist nets in China accounts for the loss of many millions of birds each year. Not least of which is their unnecessary official use on some China's airports as a way of reducing bird-strike danger on runways (see section 2.4.10).

Regulating and controlling the use of mist nets, rocket nets and other forms of animal capture is an urgent reform needed in China and should be included in revised wildlife regulations and site-level regulations.

8.3.2 Use of rocket nets

Rocket nets (canon nets) use multiple small rockets to propel a large net over a group of feeding or resting birds. Such devices are used internationally for the capture of large flocks of waterfowl for purposes of monitoring or ringing. Such devices are dangerous to both operators and birds so use should be strictly controlled and limited to only essential research programs.

8.3.3 Bird ringing programs

Bird ringing has been a founding block of our knowledge of ornithology. Ringing has a history of several decades and provides invaluable data on the movements, dispersal patterns, range sizes, migration patterns, longevity and health of many bird species. Bird ringing or bird banding is the attachment of small, individually numbered metal or plastic tag to the leg or wing of a wild bird to enable individual identification.

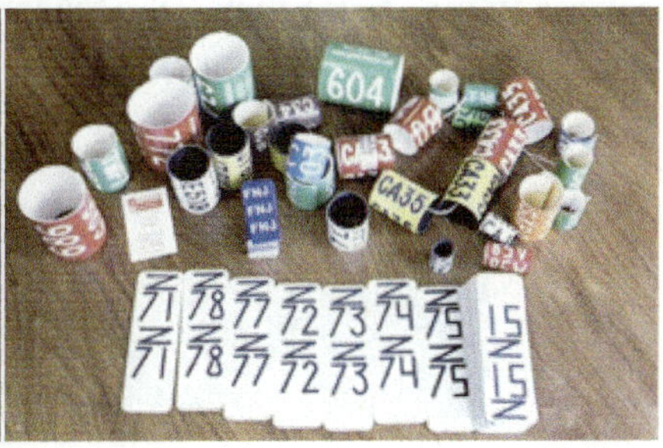

Most countries have their own national bird ringing (banding) programmes and most now conform with a smaller number of international programmes.

Birds are banded as fledglings in their nests or caught during their daily lives or on migration.

Original rings were made of aluminium stamped with a unique numeric code. The bird needed to be caught again to check the code number and identify the ringing record in a database. Numbered rings were issued in several different sizes appropriate to different sized species and fitted onto the lower tarsus by means of special pliers equipped with different sized ring holes.

Coloured plastic rings have been used in bird studies over small areas where a limited number of ringed birds could be individually identified by the colour combinations via binoculars or spotting scope without the need to recapture the individual.

Many studies now involve the use of numbered and coloured tags on the leg-rings or attached to wings to allow for easier identification at distance.

Some studies also now employ the use of small radio transmitters attached to wings of birds to monitor their movements.

Birds caught fro ringing or checking of rings can also be usefully weighed, measured and checked for health. Blood specimens may be taken for DNA analysis etc. All banding research should only be undertaken by fully trained and authorised personnel. All stages of capture, ringing, handling and release carry risks of injury or disease transmission to birds and these risks must be minimized and well below the value of the research findings. In China such bird banding is regulated through the National Bird Banding Center of China, founded in 1981 to administer bird banding issues in China. It is under the supervision of Research Institute of Forestry, Chinese Academy of Forestry, and National Bird Banding Office of the State Forestry Administration.

8.3.4 Radio tracking and use of GPS trackers

Radio tracking has been used to monitor larger mammals for some decades now. A mammal is trapped or immobilized by tranquilizer allowing biologists to fit a radio transmission collar that can then be followed by triangulation from two or more receivers over months or years. This data allows biologists to study movements, timings, home range, social interactions etc. of many species.

It is only recently that GPS radio transmitters are now made very small, some weighing less than 5 grams. Ornithologists have been fitting these to small harnesses to caught birds before release to see where they go. This is especially useful in trying to identify the routes, threats danger areas for migrating species.

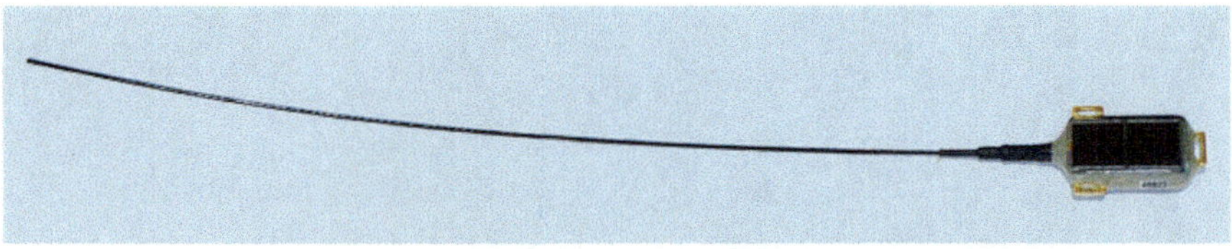

GPS tracking has been done with godwits, some other waders and waterbirds, cranes and cuckoos. Beijing Birdwatching Society is undertaking a programme to track cuckoos in this way. These birds are becoming increasing rare in Europe with problems clearly identified in the African wintering grounds. Numbers in some parts of China are also declining but we still do not know where the main problems are.

8.3.5 Use of automatic cameras

Automatic cameras are now cheap and effective. Use of multiple cameras in a grid pattern can be undertaken to estimate density and range size of individually recognisable species. Otherwise such cameras may be used to simply document presence of shy and rare species or to get decent photos for use in publications, websites etc.

Typically such cameras can operate in colour by daytime and black and white at night and can be set to still, video clips, wide angle or narrow focus. Batteries last several weeks depending on how many pictures or video shots are made, quality of batteries and temperature and humidity conditions.

Such cameras are usually set to cover terrestrial animals crossing an open target area, but they can be used to cover arboreal roosts, nest sites etc.

Setting camera traps needs skill and understanding of the behaviour of the target animal. Ex-hunters are already well skilled in identifying ideal spots to place such cameras overlooking frequently used animal trails, feeding or watering sites, scent marking points or resting sites.

Success can be increased by using bait, scent attractants or making guide fences that steer animals towards the camera.

Cameras may be stolen or destroyed by poachers, curious villagers, irritated elephants, bears etc. So it is worth setting in places less frequented by humans, concealing the camera as well as possible.

8.3.6 Use of camera drones

Camera drones are becoming increasingly popular and more cheaply available. These gadgets can be very useful in obtaining views of wetlands that are difficult to survey by land or boat. However, they do appear strange and potentially threatening to many waterbirds and care must be taken not to distress or frighten birds away.

8.3.7 Identification of footprints

It is nice to sit back home and let the automatic cameras monitor the wildlife but however many cameras you use, you are limited to the sites you have chosen and such cameras will never replace the need for persons on the ground undertaking routine patrolling and monitoring. A great deal of wildlife sightings information is gathered in this way and much of it is indirect. Mammals for instance leave many footprints in mud or sand that enable staff to know what animals are about and where. It is useful to train staff in the identification of local mammal identification of local

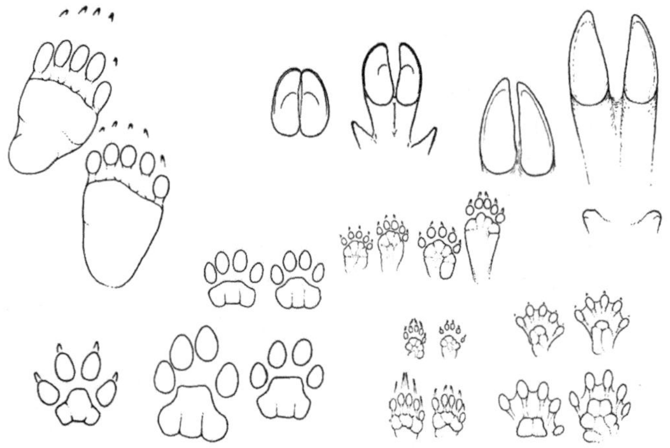

mammal tracks and even identification of their faeces or characteristic feeding signs, nests etc.

Where footprints cannot be identified or a permanent record is needed, it is common to make plaster casts. The following box indicates the steps in this procedure.

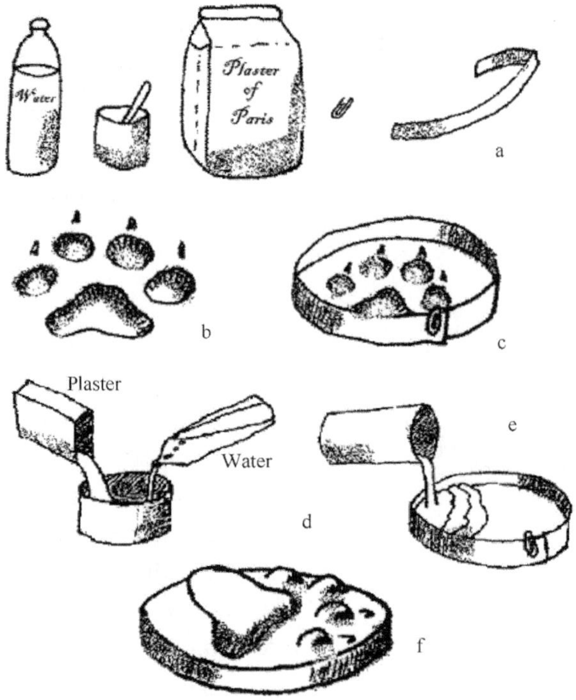

Making plaster casts of animal tracks

a. You need—Plaster of Paris, mixing container, water, paper clip and cardboard strip. b. Find a good clear track to cast in soft mud or damp sand. Carefully remove any loose leaves or sticks that have fallen into it. c. Shape the cardboard strip to build a wall around the track, held in place with the paper clip, taking care not to damage the track. Gently press the strip into the surrounding soil so the plaster will not run out from under it. d. Now mix the plaster. You should use about two parts plaster to one part water. Stir quickly to get rid of all the lumps. Tap the mixing container on the ground to remove any bubbles. e. Pour the plaster into your pre-prepared mould. Start with the finer details, such as claw marks, first. Pour it relatively thick to make a good strong cast. Let the cast set for at least 1/2 hour until matt and hard to the touch. f. Pick it up by reaching underneath it and lifting it. Do not lift by prying under it with a stick. This could crack it. Allow it to dry for several days before carefully cleaning or painting it.

8.4 Analyzing survey and monitoring data

Surveys and patrol reports generate lists, numbers and measurements but these data are meaningless in their raw form. They need to be analyzed and presented in ways that managers, leaders, public and media can understand, appreciate and act upon.

Numbers are meaningless without a yardstick by which to evaluate them. For example, the report that a survey team saw 200 green headed duck may be a very accurate fact but is meaningless without scale or context. Is that a lot or a little/What are we supposed to make of this information.

Compare the statement 'the team recorded 200 green headed duck in 2 hours from 3 stations. This is double the number counted last year and included 50 seen an area where never recorded before'. Suddenly the statistic is given some comparative meaning.

Consider the report that ' we say 200 green headed duck in sector A but only 105 in sector B.' Are we to infer there are more ducks in area A? We cannot make this conclusion without an idea of search effort. If we knew that the 200 ducks in area A took 50 hours of search at 10 stations whilst the 105 in area B only took 5 hours at 3 stations, we must conclude that in fact there are probably far more ducks in area B. The raw data cannot be fairly compared. The sample for area A is greatly biased by a much greater search effort. If we want to compare numbers from one sample to another we have to express everything in comparable units and eliminate as much bias as possible.

8.4.1 Minimize bias

Try to minimize bias by:

- Replicate conditions of method, sample sites, season, weather conditions, time of day, observer ability
- Use robust methodology
- Select commoner species to achieve large sample size
- Standardise search effort
- Use appropriate statistics

You need to express quantities in terms of comparable search effort such as mean rate of encounter, number seen per km walked, number seen per hour of observation, number caught in 10 net days etc. Even these units may be biased if observation conditions in one site are quite different from another. In the latter case you may need to use proportional figures like 60% of all duck sightings in area A were green headed duck, compared to only 40% in area B.

8.4.2 Use clear graphics

Expressing results in spreadsheets of numbers is also quite meaningless to most readers. Try to represent data in clear graphics. Most managers and leaders are familiar with histograms, pie-charts, time graphs and map insets. These can be used to effectively reveal the results of much survey data.

Provided the scores are comparable, you can show differences between different time periods, locations or species with simple histograms.

The following figure show such representations (dummy data).

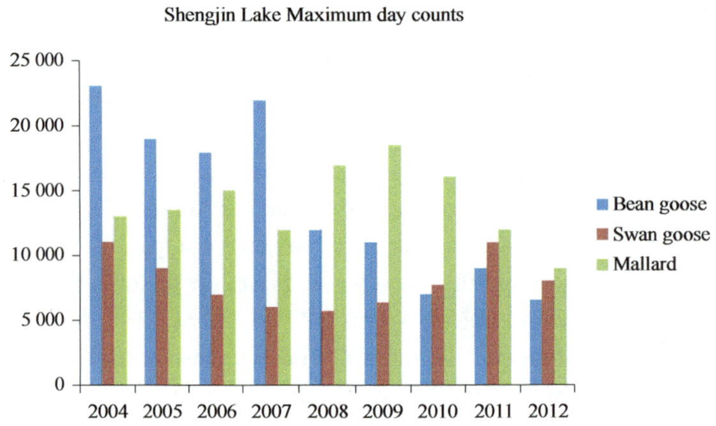

The following data (real) from Dongzhaigang shows many species combined through the year. Notice however the curious spike in numbers in mud summer. This is the presence of nesting herons and egrets and quite different from the peak of migrant waders in winter months.

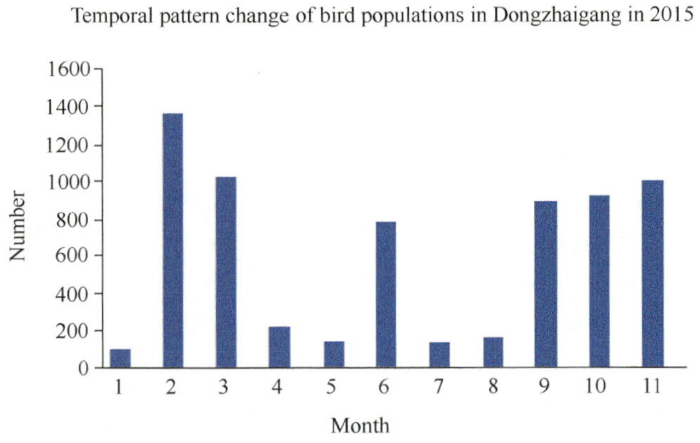

This shows the need to be careful when combining data from different species. Consider the following real data from annual estimates of the total birds (top line) in United Kingdom compiled over many years.

The data show a bit of variation but generally one can conclude that the overall numbers have not changed much. But look at the lower lines and we see that hidden in the overall

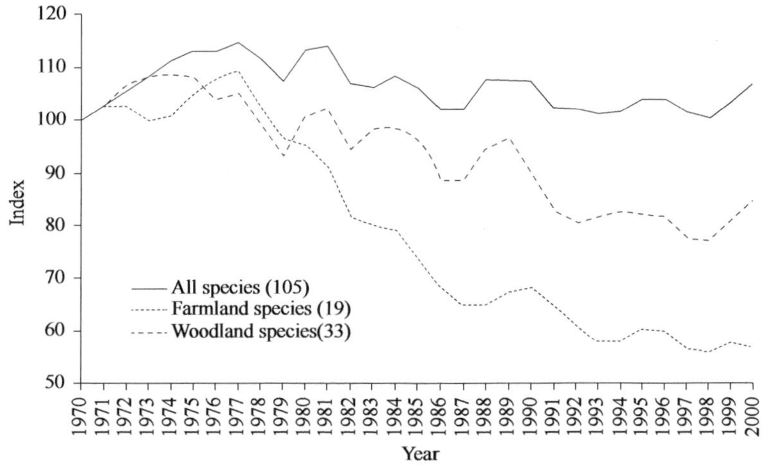

The UK Government's *Quality of Life* indicator showing population trends among common native breeding birds

satisfactory picture are serious declines in the numbers of woodland and farmland birds. The data needs to be carefully separated to see these details.

Spatial comparisons can be shown on a map. The following examples show how to display comparative data of different species or temporal distributions over large areas or a single lake. The last example shows real presentation of levels of human disturbance.

The latter graph raises the issue of correlation. Can we show that the levels of human disturbance affects the seasonal distribution of geese on the lake?

Here we need to enter the daunting realm of statistical tests.

8.4.3 What is a correlation?

Correlation tests whether there is a relationship between the values for two or more different factors. The correlation can be positive or negative or neutral and it can be very strong or weak. Use of appropriate statistics can give the answers and tell you whether the correlation is significant or not.

By significant the statistician means that the chances of such a degree of correlation being coincidental and the relationship being bogus are less than 55.

In the case of a direct comparison between two samples (e.g. seasonal use of space by two different species) a significant difference signifies that the chances of the two samples being drawn from the same statistical data pool is less than 5%.

Big caveat: A correlation may support the argument that one factor affects another but does not prove causation. Both factors may be varying in response to a third unidentified factor or combination of factors.

Bean goose, Swan goose, 2012 (dummy data)
Maximum counts per lake

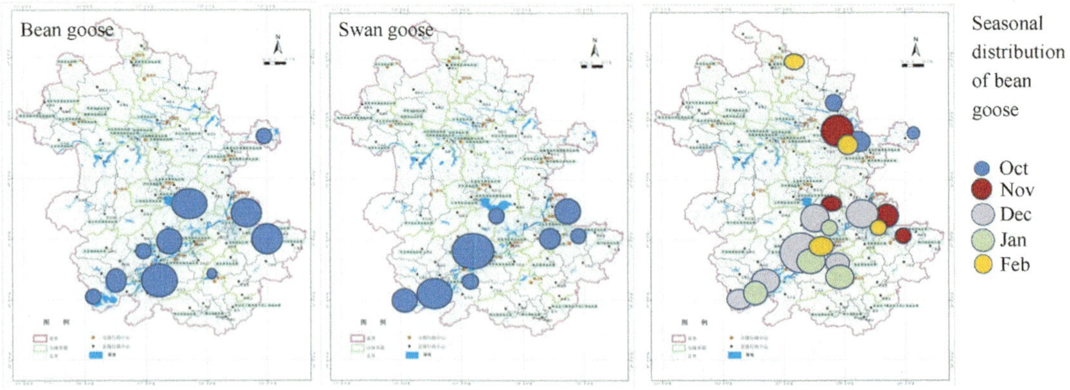

Seasonal use of single lake

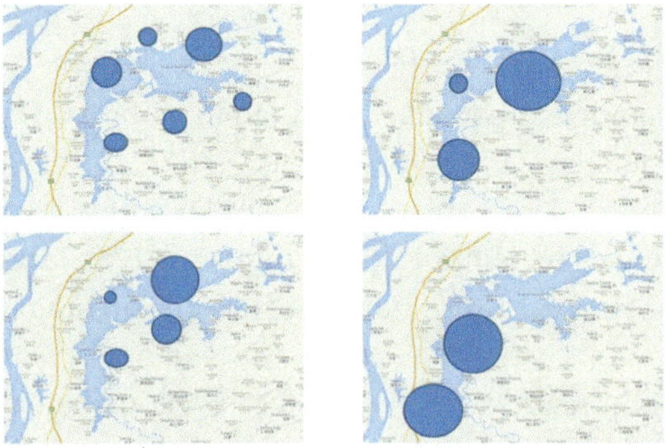

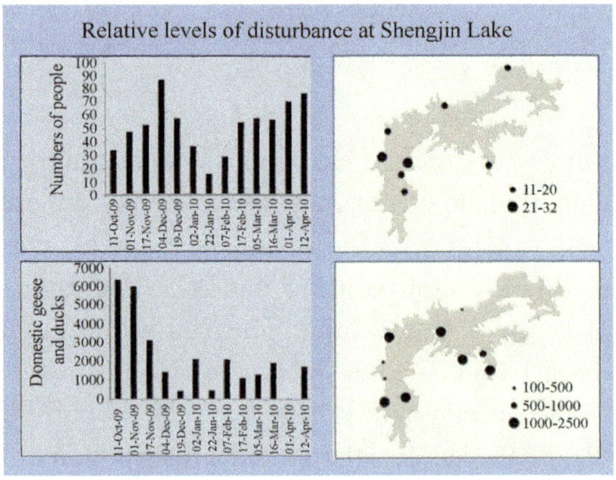

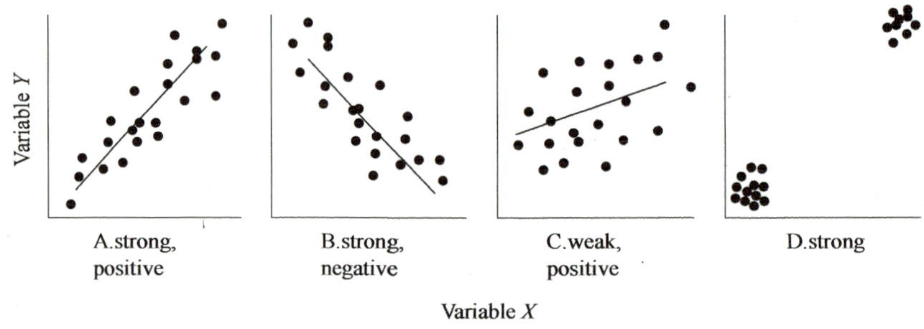

8.4.4 Longer-term data histories

Whilst the wetland manager is mostly dealing with documenting numbers, timings and spatial distributions, there are other important measurements that can reveal interesting data over much longer time periods such as growth rings on trees or age class structures that indicate whether a resident population is stable, increasing or declining.

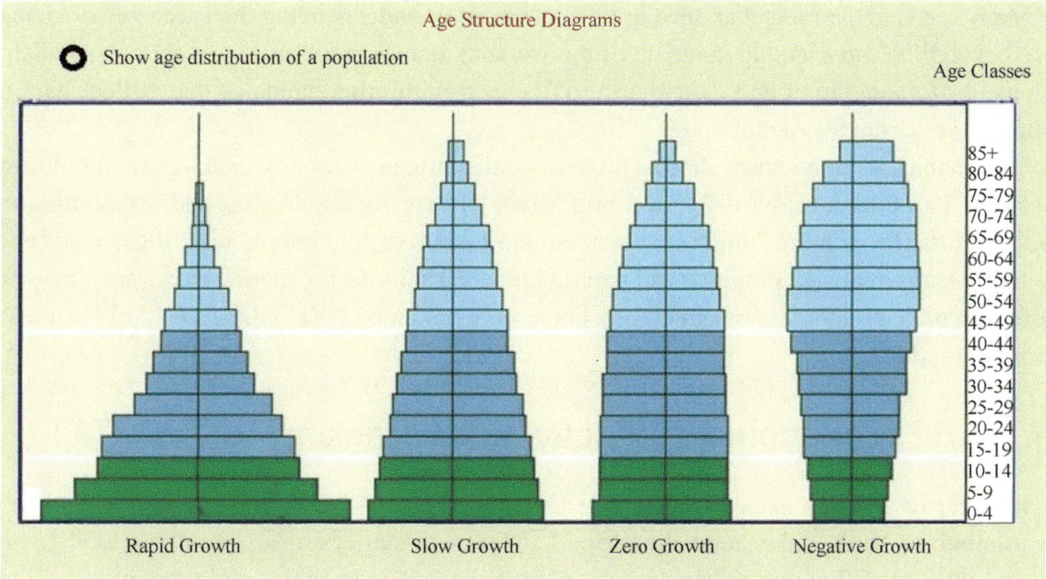

8.5 Data management, sharing and reporting

There is no point in expending great effort on monitoring unless the resulting data are used, analyzed, reported and can feed into further planning or management actions.

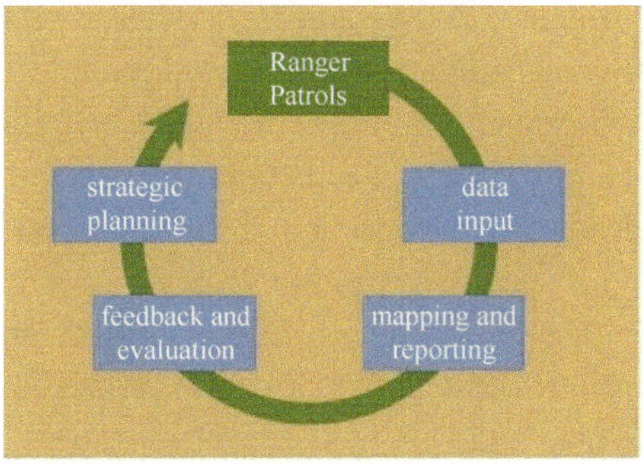

Raw data (lists, numbers, maps, forms, photos) cannot be used by managers, media, decisions makers, politicians. Data must be analyzed aggregated and reported in useful formats.

Direct relationships can usually be determined by standard statistics testing the correlation between the affected and causative factors. But ecology is complex. Variables are not independent, so do not conform to the requisites of standard statistics. There are many variables operating and few can be controlled so the process of multivariate analysis is required. If variables depart significantly from normal distribution then non-parametric statistics are required. In all cases data must be collected in an unbiased systematic manner which required strict protocols for data collection or clever corrections to counter the bias introduced by existing mode of data collection.

A good ecologist is required to design the data collection protocols, collate the resultant data and analyse the results in a revealing and convincing manner. Most wetland management bureaus lack such ecologists so it is usually important to establish relationship with a local university or technical institute to help in research and monitoring issues.

Many relevant data sets that need to be considered in understanding the processes of change are beyond the capacity and mandate of the wetland management bureau to collect, so a large element of cooperation and data sharing is needed with other agencies that collect data on climate, water quality, pollution etc.

Data management and data sharing are essential components of any biodiversity monitoring system. The internet now offers a marvelous platform for organising and assessing such information. Programmes must invest adequate resources in strengthening these aspects of communications at site, province and national levels. Facilities for sharing experience between different projects should be planned e.g. being undertaken by GEF, GIZ, ADB, WWF and WI support (and others).

8.6 Management Effectiveness Tracking Tool (METT)

In order to assist managers of protected areas, including those responsible for wetlands, a number of tools have been developed to help managers assess and respond to the effectiveness of their management planning processes and their implementation. Given that the wide range of situations and needs require different methods of assessment, the IUCN World Commission on Protected Areas (WCPA) has developed a 'framework' for assessment (M Hockings, S Stolton, F Leverington, N Dudley and J Courrau (2006); Assessing Effectiveness—A Framework for Assessing Management Effectiveness of Protected Areas, 2nd Ed. IUCN, Switzerland). This Framework aims both to provide some overall guidance in the development of assessment systems and to encourage standards for assessment and reporting. This and the more specific tracking tools provide managers and countries with mechanisms to assess their progress in meeting commitments under both the Ramsar Convention and the Convention on Biological Diversity's Programme of Work on Protected Areas and its targets.

The Management Effectiveness Tracking Tool (METT) is designed to track and monitor progress towards worldwide protected area management effectiveness. It is aimed at being cheap and simple site level tool to use by park staff, while supplying consistent data about protected areas and management progress over time. The Tracking Tool aims to:

- Identify progress on management effectiveness of protected areas;
- Provide baseline data on a protected area portfolio and assist with reporting and accountability;
- Identify portfolio trends and priorities for the development of appropriate tools and policies;
- Identify key management issues in a specific protected area and how to resolve these issues; and
- Identify appropriate follow-up steps, particularly at the site level.

The METT has been developed to help track and monitor progress in the achievement of the World Bank/WWF Alliance portfolio. It is now obligatory for all Global Environment Facility (GEF) protected area projects and has been used to develop a basic management effectiveness evaluation tool for several national protected area systems. It has been applied for many 'terrestrial' protected areas worldwide including some wetlands and Ramsar Sites.

8.7 Ecosystem Health Index (EHI)

Definition: Ecosystem Health is taken to be the suitability of a site to continue to provide secure conditions for survival of component species and delivery of key ecological services, including resilience to climate and other changes.

Objective: EHI is a not an evaluation. It is a dynamic, constantly varying index that reflects biodiversity health, just as a financial index reflects economic performance.

- EHI provides a baseline against which targets for maintaining or achieving a given level of health can be set
- EHI can be used as a results based indicator of project achievement and impacts
- EHI can indicate where the project is succeeding or failing and allow revision of activity efforts throughout the project
- EHI is complimentary to the Management Effectiveness scorecard in project M & E.

Introduction: Ecosystem health is reflected in the ability of a site to maintain its biodiversity values and ecological functions. These will vary significantly from site to site. Any index should include three components: 1) score of habitat suitability for important biodiversity; 2) status of important biodiversity and 3) the broader environmental context. The score does not necessarily indicate stability. Many wetland sites are very dynamic but what we are interested in is the ability of the biota to adapt to or even thrive with the changes. This will become increasingly important as climate and water flow patterns change. Each site using this index should undertake a baseline survey which also selects indicators and target species for subsequent surveys. Indicators should include key wetland birds, important aquatic fauna—fish, mollusks; selected indicator insects; endangered mammals; major components of vegetation; incidence of AIS.

The index establishes a snapshot value at the time of surveying; can relate present scores against baseline established at an earlier date, identifying trends in the different indicators; and can establish reasonable targets for improvement for each different indicator, and compare current state against identified targets.

Although a human body may not yet show much physical deterioration, we can identify several indicators of lifestyle that constitute health threats (excessive drinking and smoking habits, lack of sleep, lack of inoculation, living in region of known diseases, poor hygienic habits, lack of medical facilities etc.). In the same way we can recognize several threats to ecosystem health in the external context that may not be immediately reflected in condition of habitat or status of species. Such indicators include the levels of external development threats, the level of secure legal protection enjoyed, the level of human use pressures being applied or expected in the future.

8.7.1 Example of indicator species for Dongting Lake with rationale

Species	Indicates	Reasons for decline	Incidence of sampling
Common Kingfisher	Water clarity, small fish density, low human disturbance in shallow water	Water turbidity, lack of fish, high disturbance, some seasonal changes expected	Monthly counts along sample waterways
Pied Kingfisher	Small fish density in deeper water	Poor water quality and low fish density	Monthly counts along sample transects by boat
Cormorant	Large fish density and low human disturbance	Low fish stocks and human disturbance, lack of roost trees	Monthly counts at known roost sites
Grey heron	Large fish density at edges of waterways	Reduction of prey and high human disturbance	Monthly counts at sample points and transects
Night Heron	Ecological health of agricultural fields and wetlands	Lack of prey in fields due to overuse of insecticides	Monthly estimate at known roost site/sites
Egrets	Ecological health of agricultural fields and wetlands	Lack of prey in fields due to overuse of insecticides and pollution	Monthly counts at sample roosts yearround
Black drongo	Ecological health of agricultural lands, seasonality	Lack of prey in fields due to overuse of insecticides	Weekly counts along sample routes during winter
Long-tailed Shrike	Ecological health of agricultural lands yearround	Lack of prey in fields due to overuse of insecticides	Monthly counts along sample counts yearround
Coot	Vegetation quality in lake	Pollution of lake water	Monthly counts yearround
Pintail duck	General water fowl suitability	Human disturbance and reduced lake condition	Monthly counts through winter
Dragonflies	Yearround water quality and ecological health	Overuse of insecticides or pollution	Weekly counts on sample ponds and banks
Moths	Diversity and health of terrestrial flora	Low plant diversity and levels of insecticides	Moth traps, weekly

8.8 Indicators of water quality

Benthic macroinvertebrates are frequently used as a biological water quality indicator because they are abundant, easier to capture than fish, and because they are easier to identify than algae or protozoans. Macroinvertebrate samples can be collected using a Hess sampler in larger (fifth and sixth-order) streams or a Surber sampler in smaller streams. Macroinvertebrates are identified and enumerated, and the number of organisms at each site is estimated from the average of three same size sample areas. Benthic macroinvertebrate densities are reported as the total number of organisms per square meter of stream bottom. In addition to the total number of those organisms, measures of diversity particularly at the taxonomic level of order such as mayflies, stoneflies, beetles, and other organisms should also be noted. The Shannon index and the EPT index (= total species of Ephemeroptera (mayflies), Plecoptera (stoneflies), and Trichoptera (caddisflies) found in a site) measure the diversity and quality of an invertebrate community respectively.

Some bacteria, viruses and protozoa can harm humans and wildlife, from gastro-intestinal disease to minor respiratory and skin diseases. These organisms may enter waters through sewage, drains, septic tanks, farm runoff, animal processing plants and from wild life living in and around water bodies. Micro-organisms can being leached from the land into groundwater.

Because it is impossible to test waters for every possible disease-causing organism, it is usual to measure indicator very abundant species such as faecal coliforms such as *Escherichia coli*. For marine waters, a group of bacteria known as Enterococci is now commonly used. The presence of such bacteria indicates the possible presence of faecal material and, with it, the possibility that other, disease-causing organisms may be present.

Indicator bacteria are measured as a concentration, usually expressed as an estimate of the number of individual organisms per 100 ml of water. Water quality managers are interested in both the concentration of single samples, and in the "average" concentration of a series of samples taken over a period of time.

Dissolved oxygen

This is measured either as the concentration of oxygen dissolved in the water (expressed as grams of oxygen per cubic metre of water), or as the proportion of oxygen actually present relative to the theoretical oxygen-holding capacity of the water (expressed as "percentage saturation"). The latter measure is sometimes preferred because the ability of water to hold oxygen varies with temperature.

Dissolved oxygen is a basic requirement for a healthy aquatic ecosystem. Most desirable fish species (such as trout and salmon) suffer if dissolved oxygen concentrations fall below 3 to $4g/m^3$. Larvae and juvenile fish are more sensitive, requiring even higher concentrations of dissolved oxygen. Prolonged exposure to low dissolved oxygen conditions can suffocate adult fish or reduce their reproductive survival by suffocating sensitive eggs and larvae. Fish can starve when aquatic insect larvae and other prey die in response to the altered conditions. Low dissolved oxygen concentrations also favour anaerobic (without oxygen) bacterial activity that produces noxious gases or foul odours often associated with polluted water bodies.

Introducing large quantities of biodegradable organic materials, such as sewage or food processing wastes, into surface waters can rapidly consume available oxygen.

Bacteria use oxygen to decompose organic materials. Pollution containing organic wastes provides a continuous supply of food for the bacteria, which accelerates bacterial activity and bacterial population growth. In polluted waters, bacterial consumption of oxygen can rapidly outpace oxygen replenishment resulting in a net decline in oxygen concentrations in the water.

Other factors, such as temperature and salinity influence the amount of oxygen dissolved in water. Prolonged hot weather will depress oxygen concentrations and may cause fish to die even in clean waters, because warm water cannot hold as much oxygen. Warm conditions further cause oxygen depletion by stimulating bacterial activity, which consumes oxygen.

Temperature

Temperature is a fundamental factor in water quality, and the temperature exerts an enormous influence over aquatic organisms. If the overall temperature of an aquatic system is altered, a shift in community composition can be expected. Cold water fish such as salmon are very sensitive to temperature change, and as temperatures increase above about 20 ℃ suffer physiological stress. Fish are also potentially affected when swimming into localised areas of warm water.

Many factors affect water temperature. These include large fluctuations in air temperature, changes in the shape of stream channel and lake margins, reductions in overhanging vegetation, cloudiness, and most importantly, varying water flow. Wastes discharged into water can also affect temperature, if the effluent processing or treatment temperature is substantially different to the background water temperature.

pH

Acidity and alkalinity, the concentration of hydrogen and hydroxyl ions in water, drive many chemical reactions in living organisms. The standard measure of acidity and alkalinity is pH, and a pH value of 7 represents a neutral condition. A low pH value indicates acidic conditions; a high pH indicates alkaline conditions.

In tropical mangrove areas, high levels of acidity may accumulate to dangerous levels when sulphate soils are oxidized through exposure. It is safest to keep such soils wet at all times.

Shallow groundwater (usually from wells less than 30 m deep) can be slightly acidic—down to 6.0. This is because rainwater (itself slightly acidic) carries carbon dioxide (produced by plant roots and micro-organisms) into the underlying groundwater where carbonic acid is produced. Slightly acidic water is corrosive and can dissolve metals, especially copper, from pipes and pumps, into water bodies, affecting species composition.

Conductivity

This measures how electrically conductive the water is. Because conductivity increases with the number of ions (electrically charged particles) in the water, it indicates the presence of dissolved substances. Such substances may be naturally occurring minerals (see cations, and anions), or could be contaminants that are in the water as a result of human activities. Conductivity is usually measured in the field using a hand-held meter. A common unit of measurement is millisiemens per square metre, or mS/m^3.

Nutrients

The nutrients most often responsible for water quality degradation are nitrogen and phosphorus. Because these are found in the environment in a number of forms, water quality scientists measure them in different ways. For example, nitrogen present in water may be bound up in plant or animal tissue, in which case it is referred to as "organic" nitrogen. Such nitrogen eventually breaks down into "inorganic" forms; nitrate (NO_3), nitrite (NO_2) or ammonia (NH_3). Sources of nutrients include plant fertilisers, sewage effluents, animal and food-processing wastes, and urban stormwater.

Nutrients are essential building blocks for healthy aquatic communities, but excess nutrients (especially nitrogen and phosphorus compounds) over-stimulate the growth of aquatic weeds and algae. Aquatic weeds and algae out-compete the native submerged aquatic vegetation and can smother the habitat used by the aquatic fauna. Decomposition of excess weeds and algae can lead to oxygen depletion. Surface waters that have high concentrations of nutrients are referred to as "eutrophic". The adverse effects of high nutrient concentrations are particularly noticeable in lakes, where the nutrients are recycled through the same water, and tend to gradually accumulate.

Ammonia

One form of nitrogen found in water is ammonia (NH_3). Ammonia is a nutrient required for life, but, above certain concentrations it can be highly toxic to aquatic life. At very high concentrations, fish die. The chemistry of ammonia in water is quite complicated; toxicity depends on the proportion of the molecule which is un-ionised (i.e., exists as NH_3 rather than NH_4^+), which in turn is related to the pH and temperature of the water. Water samples are usually analysed for the concentration of nitrogen from total ammonia (as NH_4^+). Unionised ammonia (NH_3) is calculated from the concentration of total ammonia relative to the pH and temperature. Generally, the concentration of ammonia nitrogen in surface waters should not exceed about 1.5g/m^3. Sources of ammonia in waters include runoff of animal wastes and fertilisers from farm land, and the discharge of sewage effluents and food-processing wastes such as those from freezing plants.

Nitrates

Sources of nitrate are the same as those for ammonia. This is because an important mechanism for the formation of nitrate in soil and water is the breakdown (or "mineralisation") of ammonia. Nitrate occurs in groundwater sometimes from the decomposition of crop residues that leach down from the soil. This occurs when more nitrate is present than is required as a nutrient for plant growth. An important source of nitrate in agricultural systems is derived from nitrogen fixing plants such as legumes that capture nitrogen from the air. Also important as a source of nitrate in groundwater is urine from grazing stock.

Nitrate is measured as nitrate (NO_3), or as nitrate in the form of nitrogen (NO_3-N). Typical drinking-water standards maximum acceptable value (MAV) is 50mg/L of nitrate but may be exceeded in some areas where there is little dilution from rivers, and rainfall carries nitrogen from the soil into groundwater. Nitrogen content also increases in groundwater when the water table rises, and flushes nitrogen from the base of the soil profile.

Suspended sediment

Sediment consists of particles of all sizes, including fine clay particles, silt, sand, and gravel. In a water quality context the particles of greatest concern are the fine clays and silts. Sediment in the water column is usually referred to as suspended sediment, and measured as a concentration in g/m^3.

When sediment settles out it can severely alter aquatic communities. Sediment may clog and damage fish gills, suffocate eggs and aquatic insect larvae on the bottom, and fill in the spaces between gravel where fish lay eggs. Suspended silt and sediment interfere with aquatic pant growth by reducing water clarity

Sediment may also carry other nutrients and toxic heavy metals which attach to sediment and are then carried into surface waters. There, the pollutants may settle with the sediment or detach and become soluble in the water body.

Rain washes silt and other soil particles off all surfaces, but particularly those where the vegetative cover has been disturbed. Consequently, soil erosion, and activities such as earthworks, vegetation clearance, and cultivation can result in sediment movement into surface water, particularly after heavy rainfall. Stock trampling in the bed of a stream or trampling the margins and banks can release large amounts of sediment into the water.

Turbidity

A water quality measure that is related to suspended sediment is turbidity. This quantifies the degree to which light travelling through water is scattered by the suspended particles present. The greater the amount of suspended material, the greater the light scattering and the higher the turbidity. The light-scattering particles may be both organic (e.g., algae and other plant or animal debris) or inorganic (e.g., fine silts or clays). Reduction in light penetration reduces plant growth, which in turn reduces the food source for invertebrates and ultimately fish.

Turbidity is measured in a special type of light meter, and is generally expressed in Nephelometric Turbidity Units (NTU). An NTU less than 25 NTU is considered acceptable for aquatic life, but the appearance of the water is affected at much lower values than this.

Cations

Cations are a group of substances which, when dissolved in water have a positive electric charge. The main cations found in both surface and groundwaters are calcium (Ca^{2+}), magnesium (Mg^{2+}), iron (Fe^{3+}) and manganese (Mn^{2+}). Other cations include potassium (K^{+}) and sodium (Na^{+}). The presence of these ions is primarily a consequence of the geology of the area through which the water has flowed. For example, waters moving through limestone will contain high concentrations of calcium, magnesium and bicarbonate. These ions contribute to the hardness of water. Groundwater moving through dark coloured (mafic) volcanic rocks will generally contain lots of iron and manganese.

High iron and manganese concentrations can also occur in water from peaty deposits. Any iron or manganese present will be in solution in the ground, but will precipitate out of solution when oxygen is added as the groundwater is drawn up to the surface. High concentrations of manganese can pose a health risks to some wildlife

Anions

Anions are a group of substances which, when dissolved in water have a negative electric charge. The main anions of interest from a water quality perspective are sulphate (SO_4^{2-}), chloride (Cl^{-}), nitrate (NO_3^{-}) and bicarbonate (HCO_3^{-}). Anions may indicate the source of the water, or factors that influence its quality. For example, high concentrations of chloride in groundwater may indicate that saltwater is entering the system.

Toxic organic chemicals

Toxic organic chemicals are synthetic compounds that contain carbon, such as polychlorinated biphenyls (PCBs), dioxins, and the pesticide DDT. These compounds often persist and accumulate in the environment because they do not readily break down in natural ecosystems. Thus, they are often referred to as persistent bio-accumulative toxins. Many of these compounds cause cancer in people and birth defects in other predators near the top of the food chain, such as birds and fish.

Toxic hydrocarbons are used in petroleum products, refrigerators, insecticides, solvents, propellants, and cleaners. They can contaminate water as a result of spillage or disposal. Because hydrocarbons are frequently stored in underground tanks, they pose a potential risk to groundwater in the event of tank rupture.

Pesticides can contaminate water, including fungicides, insecticides, herbicides and growth regulators. Generally pesticides are transported into surface waters via rainfall runoff from areas such as road surfaces or farmland that have been sprayed. Occasionally they may be found in streams as a result of over-spraying or spray drift, or when they have been applied directly to stream channels to control aquatic weeds or kill snails (hosts of Schistosomiasis disease). Some pesticides are very mobile, and can leach through the soil into groundwater.

Heavy metals and metalloids

High density metals such as lead, zinc, copper and chromium and metalloids such as arsenic occur naturally in the environment. However, human activities (such as industrial processes and mining) have altered the distribution of such metals leading to both surface and groundwater contamination. In surface waters metals are usually found in association with sediments, to which they attach readily. Animals higher up the food chain, such as fish are likely to accumulate metals in increasing concentrations.

8.8.1 Monitoring water quality

Measuring and monitoring the water quality for human safety is the responsibility of the Ministry of Environmental Protection (MEP) and the work is highly technical requiring a wide range of instruments and laboratory facilities. However, ecologists and wetland mangers may want to undertake some basic and simpler tests independently. The following table indicates how this can be done.

Factor	Method	Frequency	Equipment	Note	Safety level
O_2	Smell for methane or sulphur. Collect water in sample bottle	Monthly	Clean collecting bottles; standard chemistry lab.	Key to fish and other animal health	> 3 to 4g/m^3
Temperature	Measure at different depths	Monthly or in extreme weather events	Thermometer	Water holds more oxygen when it is cold	Temperate < 20℃ Tropical < 32℃
pH	Sample from several points of a large wetland	Monthly	pH meter, Litmus paper	pH rising due to excess CO_2 or lime spillage	5-8
Turbidity	Visual observation or depth of sediment resulting from standard sample	Monthly	Sampling bottle		
E. coli (bacteria)	Collect water samples in sterile bottles from several points	Monthly	Send to laboratory	General measure of biotic pollution. Low levels are normal	
N				Best indicator of eutrophication in fresh water	>10mg/L
P				Best indicator of eutrophication in fresh water	
Hg, S, Pb,Cu		Annual or monthly if close to discharge source	Send to laboratory	Heavy metals are very toxic	
Organic compounds	Visual oil on surface. Collect sample in clean bottle	Annual unless you suspect a new spillage	Send to laboratory	Most organic compounds are highly toxic	
Cl and other halides				Disinfectant and toxic	
Na, K	Marine salinity can be roughly estimated by tasting	Monthly from several points	Simple voltmeter can measure electric conductivity	Salinity can be a problem in some freshwater systems	

Appendix

Appendix 1 International wetlands organizations

Title	Scope	Website
Wetlands International	Wetlands International is a global organisation that works to sustain and restore wetlands and their resources for people and biodiversity	www.wetlands.org/
Royal Society for Protection of Birds, UK	The RSPB is a UK-based charity working to secure a healthy environment for birds and all wildlife, helping to create a better world for everyone.	www.rspb.org.uk/
British Trust of Ornithology	Independent charitable research institute combining professional and citizen science aimed at using evidence of change in wildlife populations, particularly birds, to inform the public, opinion-formers and environmental policy- and decision-makers. UK focus but participates in many global projects.	www.bto.org/
National Audubon Society	Mission is to conserve and restore natural ecosystems, focusing on birds and other wildlife for the benefit of humanity and the earth's ecosystems. Has many global partners to help birds that migrate beyond USA.	www.audubon.org/
Wildfowl and Wetlands Trust, Slimbridge	One of the world's largest and most respected wetland conservation organisations working globally to safeguard and improve wetlands for wildlife and waterfowl with global outreach.	www.wwt.org.uk/
International Crane Foundation	Foundation specifically dedicated to conserving all 15 global species of cranes through broad commitment to the people and places essential to cranes.	www.savingcranes.org/
East Asian Australasian Flyway Partnership	The Partnership (EAAFP) aims to protect migratory waterbirds, their habitat and the livelihoods of people dependent on them.	www.eaaflyway.net/
Hong Kong Bird Watching Society, China	Aims at appreciation and conservation of Hong Kong birds and natural environment and set up the *HKBWS China Conservation Fund* in 1999 so as to support the birdwatching promotion and research works of birdwatchers and ornithologists in China mainland.	www.hkbws.org.hk/

Appendix 2 Wetland biodiversity conventions/programmes

Title	Date	Scope	Website
Convention on Biological Diversity	1992	CBD is a global agreement addressing all aspects of biological diversity: genetic resources, species, and ecosystems.	www.cbd.int/

Continued

Title	Date	Scope	Website
The United Nations Framework Convention on Climate Change (UNFCCC)	1994	International environmental treaty negotiated at the United Nations Conference on Environment and Development (UNCED)	newsroom.unfccc.int/
Convention on combating desertification	1996	to forge a global partnership to reverse and prevent desertification/ land degradation and to mitigate the effects of drought in affected areas in order to support poverty reduction and environmental sustainability	www.unccd.int/en/Pages/default.aspx
Convention on Wetlands	Ramsar, Iran 1971	conservation and wise use of all wetlands through national actions and international cooperation, as a contribution towards achieving sustainable development.	By 2010, 160 nations have joined the Convention and more than 1900 wetlands, covering more than 186 million hectares, are designated for inclusion on the Ramsar List of Wetlands of International Importance. www.ramsar.org/
Convention on the conservation of migratory species of wild animals CMS	1979	a global platform for the conservation and sustainable use of migratory animals and their habitats	www.cms.int/
East Asian Australasian Flyway Partnership EAAFP	2006	protect migratory water-birds, their habitat and the livelihoods of people dependent on them.	www.eaaflyway.net/
Convention on international trade in endangered species of wild fauna and flora CITES	1975	International agreement between governments. Its aim is to ensure that international trade in specimens of wild animals and plants does not threaten their survival.	www.cites.org/
Global Invasive Species Programme GISP	1997	A project of DIVERSITAS aiming at preventing and managing invasive species	www.diversitas-international.org/activities/past-projects/global-invasive-species-programme-gisp/

Appendix 3 Best practices guidelines

Appendix 3.1 IUCN World Commission on Protected Areas (WCPA) best practice guidelines series

BP1: National Systems Planning for Protected Areas

BP2: Economic Values of Protected Areas

BP3: Guidelines for Marine Protected Areas

BP4: Indigenous Peoples and Protected Areas

BP5: Financing Protected Areas

BP6/BP14: Evaluating Effectiveness: A Framework for Assessing the Management of Protected Areas

BP7: Transboundary Protected Areas for Peace and Cooperation

BP8: Sustainable Tourism in Protected Areas

BP9: Management Guidelines for IUCN Category V Protected Areas Protected Landscapes/ Seascapes

BP10: Guidelines for Management Planning of Protected Areas

BP11: Indigenous and Local Communities and Protected Areas: Towards Equity and Enhanced Conservation

BP12: Forests and Protected Areas: Guidance on the use of the IUCN Protected Area Management Categories

BP13: Sustainable Financing of Protected Areas

BP14 (see BP6)

BP15: Identification and Gap Analysis of Key Biodiversity Areas: Targets for Comprehensive Protected Area Systems

BP16: Sacred Natural Sites: Guidelines for Protected Area Managers

BP17: Protected Area Staff Training: Guidelines for Planning and Management

Appendix 3.2 Ramsar Handbook series for the wise use of wetlands

1. Concepts and approaches for the wise use of wetlands
2. National Wetland policies
3. Laws and institutions
4. Avian influenza and wetlands
5. Partnerships
6. Wetland CEPA
7. Participatory skills
8. Water related guidance
9. River basin management
10. Water allocation and management
11. Managing groundwater
12. Coastal Management
13. Inventory, Assessment and Monitoring
14. Data and information needs
15. Wetland Inventory http://www.ramsar.org/sites/default/files/documents/pdf/lib/hbk4-15.pdf
16. Impact Assessment
17. Designating Ramsar Sites
18. Managing Wetlands
19. Addressing change in wetland ecological character
20. International Cooperation
21. The Ramsar Strategic Plan 2009-2015

Ramsar Technical Report No. 3 CBD Technical Series No. 27 Valuing wetlands https://www.cbd.int/doc/publications/cbd-ts-27.pdf

Appendix 4 Some tips on field-work in wetland sites

Appendix 4.1 The bare-foot doctor

All field staff should be given some minimal instruction about health, safety and how to treat common wounds and problems. The following rules apply.

Keep dry

The weather may change, you may get wet and it may get cold. Wet clothes greatly increase heat loss. So stay dry. Always take a waterproof cape or jacket. Take a dry set of clothes wrapped in plastic bag in case you get wet on a longer trip.

Keep warm

Always wear enough clothes. You will feel colder in a wind or if you get wet and colder in evening or if you have to stay out at night. Useful to keep some dry clothes in a plastic bag. Use shelter to reduce exposure and huddle together for warmth if you start shivering.

Wear life jacket

No need to take risks. Even if you are a strong swimmer you may get knocked into water whilst boating. Wear a life jacket and make sure anyone you are taking on a boat wears on also.

Mind the gap

Many boating accidents happen when someone gets trapped or crushed between a boat and a solid jetty or other larger boat. Crushed bones are hard to heal. Take great care getting in and out of boats.

Take a water bottle of safe water

Take enough water to last a whole day. You may trust some clear upland streams for drinking or topping up your water bottle but be aware most streams and rivers 9in China are very polluted, especially near mining areas or downstream of industry or settlement. In the lower Yangtze valley the water is also infested by Eastern Schistosomiasis parasites that are hosted by water snails and can infect humans causing illness.

Use a map

Make sure you have a clear route in mind and if far from main roads or clear physical features, take a map with you. Maps are rather essential for off trail walking, surveys, monitoring etc. You may be able to have a digital map on a cell phone (see below) but not all areas of off-road terrain are covered by cell phone signal.

Cellphone

If there is coverage of cell phone signals, It is wise to carry a charged cell phone. You can call someone if you get into trouble, get lost or have an accident. It is increasingly common that you can also have GPS signal on your phone and have mapping app to help see where you are.

Cuts and scrapes

Clean wounded area with water or alcohol. Dry. Apply iodine or antiseptic powder if available. Cover wound with clean dry dressing. Keep dressing not too tight and allow breathing of would. Dry out wound e.g. in sunshine as soon as possible.

Insect bites, stings

Resist scratching. This only increases the itch and chances of infection. Apply alcohol, eucalyptus oil or antihistamine cream. In areas with many biting flies wear a netted hat.

Snakebite

Do not panic. Move very slowly. Keep heartbeat minimal. If snake thought to poisonous you can open puncture holes with a knife blade to promote faster bleeding and may be able to remove some of the poison by sucking the wound or applying vacuum from a half rubber ball or absorbant material. There is no need to tie a tourniquet.

Photograph or collect the snake for identification. Use phone to notify others of your locality. Make your way slowly to nearest road, river junction etc. to find transport to medical help.

Drowning and artificial respiration

If a patrol team member stops breathing as a result of drowning, coma after accident, ingestion of drugs or poison (e.g. from snake-bite), it is more urgent to restore breathing before attending to any other aspect of injury.

There are several different techniques for giving artificial respiration. The most widely taught and accepted technique goes by a variety of names, such as "mouth to mouth" or "rescue breathing". Air is expelled from the rescuers lungs directly into the victim's mouth. Assuming a tight air seal, the air is forced into the victim's lungs, and the rescuer watches to see the victim's chest rise with each breath.

To perform rescue breathing perform the following steps:

1. Check the mouth for obstructions (tongue, vomit etc.), lift the neck and tilt the head back.
2. Pinch the nostrils and seal the mouth, and exhale directly into the victim's mouth.
3. Release the nostrils and the seal around the mouth.
4. Watch for the victim's chest to rise by itself.
5. Feel for a pulse on the victim's neck.
6. If the victim's chest does not start to rise on its own, repeat this process from number 1, until professional help arrives.

Rescue breathing should continue at a normal breathing pace of about 12 times per minute, until the victim is fully able to breathe on his or her own. Even if the victim appears able to breathe on their own, be sure to keep a close watch since relapses are common.

Broken limbs

If possible call for medical help. For simple fracture, pull limb into normal extended position and make a splint by binding both sides of the break to a stiff piece of wood.

For more complex crushed or twisted bone fractures it is best to stay still and wait for medical team to arrive. Keep warm.

Making a splint

A splint is a rigid support used to keep an injured body part from moving and to protect it from any further damage. It is often used to stabilize a broken bone while the injured person is taken to the hospital for more advanced treatment. It can also be used if you have a severe strain or sprain in one of your limbs. If you or a partner is injured in the field, you can create a temporary splint from materials around you such as a stick bound in place by string, bandage, shredded clothing or fibrous tree bark.

- Attend to any bleeding before you attempt to place the splint by putting pressure directly on the wound, then, apply a bandage, a square of gauze, or a piece of cloth.
- Do not try to move the body part that needs to be splinted—you may accidentally cause more damage.

- Position the splint so that it rests on the joint above the injury and the joint below it. Then, tie or tape it to the limb on either side of the injury, avoiding placing ties directly over the wound. Fasten tightly enough to hold the body part still, but not so tightly that the ties will cut off the victim's circulation.
- Check around the splint every few minutes for signs of decreased blood circulation. If the extremities begin to appear pale, swollen, or tinged with blue, loosen the ties.
- The injured person may be suffering from shock if they are feeling faint or taking only short, rapid breaths. In this case, try to lay the person down without affecting the injured body part. If possible, you should elevate the legs and position the head slightly below heart level.

Applying a Tourniquet

Tourniquet is a temporary tight binding that limits blood flow to and from a part of the body. It can be used to prevent excessive bleeding from a deep cut or can be used to prevent poison from a snakebite spreading to the rest of the body. It's important to learn how to apply a tourniquet because improper technique (or leaving it on too long) can actually lead to dire complications, such as tissue death and amputation. The idea is to cut off the strong blood flow within arteries leaving the heart, not the more superficial veins returning blood back to the heart.

However, if the wound is too severe and the bleeding cannot be stopped with applied pressure, then (and only then) should you consider a tourniquet.

- If left uncontrolled, bleeding will eventually lead to shock, then death.
- Even if you have to use a tourniquet, leave the make-shift bandage on the wound because it will help promote clotting when the blood flow slows down.
- Elevate the wound if possible. Often the combination of pressure and reducing the pull of gravity on the flow of blood in the vessels will be enough to stop the bleeding and allow clot formation.

Carrying an injured person

Injured persons should be taken to clinic or hospital as soon as possible but if you are far from a road the injured may need to be carried. The easiest way to carry a person is the so-called fireman's lift but before attempting this you must first determine if any limbs are broken and the spine is not damaged. Broken limbs may need splinting before the injured person is carried. In the case that the injured person experiences extreme pain in moving the neck or back a stretcher should be used to prevent further movement of those parts, which should also be well cushioned to prevent lateral movement during carriage.

A improvised stretcher can be made by finding two strong poles approximately 2.5 m long and slipping these through the arms of two jackets. Tie the bottom corners of the jackets together to form a strong sling. Place the patient on the sling taking great care not to bend or jolt the body further and giving particular support to the neck. If you have a blanket, this is better than the two jackets. Place the poles on top of the blanket across its narrow width about 70 cm apart. Fold the longer ends of the blanket over and round as many times as it will go. Place the injured on the blanket. The persons weight will hold the blanket in place so it does not unroll during carriage.

Appendix 5 Ecological principles for wetlands management

Reproduced from *Biodiversity Principles for Developers and Planners*. 2002. John MacKinnon, (ed). BWG/CCICED, Beijing, China.

建立具代表性的保护区系统

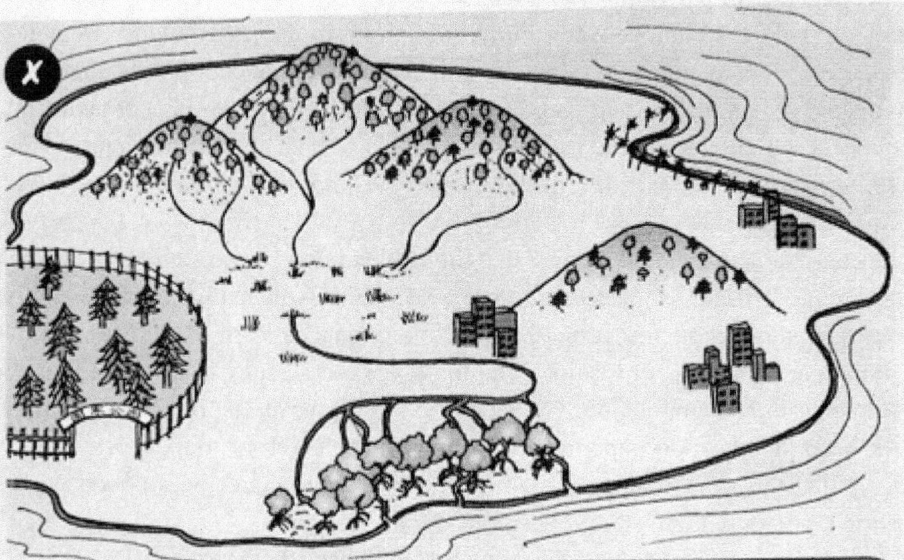

应该建立一个保护区体系，它应该具有代表性，保护着中国所有自然生态系统、栖息地和各种海拔范围内的典范地区。地点应选择在物种丰富或当地特有种集中分布的地区。面积要足够大，以实现可持续性发展。关键物种的种群要达到能够自我维持的数目。大的、连通的、变化多的地点，比小的、孤立的或分散的地点要好。

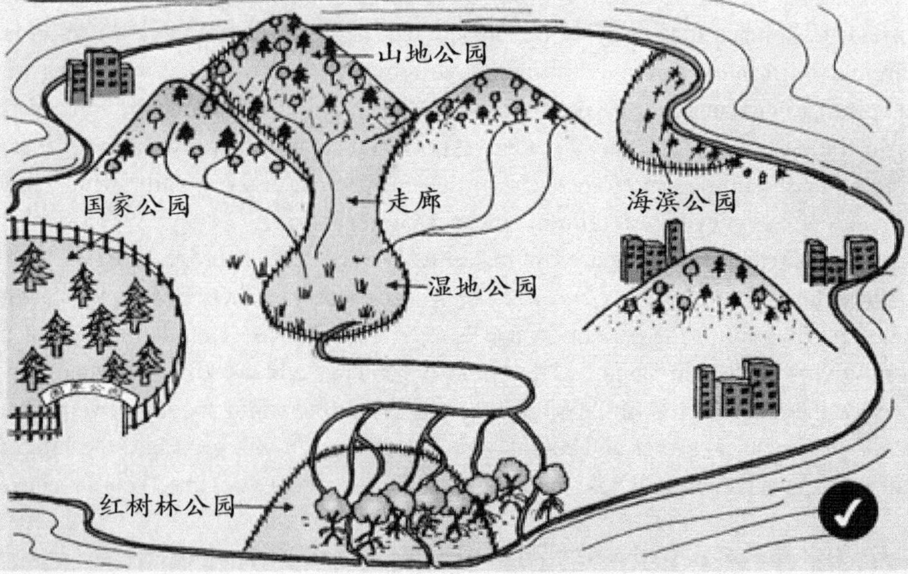

禁止在重点湿地内开垦农田

围湖造田和开发破坏了湖泊的生态系统，降低其蓄水（海绵）功能，使重要的生物多样性丧失，破坏微气候和当地湖泊河流连通体系的生态平衡。为了改善人类的健康水平、提高生产力和环境质量，就必须维护湖泊和天然水系的连通渠道，不要在重要地区开垦农田。保护中国健康的水文功能比生产更多的粮食作物要重要得多。

保持淡水系统的连通性

在水系通道设立人为屏障将影响当地鱼类和其他水生动物的迁徙。必须保持堰坝和其他屏障水体间的连通性，才能保护水生生态系统，使成年鱼类能够到达繁殖区，以及鱼苗能够到达采食栖息地。

维护水系统的排蓄水功能

湖泊在干旱季节是水的来源地，而河流却是雨季的排水渠道。保护湖泊河流的这种功能和防止淤塞是相当重要的。必须在集水区和所有必要的地方保持良好的森林覆盖以便减少河流上游土壤侵蚀，挖走多余的泥土和沙砾，避免河道淤塞。

保护水域和湿地的生物多样性

提高湿地生态系统的连通性，降低污染，避免过度捕鱼和水流量减少的威胁。管理好湿地生态系统以提高生产力、生物多样性、娱乐和旅游潜力，也有助于调节小气候。通过建立保护区或控制捕鱼、狩猎和植被采集来加强保护这些生态系统的重要组成成分。

把捕鱼量限制在可持续性范围内

过度捕鱼减少了单位捕鱼量，破坏了鱼类的多样性，降低了鱼的平均大小和使不受喜欢的鱼类扩散。如果传统的捕鱼方法已经使捕鱼量过高，那么希望通过更大面积和更现代的捕捞方法来达到增加捕鱼量的目的是不可能的。小规模的传统捕鱼方法能在社会上更加公平地分享利益。

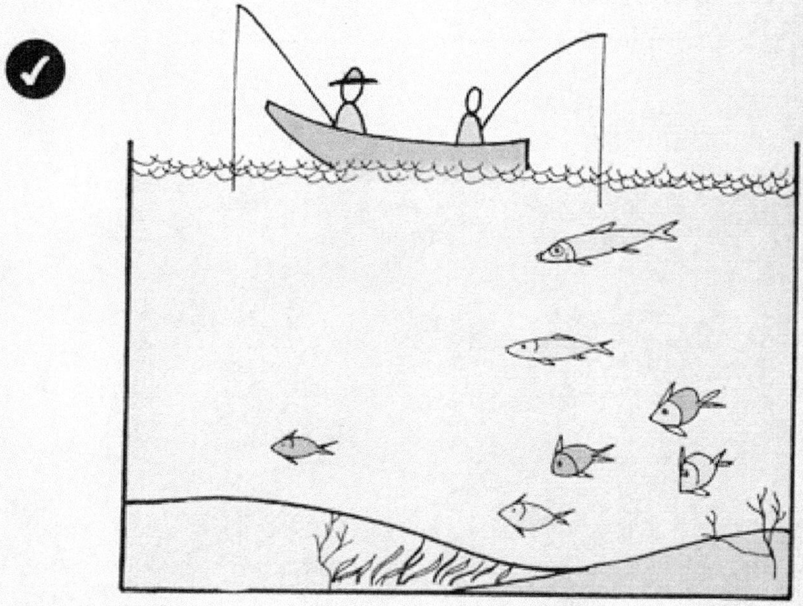

保持水系统的平衡

不要提倡通过引入或过度养殖，使某些物种形成优势。这会导致某些食物类型的过度消费，其他物种的丧失，生态平衡的破坏，最终导致产量的降低。可以通过保护多样性，来达到平衡的和可持续性的渔业管理。生态系统中物种越多，系统越稳定。

合理利用水资源

在季节性干旱地区，过度使用水资源将降低水位、破坏植被和加剧荒漠化。在任何地区必须确定可持续性发展的利用限度，发展只能在允许范围内进行。特别注意控制使用深井。

保护风景区的自然景观

有植被环绕的湖泊和瀑布不仅能吸引游客，而且对生物多样性也有着重要的意义。它们既是室外娱乐、踏青、旅游的绝好去处，也是人们陶冶情操的地方。一定不要以牺牲这些不断增值的宝贵财富为代价，制定那些会造成污染的发展计划。中国有大量并不十分脆弱的地区能为这类发展提供广阔的空间。

需要正确处理废水

城市生活和工业废水对当地的生物多样性有相当大的负面影响，不仅破坏了河流和湖泊，还使我们的生活环境质量降低。要设计和使用足够的废水处理设施，使城镇居民享受美丽、富于生物多样性和健康的环境。必须严格设置污染标准，对造成污染的单位和个人处以高额罚款。

保护城镇内湖泊和水道周围的天然植被

湖泊河流周围的植被有助于滋养和净化生态系统，并可创造出优美、清洁、健康的环境。但这些水系统若被不能生长植被的混凝土填塞，则会造成水体超营养作用，形成无生命的、不健康的和没有吸引力的厌氧性水体。

尽量减少挖泥所带来的负面影响

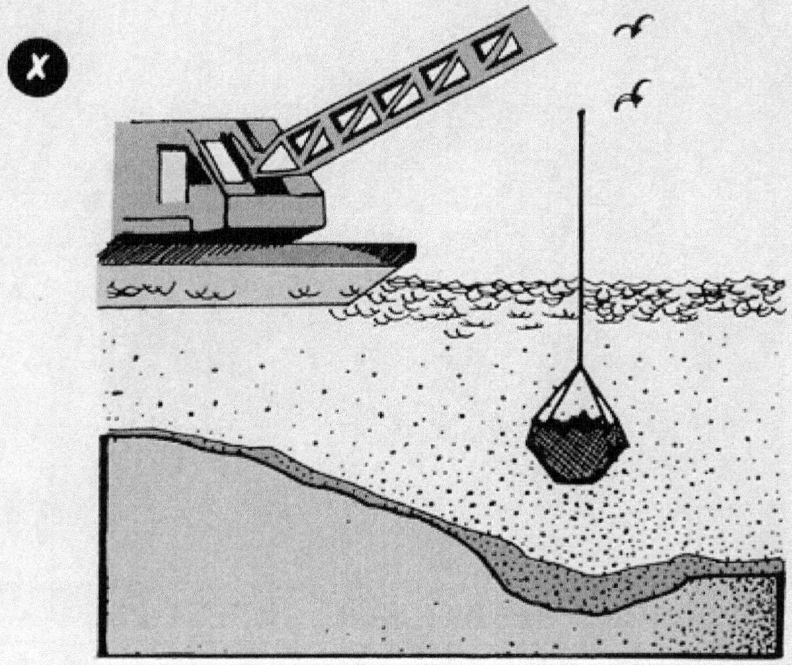

河床湖泊淤塞会降低其蓄水、排水的功能，因此有必要进行挖泥。但挖泥也会带来负面影响，包括降低水质、改变水的流速和流向以及导致洪水和侵蚀等。挖泥量应最小化，弃泥应安置在适宜的地方，使其不会加大侵蚀程度。

将泛滥平原保留为开放地区

随着气候不稳定性的增强，在河流易泛滥的平原开发建设会造成生命和财产的损失，甚至会影响到多年未遭受洪水的地区。易遭洪水的地区应该保留作为开放地区、娱乐地区、野生动植物避难所或农田。

创建人造池塘或湖泊

很多废弃的采石场和矿井都可以转化为宝贵的湖泊或池塘，其他池塘也可以很容易地纳入当地周密的发展规划中。如果这些池塘适合于鱼类生长，其娱乐和生物多样性价值更会提高。深水与浅水区应该结合在一起。浅水区有利于鱼类和两栖动物的生活，芦苇生长区可为秧鸡和鸣禽所用。如果在安全的岛屿上保留矮树丛，则可作为白鹭和其他水鸟的栖息所和繁殖地。

适度开发湖滨地带或海岸线

不正确的开发容易毁坏湖滨地带或海岸线的自然环境质量和生物多样性价值。大量建设码头区会破坏植被和生物多样性，增加腐烂物的污染危险。房屋应集中建设并远离这些地区并使负面影响降至最低。应该建设少量的公用码头和入口处，而不是建设大量的码头和上岸平台。

保护河边和沼泽森林

沼泽和河边森林在蓄留洪水、避免河口淤积、阻拦污染源、维持干旱季节水流量、保持船只畅通以及保护生物多样性等方面起着决定性的作用。伐除这一地带的森林会导致洪灾、河堤倒塌、河流通道淤积以及生物多样性损失。

正确使用桥梁和堤道

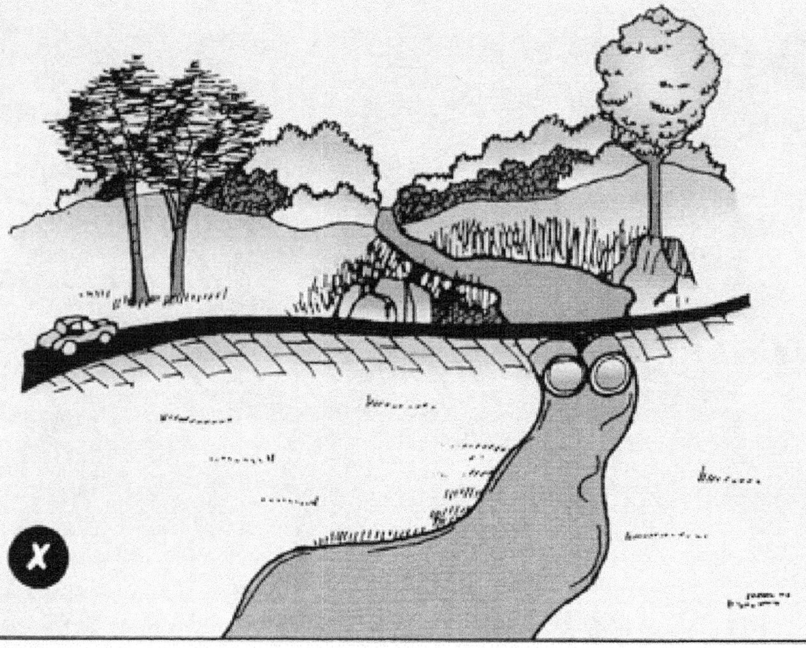

应在湿地和潮水地带建立桥梁和堤道，使水循环不受阻碍。桥梁比管道的作用大得多。途经沼泽和潮汐地的堤道可以保护天然栖息地和生物多样性，这比填充阻塞的建筑方式破坏性要小得多。

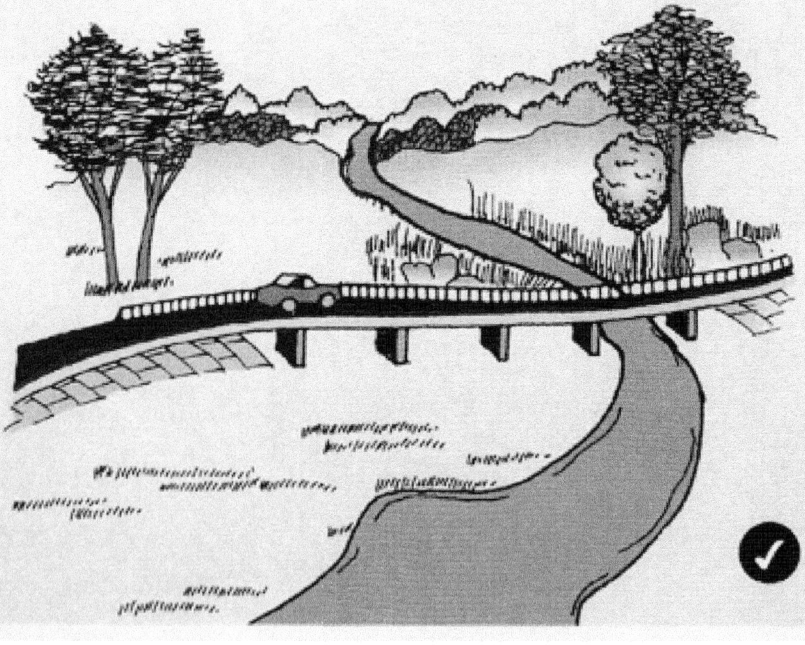

限制沿码头的建设

水道的堤岸非常富于变化。天然植被对于保护堤岸的结构有十分重要的作用。房屋建设要远离堤岸，以免造成堤岸的侵蚀、下沉、洪水破坏等危险，以及对环境和生物多样性的破坏。

保护红树林和沿海植被

沿海植被对稳定泥泞多沙的地层起着重要作用。红树林、番薯属（*Ipomoea*）和鬣刺属（*Spinifex*）沙丘植被，以及浓密的露兜树属（*Pandanus*）植被和海滩森林不仅促进了新农田的开垦，还保护着沿海地区不受海水侵蚀、台风和移动沙丘的影响。它们还是野生动植物的庇护所。红树林是鱼虾的繁殖地，而且能够固着沉积物和防止附近珊瑚礁被遮盖。必须防止不必要地清除沿海植被。如果已经被破坏，就应鼓励人们恢复之。

限制潮间带的开发

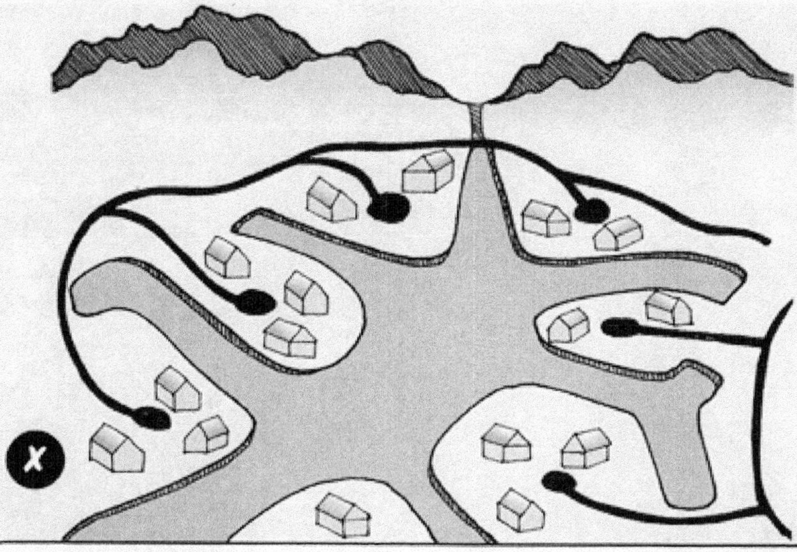

潮间带非常富饶，是重要的净化区、鱼虾繁殖地和大量食用甲壳类的栖息地，也是娱乐休闲场所和重要的生物多样性来源地。同时它也很脆弱，污染、土壤流失或潮汐运动都会毁坏该地带。该地带一旦被污染，采自这里的食物会威胁人类的健康。要严格限制在潮间带内或可能影响该地带的开发，这些活动应该在内陆进行。

建立综合的海洋地区发展计划

由于不同部门在海洋和沿海地区的利益有很大的交叉，因此有必要建立综合的发展计划，确保由某个部门发起的开发不会危及其他部门的需要和利益。

保护珊瑚礁

珊瑚礁是高生产力的生态系统，有很多有价值的功能，如鱼苗的育苗地、旅游胜地、水资源净化、海岸保护等。很多珊瑚礁有机体富含大量蛋白质，正在被越来越多地制成药物。这些利益不应被没有必要的破坏和开发所损害。必须保护珊瑚礁，禁止进行珊瑚礁开采、炸鱼、毒鱼、违规的船只停泊、淤泥或其他污染物排放等活动。

为海洋哺乳动物提供特殊保护

海豹、鲸和儒艮等珍稀海洋哺乳动物的觅食和繁殖栖息地极易遭到污染、人为干扰和非法猎捕等的影响。开发计划必须保证不给海洋动物种群集中地增加更多的威胁。

避免在沿岸和潮汐海域进行不必要的开垦

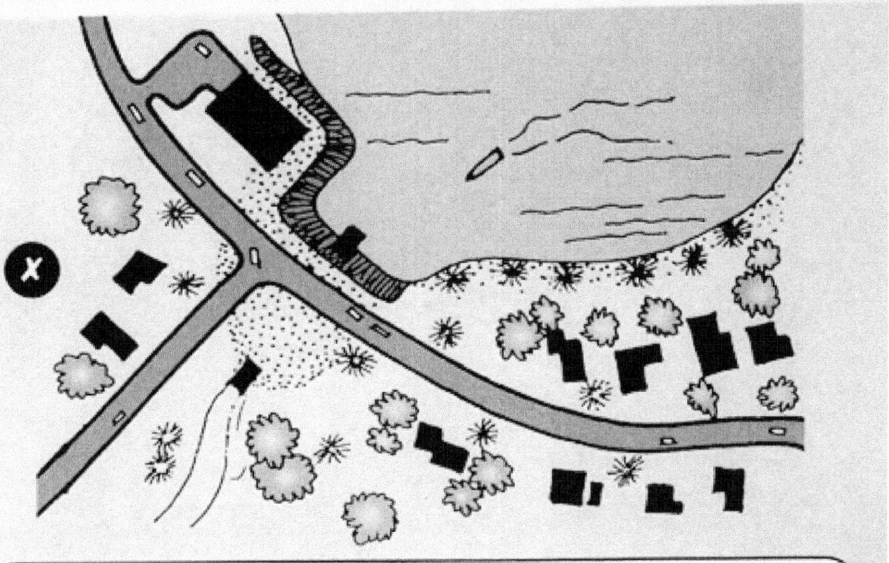

一般来说，沿岸和潮汐海域不应该被填起来或者用其他方式进行开垦。被填起来的潮汐海域往往会受到洪水的侵袭，也可能导致侵蚀问题。而且，填塞会改变水和沉积物的流动，破坏野生动植物的栖息地和高产的浅水区。一旦这些情况确实发生了，我们就必须建造人工暗礁或栖息地来弥补对生物多样性和其他功能造成的损失。这些工作既困难又花费昂贵。

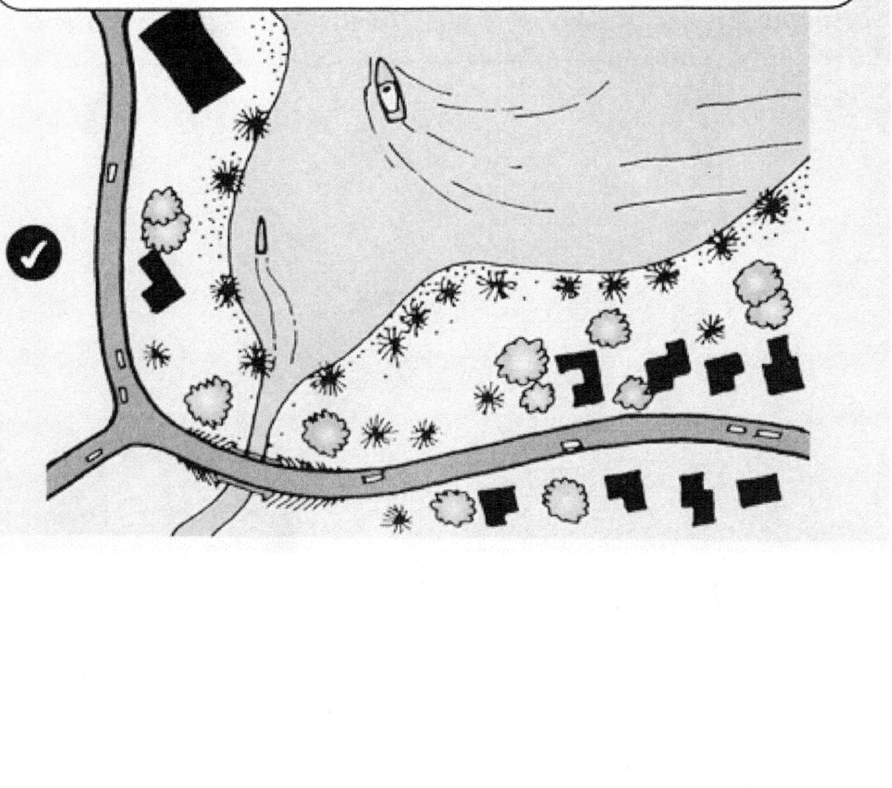

防止赤潮和其他污染的影响

海洋经常被当作巨大的垃圾存储地，但是这些污染物会对海洋生态系统造成严重的破坏。珊瑚在淤泥中会窒息。由于重金属的污染，食用这些鱼类、贝类和虾类对人类的健康构成了威胁。其他废弃的化学物质则导致有毒藻类的大爆发，这些藻类对食物链中的鱼类和软体动物释放出大量的神经毒素，甚至威胁到人类的健康，这种现象被称作“赤潮”。必须对河流和沿岸工厂的污染源头进行净化处理，减少污染的扩散。

消除防水壁

建设开发应远离海岸线，以免沿着被冲蚀的海岸线形成防水壁。在防水壁不可避免的地方，可以在防水壁和海水之间保留由天然植被组成的缓冲带，这将有助于保护防水壁和野生动植物的栖息地，并提高生物生产力。

注重风景质量

沿海地区发展规划应注意保护景观的价值。尽可能地使这些景观对大众开放。可以通过建立缓冲地带或仔细规划建筑物的规模和地点，来减少建筑物对景观的视觉破坏，保护沿河口和沿海地区的景色。

使用桩子

在潮汐带建立码头时，打桩比填方好。沉积物和野生动植物可在桩子底下自由活动，而填方只会成为阻碍，并改变流速和侵蚀过程。另外，打桩的建筑和保养成本低，而且可以被生物分解。

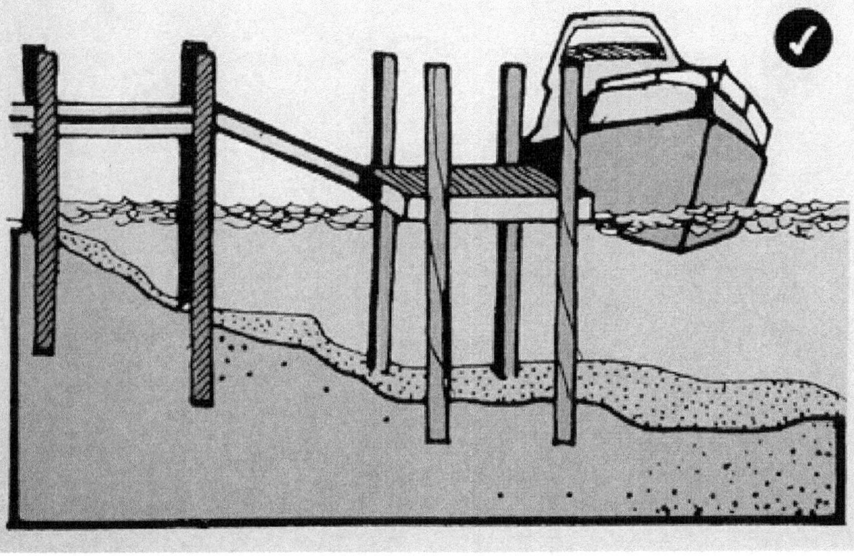

保护天然排水模式

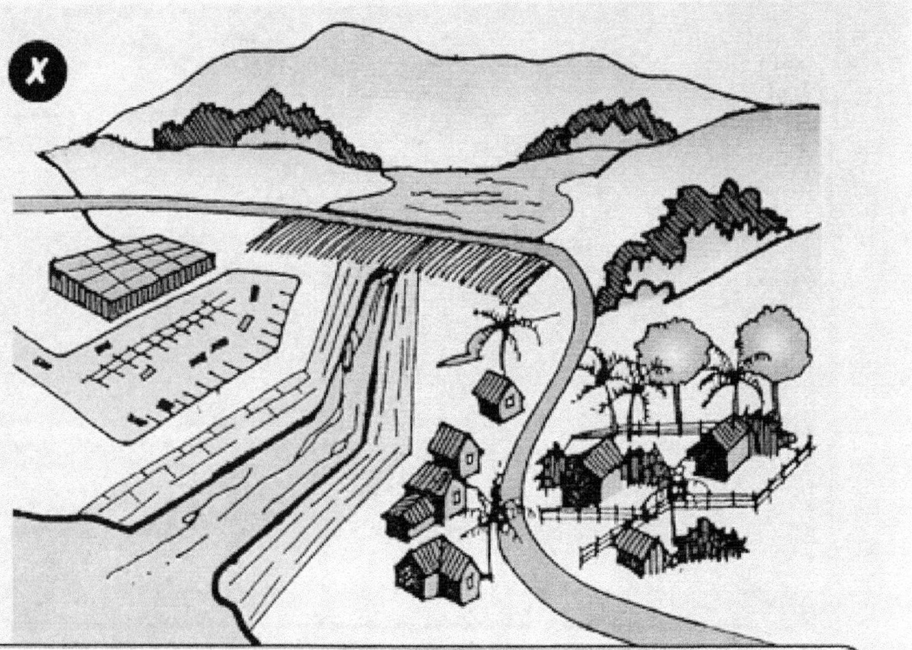

保护沿岸地区的天然排水模式。开凿运河和改变沿海水流的流向会改变或转移沼泽、潮汐带和其他浅水区的水流流向，从而增加污染、改变盐度、减少河口的生物活动。同样的道理，沿岸地区也不应建设堤坝和水电站。

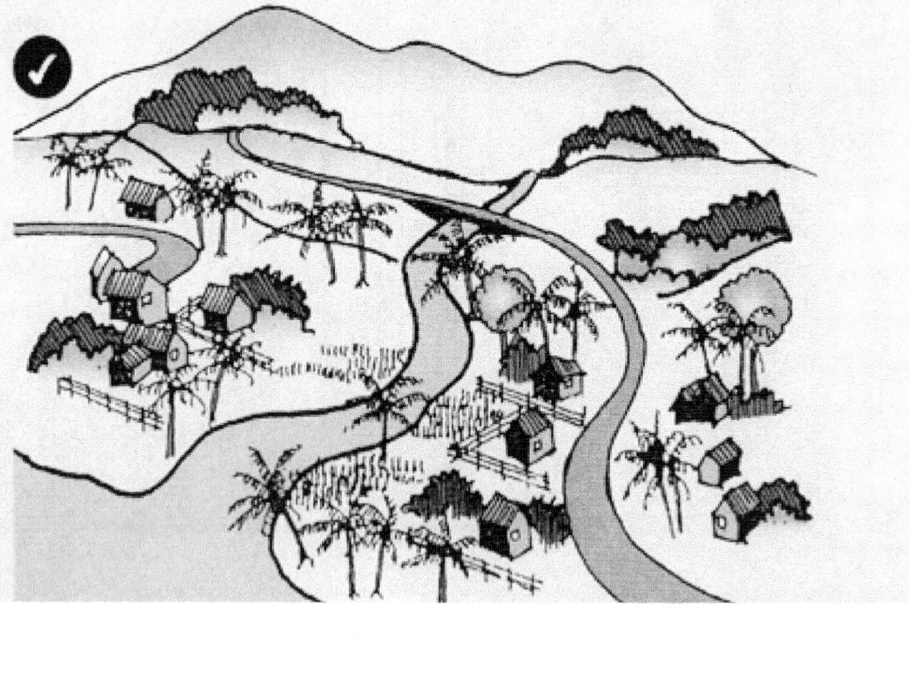

避免沿沙质海岸的开发

沙质海岸的建设要十分小心。这些地区十分脆弱，如果失去天然植被的覆盖，很容易被风暴侵蚀而破坏。人类和机械的过度使用会增加侵蚀的严重程度。洪水、坍塌、水供应和废物排放等问题随之而来。开发应该在远离海岸的更坚实的地表进行。

保护被保护地区的“荒野”价值

建设、分区和制定规章制度要以保护被保护地区的“荒野”价值为前提。这就是说游客们在这里应感觉到其身处真正的大自然中，而不是在人造环境之中。应特别注意要尽量减少噪音和影响视觉效果的建筑，严格禁止采集植物，损伤树木，以及乱扔垃圾和乱涂乱画等。

限制建设进入保护区的公路

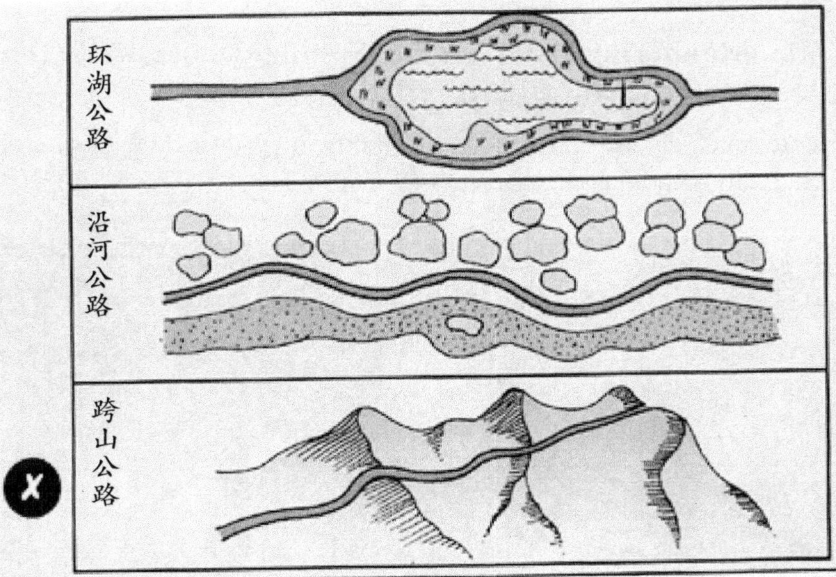

进入保护区的公路虽然便于管理人员工作，但更为滥用和掠夺保护区内的树木及野生动植物资源提供方便。此外，还具有经济效益的诱惑力，使人们想在保护区内开垦农田，所以必须抵制这种诱惑，不要开设进入保护区的公路。难以进入保护区可以避免不适宜的发展，是保护自然最好的方式。公路要建在负面影响最小的地方。

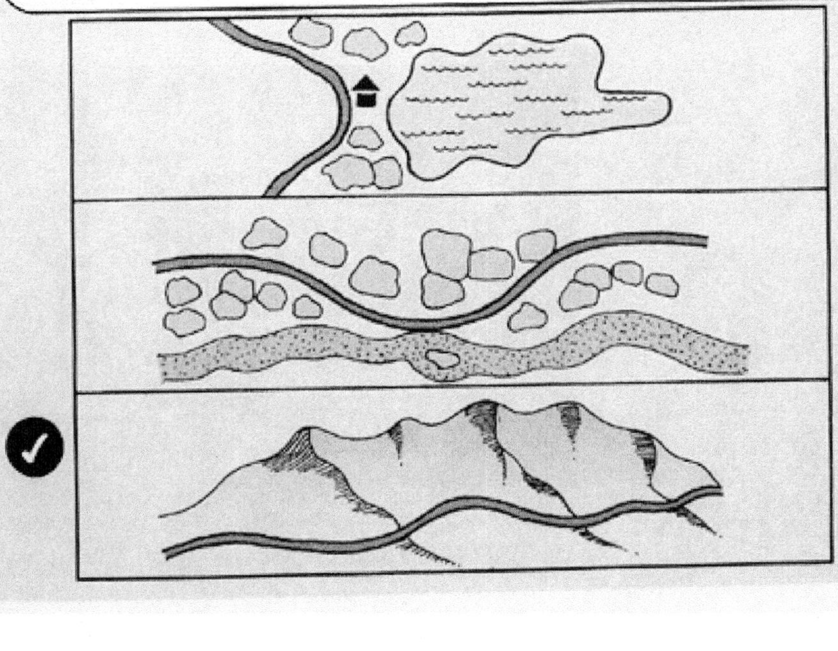

Appendix 6　Some more specific details for specific indicator monitoring

Appendix 6.1　Monitoring Hawk moths (Sphingidae) and Silk (Emperor) moths (Saturniidae)

Many moths are small, diverse and difficult to identify. It is adequate to monitor these two families of large, easily identified and characteristic moths.

Sphingidae are shaped like fighter jets, wings pointed and held in backward position.

Silk moths are large with broad rounded wings decorated with eyespots or crescent patterns.

A simple moth trap with a mercury vapour bulb can be set one night per week throughout the year in southern sites or in summer months only in northern sites. The trap should be in a fixed, well visible position, away from other bright lights. Moths caught can be examined in daytime but should be released only at night as birds will learn to come and eat them otherwise. Old egg boxes can be used for the moths to hide under at bottom of trap.

Typical home-made moth trap. Commercial suppliers can supply professional model.

Appendix 6.2 Monitoring Dragonflies

Dragonflies make good indicators of wetlands conditions as both larvae and adults live in or around water. Select a few good sampling points or small ponds or lengths of ditch or stream for repeated sampling. Count total number of individuals and species seen on each visit (set time limit for sample points). It is enough at first to use para-taxonomic names for different species (species A, species B etc.). Over time you will get to recognize the main species and be able to apply proper names. Note males may be different colour than females so watch for mating pairs to learn which are same species. Learn the main families first.

Anisoptera—Dragonflies (powerful, hold wings flat)

Family **Aeshnoidea** Large dragonflies with long straight bodies. Eyes touching down entire head line.

Family **Cordulegastridae** Very large, yellow striped, blackish dragonflies whose green eyes touch marginally on top of head only

Family **Libellulidae** Shorter dragonflies with sometimes broad flattened abdomen. Usually sit with wings held in forwards position (This is largest family). Some species have tinted wings.

Family **Chlorogomphidae** Very large, yellow striped, blackish dragonflies but eyes do not touch at all. Hind wing curved rather than angled back edge. Females may have coloured hind wings.

Family **Gomphidae** Smaller, yellow striped, blackish dragonflies whose eyes do not touch. Hind wing angled rather than curved back edge. End of abdomen expanded.

Zygoptera—Damselflies (delicate, eyes wide apart, usually hold wings vertically)

Family **Calopterygidae** Broad-winged damselflies with metallic coloured bodies and wings.

Family **Chlorocyphidae** Small damselflies with distinct 'nose'

Family **Coenagrionidae** Slender damselflies with narrow short wings (the largest family)

Family **Platycnemididae** Similar to Coenagrionidae but males have somewhat expanded legs

Family **Protoneuridae** Dark slender damselflies with short wings

Family **Diphlebiidae** Big stocky damselflies with shortish body and holds wings in X

Family **Euphaeidae** Medium-sized thickset damselflies that hold wings slightly separated

Family **Megapodagrionidae** Larger robust damselflies that hold their wings half open

Family **Synlestidae** Large metallic damselflies with broad tip to abdomen and very small wings

Family **Lestidae** Aberrant group of small damselflies that hold wings flattened over abdomen

Appendix 7 Selecting a suitable statistical test (adapted from www.graphpad.com)

	Type of Data			
Reason for testing	Measurement (from Normal Population)	Rank, Score, or Measurement (from Non-Normal Population)	Binomial (Two Possible Outcomes)	Survival Time
Describe one group	Mean, SD	Median, interquartile range	Proportion	Kaplan Meier survival curve
Compare one group to a hypothetical value	One-sample *t* test	Wilcoxon test	Chi-square or Binomial test **	
Compare two unpaired groups	Unpaired *t* test	Mann-Whitney U test, Kolmogorov-Smirnov test	Fisher's test (chi-square for large samples)	Log-rank test or Mantel-Haenszel*
Compare two paired groups	Paired *t* test	Wilcoxon test	McNemar's test	Conditional proportional hazards regression*
Compare three or more unmatched groups	One-way ANOVA	Kruskal-Wallis test	Chi-square test	Cox proportional hazard regression**
Compare three or more matched groups	Repeated-measures ANOVA	Friedman test	Cochrane Q**	Conditional proportional hazards regression**
Quantify association between two variables	Pearson correlation	Spearman correlation	Contingency coefficients**	
Predict value from another measured variable	Simple linear regression or Nonlinear regression	Nonparametric regression**	Simple logistic regression*	Cox proportional hazard regression*
Predict value from several measured or binomial variables	Multiple linear regression* or Multiple nonlinear regression**		Multiple logistic regression*	Cox proportional hazard regression*

Review of Nonparametric Tests

Choosing the right test to compare measurements involves choosing between two families of tests: parametric and nonparametric. Many-statistical test are based upon the assumption that the data are sampled from a Normal distribution. These tests are referred to as parametric tests. Commonly used parametric tests are listed in the second column of the table above and include the *t* test and analysis of variance.

Tests that do not make assumptions about the population distribution are referred to as nonparametric-tests. Most nonparametric tests rank the outcome variable from low to high and then analyze the ranks. These tests are listed in the third column of the table and include the Wilcoxon, Mann-Whitney U test, and Kruskal-Wallis tests. These tests are also called distribution-free tests. The Kolmogorov-Smirnov test can compare scores for totally non-arithmetic variables such as habitat preference.

Choosing Between Parametric and Nonparametric Tests: The Easy Cases

Choosing between parametric and nonparametric tests is sometimes easy. You should definitely choose a parametric test if you are sure that your data are sampled from a population that follows a Normal distribution (at least approximately). You should definitely select a nonparametric test in three situations:

The outcome is a rank or a score and the population is clearly not normal. Examples include class ranking of students, the Apgar score for the health of newborn babies (measured on a scale of 0 to 10 and where all scores are integers), the visual analogue score for pain (measured on a continuous scale where 0 is no pain and 10 is unbearable pain), and the star scale commonly used by movie and restaurant critics (* is OK, ***** is fantastic).

Some values are "off the scale", that is, too high or too low to measure. Even if the population is normal, it is impossible to analyze such data with a parametric test since you don't know all of the values. Using a nonparametric test with these data is simple. Assign values too low to measure an arbitrary very low value and assign values too high to measure an arbitrary very high value. Then perform a nonparametric test. Since the nonparametric test only knows about the relative ranks of the values, it won't matter that you didn't know all the values exactly.

The data are measurements, and you are sure that the population is not distributed in a normal manner. If the data are not sampled from a normal distribution, consider whether you can transform the values to make the distribution become normal. For example, you might take the logarithm or reciprocal of all values. There are often biological or chemical reasons (as well as statistical ones) for performing a particular transform.

Choosing Between Parametric and Nonparametric Tests: The Hard Cases

It is not always easy to decide whether a sample comes from a normal population. Consider these points:

If you collect many data points (over a hundred or so), you can look at the distribution of data and it will be fairly obvious whether the distribution is approximately bell shaped. A formal statistical test (Kolmogorov-Smirnoff test, not explained in this book) can be used to test whether the distribution of the data differs significantly from a Normal distribution. With few data points, it is difficult to tell whether the data are Normal by inspection, and the formal test has little power to discriminate between Normal and non-Normal distributions.

You should look at previous data as well. Remember, what matters is the distribution of the overall population, not the distribution of your sample. In deciding whether a population is Normal, look at all available data, not just data in the current experiment.

Consider the source of scatter. When the scatter comes from the sum of numerous sources (with no one source contributing most of the scatter), you expect to find a roughly Normal distribution. When in doubt, some people choose a parametric test (because they aren't sure the Normal assumption is violated), and others choose a nonparametric test (because they aren't sure the Normal assumption is met).

Choosing Between Parametric and Nonparametric Tests: Does It Matter?

Does it matter whether you choose a parametric or nonparametric test? The answer depends on sample size. There are four cases to think about:

Large sample. What happens when you use a parametric test with data from a nonnormal population? The central limit theorem ensures that parametric tests work well with large samples even if the population is non-Normal. In other words, parametric tests are robust to deviations from Normal distributions, so long as the samples are large. The snag is that it is impossible to say how large is large enough, as it depends on the nature of the particular non-Normal distribution. Unless the population distribution is really weird, you are probably safe choosing a parametric test when there are at least two dozen data points in each group.

Large sample. What happens when you use a nonparametric test with data from a Normal population? Nonparametric tests work well with large samples from Normal populations. The *P* values tend to be a bit too large, but the discrepancy is small. In other words, nonparametric tests are only slightly less powerful than parametric tests with large samples.

Small samples. What happens when you use a parametric test with data from non-normal populations? You can't rely on the central limit theorem, so the *P* value may be inaccurate.

Small samples. When you use a nonparametric test with data from a Normal population, the *P* values tend to be too high. The nonparametric tests lack statistical power with small samples.

Thus, large data sets present no problems. It is usually easy to tell if the data come from a Normal population, but it doesn't really matter because the nonparametric tests are so powerful and the parametric tests are so robust. Small data sets present a dilemma. It is difficult to tell if the data come from a Normal population, but it matters a lot. The nonparametric tests are not powerful and the parametric tests are not robust.

One- or Two-Sided *P* Value?

With many tests, you must choose whether you wish to calculate a one- or two-sided *P* value (same as one- or two-tailed *P* value). Let's review the difference in the context of a *t* test. The *P* value is calculated for the null hypothesis that the two population means are equal, and any discrepancy between the two sample means is due to chance. If this null hypothesis is true, the one-sided *P* value is the probability that two sample means would differ as much as was observed (or further) in the direction specified by the hypothesis just by chance, even though the means of the overall populations are actually equal. The two-sided *P* value also includes the probability that the sample means would differ that much in the opposite direction (i.e., the other group has the larger mean). The two-sided *P* value is twice the one-sided *P* value.

A one-sided *P* value is appropriate when you can state with certainty (and before collecting any data) that there either will be no difference between the means or that the difference will go in a direction you can specify in advance (i.e., you have specified which group will have the larger mean). If you cannot specify the direction of any difference before collecting data, then a two-sided *P* value is more appropriate. If in doubt, select a two-sided *P* value.

If you select a one-sided test, you should do so before collecting any data and you need to state the direction of your experimental hypothesis. If the data go the other way, you must be willing to attribute that difference (or association or correlation) to chance, no matter how striking the data. If you would be intrigued, even a little, by data that goes in the "wrong" direction, then you should use a two-sided *P* value.

Paired or Unpaired Test?

When comparing two groups, you need to decide whether to use a paired test. When comparing three or more groups, the term paired is not apt and the term repeated measures is used instead.

Use an unpaired test to compare groups when the individual values are not paired or matched with one another. Select a paired or repeated-measures test when values represent repeated measurements on one subject (before and after an intervention) or measurements on matched subjects. The paired or repeated-measures tests are also appropriate for repeated laboratory experiments run at different times, each with its own control.

You should select a paired test when values in one group are more closely correlated with a specific value in the other group than with random values in the other group. It is only appropriate to select a paired test when the subjects were matched or paired before the data were collected. You cannot base the pairing on the data you are analyzing.

Fisher's Test or the Chi-Square Test?

When analyzing contingency tables with two rows and two columns, you can use either Fisher's exact test or the chi-square test. The Fisher's test is the best choice as it always gives the exact *P* value. The chi-square test is simpler to calculate but yields only an approximate *P* value. If a computer is doing the calculations, you should choose Fisher's test unless you prefer the familiarity of the chi-square test. You should definitely avoid the chi-square test when the numbers in the contingency table are very small (any number less than about six). When the numbers are larger, the *P* values reported by the chi-square and Fisher's test will be very similar.

The chi-square test calculates approximate *P* values, and the Yates' continuity correction is designed to make the approximation better. Without the Yates' correction, the *P* values are too low. However, the correction goes too far, and the resulting *P* value is too high. Statisticians give different recommendations regarding Yates' correction. With large sample sizes, the Yates' correction makes little difference. If you select Fisher's test, the *P* value is exact and Yates' correction is not needed and is not available.

Regression or Correlation?

Linear regression and correlation are similar and easily confused. In some situations it makes sense to perform both calculations. Calculate linear correlation if you measured both *X* and *Y* in each subject and wish to quantity how well they are associated. Select the Pearson (parametric) correlation coefficient if you can assume that both *X* and *Y* are sampled from Normal populations. Otherwise choose the Spearman nonparametric correlation coefficient. Don't calculate the correlation coefficient (or its confidence interval) if you manipulated the *X* variable.

Calculate linear regressions only if one of the variables (*X*) is likely to precede or cause the other variable (*Y*). Definitely choose linear regression if you manipulated the *X* variable. It makes a big difference which variable is called *X* and which is called *Y*, as linear regression calculations are not symmetrical with respect to X and Y. If you swap the two variables, you will obtain a different regression line. In contrast, linear correlation calculations are symmetrical with respect to X and Y. If you swap the labels X and Y, you will still get the same correlation coefficient.

Appendix 8　Example report forms for use in patrolling and monitoring

<table>
<tr><td colspan="4">Station:</td><td colspan="8">Date:　　year:　　month:　　day:</td></tr>
<tr><td colspan="8">Name of patrollers:</td><td colspan="4">Name of reporter:</td></tr>
<tr><td colspan="7">Method of survey: 1, foot, 2, car　3, boat　4, other</td><td colspan="5">Itinerary/route:</td></tr>
<tr><td colspan="7">Locations:</td><td colspan="5">Weather:</td></tr>
<tr><td colspan="3">Start time:</td><td colspan="4">End time:</td><td colspan="5">Duration:</td></tr>
<tr><td>Sectors:</td><td colspan="3"></td><td colspan="3"></td><td colspan="3"></td><td>Habitat description</td><td>Remarks</td></tr>
<tr><td>Altitude/ Lat. Long.</td><td colspan="3"></td><td colspan="3"></td><td colspan="3"></td><td></td><td></td></tr>
<tr><td>Name</td><td>No.</td><td colspan="2">Includes</td><td>No.</td><td colspan="2">Includes</td><td>No.</td><td colspan="2">Includes</td><td rowspan="2"></td><td rowspan="2">Sex ratio</td></tr>
<tr><td></td><td></td><td>Adult</td><td>Young</td><td></td><td>Adult</td><td>Young</td><td></td><td>Adult</td><td>Young</td></tr>
<tr><td></td><td></td><td></td><td></td><td></td><td></td><td></td><td></td><td></td><td></td><td></td><td></td></tr>
<tr><td></td><td></td><td></td><td></td><td></td><td></td><td></td><td></td><td></td><td></td><td></td><td></td></tr>
<tr><td></td><td></td><td></td><td></td><td></td><td></td><td></td><td></td><td></td><td></td><td></td><td></td></tr>
<tr><td></td><td></td><td></td><td></td><td></td><td></td><td></td><td></td><td></td><td></td><td></td><td></td></tr>
<tr><td></td><td></td><td></td><td></td><td></td><td></td><td></td><td></td><td></td><td></td><td></td><td></td></tr>
<tr><td></td><td></td><td></td><td></td><td></td><td></td><td></td><td></td><td></td><td></td><td></td><td></td></tr>
<tr><td></td><td></td><td></td><td></td><td></td><td></td><td></td><td></td><td></td><td></td><td></td><td></td></tr>
<tr><td>Total</td><td></td><td></td><td></td><td></td><td></td><td></td><td></td><td></td><td></td><td></td><td></td></tr>
<tr><td>Plants</td><td colspan="3"></td><td colspan="3"></td><td colspan="3"></td><td></td><td></td></tr>
<tr><td></td><td colspan="3"></td><td colspan="3"></td><td colspan="3"></td><td></td><td></td></tr>
<tr><td></td><td colspan="3"></td><td colspan="3"></td><td colspan="3"></td><td></td><td></td></tr>
<tr><td>Human activity</td><td colspan="11"></td></tr>
</table>

Appendix 9　Summary of reforms and changed practice changes required to improve Wetlands protection in China

- Reform of PA categories and regulations
- Improved zoning within WPAs
- Red lining of wider landscape around wetlands
- Mitigation of dams and water diversions
- Better control of water levels
- Use of EHI monitoring

- Addressing pollution sources
- Maintaining network of critical migratory stopover sites
- greater co-management by local communities
- Greater awareness
- Adoption of competency standards with relevant raining
- Better mainstreaming with fisheries, power companies, water resources etc.
- Promotion of less damaging types of eco-tourism
- Planning mitigation for climate change
- Allowing more natural methods of regeneration
- Halt planting trees inside wetlands
- Improved methods for restoring mangroves
- Public sharing of data

Bibliography

Appleton M, Texon G, Uriarte M. 2003. Competence Standards for Protected Areas Jobs in South East Asia. Laguna: ASEAN Regional Centre for Biodiversity Conservation.

Athanas A, Ghersi F, Phillips A, et al. 2001. Guidelines for Financing Protected Areas in East Asia. Gland: IUCN.

Bailey JA. 1984. Principles of Wildlife Management. New York: John Wiley & Sons.

Bibby CJ, Burgess ND, Hill DA, et al. 2000. Bird Census Techniques Second Edition. San Diego: Academic Press.

BirdLife International. 2001. Threatened Birds of Asia: The BirdLife International Red Data Book. Cambridge: BirdLife International.

Bowman ME, Eagles PFJ, Tao TCH. 2001. Guidelines for Tourism in Parks and Protected Areas of East Asia. Gland: IUCN; Waterloo: University of Waterloo.

Brown J, Mitchell N, Beresford M. 2005. Protected Landscape Approach (The): Linking Nature, Culture and Community.

Christ C, Hillel O, Matus S, et al. 2003. Tourism and Biodiversity: Mapping Tourism's Global Footprint. Washington DC: Conservation International.

Davey AG. 1998. National System Planning for Protected Areas. Gland: IUCN.

de Sherbinin A, Lacko A, Jaiteh M. 2012. Evaluating the Risk to Ramsar Sites from Climate Change Induced Sea Level Rise. Ramsar Scientific and Technical Briefing Note No.5. Gland: Ramsar Convention Secretariat.

Diamond JM. 1975. The island dilemma: lessons of modern biogeographic studies for the design of natural reserves. Biological Conservation, 7: 129-145.

Ding CQ. 2010. Crested Ibis. Chinese Birds, 1(2): 156-162.

Dudley N, Hockings M, Stolton S. 2000. Evaluating Effectiveness: A Framework for Assessing Management of Protected Areas. Cambridge: IUCN.

Eagles PFJ, Haynes CD, McCool SF. 2002. Sustainable Tourism in Protected Areas: Guidelines for Planning and Management. Gland: IUCN.

Guidelines for Protected Areas Management Categories.1994. IUCN World Commission on Protected Areas and the World Conservation Monitoring Centre.

Hamilton LS, Sandwith T, Sheppard D, et al. 2001. Transboundary Protected Areas for Peace and Co-operation. Gland: IUCN.

IUCN. 2000. Financing Protected Areas: Guidelines for Protected Area Managers. Gland: IUCN.

Kelleher G. 1999. Guidelines for Marine Protected Areas. Gland: IUCN.

Li S. 1981. Studies on Zoogeographical Divisions for Freshwater Fishes of China. Beijing: Science Press.

Li S, Qu R. 1937. History of Chinese Fisheries. Shanghai: The Commercial Press.

MacKinnon J. 2002. Biodiversity Principles for Developers and Planners. Beijing: BWG/CCICED.

MacKinnon J, MacKinnon K, Child G, et al. 1987. Managing Protected Areas in the Tropics. Journal of

Applied Ecology, 25(1): 369.

MacKinnon J, Mengsha G, Cheung C, et al. 1996. A Biodiversity Review of China. Hong Kong: WWF International.

MacKinnon K, MacKinnon J. 1991. Habitat protection and re-introduction programmes//Gipps JHW. Beyond Captive Breeding: Re-introducing Endangered Mammals to the Wild. Symposia Zoological Society of London No. 62. Oxford: Clarendon Press: 173-198.

Meffe GK, Carroll CR. 1994. Principles of Conservation Biology. Sunderland: Sinauer Associates.

Middleton J, Thomas L. 2003. Guidelines for Management Planning of Protected Areas. Gland: IUCN.

Protected Areas Task Force. 2004. Using Protected Areas to Extend Economic Benefits to Rural China—Evaluation of the Protected Area System of China and Policy Recommendations for Rationalizing the System. Beijing: Protected Area Task Force Report to the China Council for International Cooperation on Environment and Development (CCICED).

Riney T. 1982. Study and Management of Large Mammals. New York: John Wiley & Sons.

Schemnitz SD. 1980. Wildlife Management Techniques Manual. Washington DC: The Wildlife Society.

She ZG, Lin JX, Peng YG, et al. 2005. A preliminary study on mangrove and aquaculture system. Chinese Journal of Ecology, 7: 837-840.

Soulé ME. 1996. Viable Populations for Conservation. Cambridge: Cambridge University Press.

Sutherland WJ. 1997. Ecological Census Techniques: A Handbook. Cambridge: Cambridge University Press.

Sutherland WJ. 2001. The Conservation Handbook: Research, Management and Policy. Oxford: Blackwell.

Wang S, MacKinnon J. 1993. Urgent Recommendations to Save China's Biological Diversity. Report to the Chinese Council for International Cooperation in Environment and Development (CCICED). *Chinese Biodiversity*, 1(1): 2-13.

Worboys G, Lockwood M, De Lacy T. 2001. Protected Area Management: Principles and Practice. Melbourne: Oxford University Press.

Xie Y, Wang S, Peter Schei. 2004. China's Protected Areas. Beijing: Tsinghua University Press.